Clinical Trials of Genetic Therapy with Antisense DNA and DNA Vectors

Clinical Trials of Genetic Therapy with Antisense DNA and DNA Vectors

edited by

ERIC WICKSTROM

Thomas Jefferson University
Philadelphia, Pennsylvania

MARCEL DEKKER, INC.   NEW YORK · BASEL · HONG KONG

Library of Congress Cataloging-in-Publication Data

Clinical trials of genetic therapy with antisense DNA and DNA vectors/edited by Eric Wickstrom.
    p. cm.
    Includes bibliographical references and index.
    ISBN 0-8247-0085-6 (alk. paper)
    1. Gene therapy—Testing.   2. Antisense DNA—Therapeutic use—Testing. 3. Genetic vectors—Therapeutic use—Testing.   4. Clinical trials.
I. Wickstrom, Eric.
    [DNLM: 1. Gene Therapy.   2. DNA, Antisense.   3. Genetic Vectors.
4. Clinical Trials. QZ 50 C641 1998]
  RB155.5.C5728   1998
  616'.042—dc21
  DNLM/DLC
  for Library of Congress   98-13317
                                        CIP

The publisher offers discounts on this book when ordered in bulk quantities. For more information, write to Special Sales/Professional Marketing at the address below.

This book is printed on acid-free paper.

MARCEL DEKKER, INC.
270 Madison Avenue, New York, New York 10016
http://www.dekker.com

Current printing (last digit):
10 9 8 7 6 5 4 3 2 1

PRINTED IN THE UNITED STATES OF AMERICA

Foreword: Genetic Therapy in the 21st Century

O. Michael Colvin
Duke University Medical Center, Durham, North Carolina

The elucidation of the nature of the genetic code by Watson and Crick in 1952 presaged that we will ultimately be able to describe the determinants of the normal structure and functions of our bodies. The rapid increase in our ability to identify genes and sequence them is bringing that goal steadily closer to reality. A logical extension of this ability will be the capability to ascertain the genetic abnormalities associated with specific diseases, including cancer, and to treat these diseases by returning the altered segments of the genetic code to normal, or in some other way utilizing genetic knowledge and techniques for treatment. At this time a number of such initiatives are in preclinical development and several have entered the clinical trial stage. Examples of these preclinical and clinical efforts are described in this volume.

Efforts in genetic therapy have taken two general approaches. The first is the introduction into the patient of a vector that can insert into the genetic code a sequence that will restore a normal function or alter an abnormal function. Probably the most readily achieved example of this approach will be to introduce a gene for a single missing enzyme. A more difficult task will be to reverse a malignant state in all the cells of a tumor. More immediately realized antitumor efforts will likely be the genetic enhancement of the function of immune cells to recognize and eradicate a tumor; the induction of a portion of the cells in a tumor to secrete an antisense molecule, a ribozyme, or other lethal entity to attack those cells and neighboring malignant cells; or introduction of an enzyme into tumor cells which will activate a potent antitumor agent selectively in the region of the tumor.

A second general approach to using the genetic code to treat disease is to deliver to the affected cells antisense molecules that enter abnormal cells and by sequence recognition selectively inhibit a pathologic process in these cells. This approach has been limited by problems of selectivity for specific genetic sequences, inadequate avidity of binding to the target sequence, and inadequate delivery of the molecules into tissues and into cells.

The delivery of agents to tumor cells has traditionally been more readily accomplished with leukemias than with solid tumors, and for this reason we are most likely to initially see beneficial clinical effects with antisense molecules in leukemias. The problems with delivery also suggest that the initial significant successes will be in local applications of antisense technology, such as local administration to stenotic vasculature, the purging of tumor cells from bone marrow and stem cells, local injection into accessible tumors, and intraocular administration, as described in these chapters.

However, it seems very likely that more successful delivery of both vectors and antisense molecules will be achieved. It is probable that at least limited evidence of effectiveness will be forthcoming from the studies discussed here and from other studies underway. Nevertheless, as with any difficult challenge, the full realization of the potential of this therapeutic approach will depend on our ability to select the appropriate targets, to improve the delivery of the agents to the critical sites, and to optimize the effector functions of the agents. In this respect it is critical in the clinical experiments to understand why the efforts do not work when they don't, and how they could work better when they do.

This volume clearly demonstrates that genetic therapy is underway, and that the efforts described are soundly based in rationale and feasibility. However, we must remember that successful genetic therapy represents a formidable technical challenge whose promise well justifies our efforts to bring this type of treatment to fruition. This promise and our increasing understanding of the mechanisms of conversion of the genetic code into function indicate that gene-based therapy will become a mainstay of disease control in the 21st century.

It is appropriate to remember that the first antitumor agents were introduced into clinical therapy a half century ago. In retrospect, most of the subsequent challenges and successes of cancer therapy were predictable from the initial clinical experiments. I believe the same will be true for the genetic therapy of cancer and other diseases.

Preface: From Theory to Practice in Thirty Years

Eric Wickstrom
Thomas Jefferson University, Philadelphia, Pennsylvania

This collection of clinical and preclinical reports on genetic therapy is designed to inform clinical and biomedical investigators in all disease areas of gene-based interventions which appear to be plausible at this time, by presenting a few examples of the wide variety of current studies in this field. Since this editor's first volume on this subject (Wickstrom, 1991), the whole field of genetic therapy, by antisense DNA and by DNA vectors, has advanced dramatically. Subsequent advances of genetic therapy in cell culture (Crooke and Lebleu, 1993) and in animals (Agrawal, 1996) have been described in later volumes.

A number of laboratories have published results of successful human clinical trials. A larger series of papers on clinical trials of genetic therapy will be published over the next year. Thus, thirty years after the first suggestion of nucleic acid therapy against naturally occurring gene sequences (Belikova, et al., 1967), and the first demonstration of antisense activity in cell culture (Zamecnik and Stephenson, 1978), successes have been achieved in clinical trials.

Traditionally, chemotherapy agents for cancer or viral diseases have been designed to inhibit the enzymes necessary for cell growth or production of new viruses. Chemotherapeutics have strong side effects, extending to fatal toxicity. The diseased cells often become resistant to existing drugs, so new approaches are needed for effective disease management.

Cloning and sequencing of pathogenic genes over the last two decades have made possible a direct genetic approach to the treatment of disease, using nucleic acid therapeutics or vectors. The ability to turn off individual disease-causing genes, or correct them, or replace them at will in a patient's cells provides a powerful tool for therapeutic intervention in genetic diseases. Gene-specific nucleic acid therapy has gone from theory to practical possibility in a very short time.

These new DNA and RNA agents are intended to stop the growth of cancerous cells, the production of HIV from infected cells, the production of faulty proteins, or the pathogenic absence of necessary proteins, or to make abnormal cells more easily recognized by the immune system. Aside from traumatic events that require surgical repair, one may now say that all diseases represent examples of genetic dysfunction.

In the Foreword, Dr. O. Michael Colvin describes the scope of the present volume, which extends from nucleic acid therapeutics acting as drugs to turn off pathogenic genes to biological vectors carrying antisense, ribozyme, corrected, replacement, or immunostimulatory genes. Questions of preparation, delivery, toxicity, mechanism, and specificity are addressed.

Following this Preface, a brief history of genetic therapy is provided by Dr. James Hawkins, followed by a section on synthesis, preparation, and production of synthetic oligonucleotides and biological vectors. Applications to treat gene deficiencies, leukemia and lymphoma, solid tumors, human immunodeficiency virus (HIV), cytomegalovirus (CMV), and coronary restenosis are described in subsequent sections.

In the preparation section, the perspective and expectations of the Food and Drug Administration towards synthetic oligonucleotides are discussed by Drs. Ahn and DeGeorge, while the practical considerations of meeting FDA requirements for large-scale cGMP production are described by Dr. Gonzalez and his colleagues. Similarly, the perspective and expectations of the Food and Drug Administration towards biological vectors are discussed by Dr. Epstein, while practical considerations of large-scale cGLP production are described by Drs. Davis and Baker.

Focusing on treatment of adenosine deaminase deficiency, Drs. Gordon and Anderson recount their clinical experiences with vector administration. For hematological malignancies, several approaches to treatment of leukemia and lymphoma are presented. Drs. Gewirtz and Sokol describe their experiences treating patients with chronic myelogenous leukemia with antisense DNA against c-*myb* oncogene. In the case of acute myelogenous leukemia, Dr. Iversen discusses the pharmacokinetics and toxicity of an oligonucleotide against the tumor suppressor gene p53. Next, a clinical trial of anti-*bcl2* DNA oligomers to treat follicular lymphoma is reported by Dr. Cotter and his colleagues. In the area of bone marrow transplantation therapy, Drs. Abonour and Cornetta describe improvements in *ex vivo* retroviral transfection of stem cells with marker genes and drug resistance genes.

Therapy of solid tumors is considered next, beginning with application by Dr. Alavi and her colleagues of an adenoviral vector capable of transforming gliomal cells to ganciclovir, in patients with brain cancer. Preclinical and phase I clinical trials of a retroviral vector encoding an antisense RNA against c-*fos* oncogene in breast, prostate, and ovarian cancer cells are then reported by Dr. Obermiller and her colleagues. In an attempt to boost immune cell responses to tumor cells, Drs. Stopeck and Hersh discuss their clinical experiences using cytolytic T cells expressing an allogeneic MHC class I antigen. Dr. Roth then presents the results of treating lung cancer with retroviral p53 tumor suppressor gene replacement, while bladder cancer is the focus of a preclinical study of adenoviral delivery of ribozyme genes against H-*ras* oncogene, described by Dr. Small and his colleagues. To elevate immune cell responses in patients with melanoma, Drs. Lattime and his colleagues then report on the efficacy of vaccinia virus for insertion of granulocyte-macrophage colony-stimulating factor gene insertion for immunotherapy using vaccinia virus vectors.

In the HIV section, Dr. Agrawal describes administration to patients of antisense DNA against HIV *gag* gene, with little or no toxicity detected. For opportunistic CMV infection of the retinas of AIDS patients, Dr. Crooke reports that intravitreal treatment with an oligonucleotide complementary to a CMV immediate early gene clearly reduced the severity of infection, with modest side effects.

Finally, two different approaches to prevention of arterial restenosis following coronary bypass or balloon angioplasty are presented. An oligonucleotide complementary to c-*myc* mRNA, the first antisense DNA applied to an oncogene (Wickstrom, 1991), successfully inhibited remodeling and regrowth of vascular smooth muscle cells in the hearts of both pigs and human patients, as described by Dr. Zalewski and his colleagues. In preclinical trials, Dr. Mann and his colleagues report that encapsulation of antisense DNA, double-stranded decoy DNA, or plasmids with hemagglutinating virus of Japan permitted excellent cellular uptake and potency of gene therapy agents against a panel of proliferative genes expressed in animal models of restenosis.

The CMV therapeutic discussed by Dr. Crooke is currently in a Phase III trial, and may plausibly be approved for prescription and sale this year, two years earlier than this editor's previous prediction (Wickstrom, 1991). Great efforts must be applied to careful and thorough development of oligonucleotides and vectors, but safety and efficacy are sufficiently well demonstrated to permit optimism that nucleic acid therapeutics will fulfill their expectations.

ACKNOWLEDGEMENTS

My wife, Lois Wickstrom, and my postdoctoral fellow, Dr. Janet B. Smith, contributed immeasurably to the editing and preparation of this volume. I am greatly indebted to them for their tireless and invaluable help.

REFERENCES

Agrawal, S. (ed.) (1996) *Antisense Therapeutics*. Humana Press, Totowa, NJ.

Belikova, A. M., Zarytova, V. F., and Grineva, N. I. (1967) Synthesis of ribonucleosides and diribonucleoside phosphates containing 2-chloro-ethylamine and nitrogen mustard residues. *Tet. Lett.* 37:3557-3562.

Crooke, S. T., and Lebleu, B. (eds.) (1993) *Antisense Research and Applications*. CRC Press, Boca Raton, FL.

Wickstrom, E. (ed.) (1991) *Prospects for Antisense Nucleic Acid Therapy of Cancer and AIDS*. Wiley-Liss, New York.

Zamecnik, P. C., and Stephenson, M. L. (1978) Inhibition of Rous sarcoma virus replication and cell transformation by a specific oligodeoxynucleotide. *Proc. Natl. Acad. Sci. USA* 75:280-284.

Contents

Contributors

Rafat Abonour, M.D., Section of Hematology/Oncology, Department of Medicine, Indiana University School of Medicine, 975 W. Walnut St., Room IB442, Indianapolis, Indiana 46202

Sudhir Agrawal, Ph.D., Hybridon, Inc., 620 Memorial Drive, Cambridge, Massachusetts 02142

Chang-Ho Ahn, Ph.D., Oncology Drug Products, HFD 150, Center for Drug Evaluation and Research, Food and Drug Administration, 5600 Fishers Lane, Rockville, Maryland 20857

Jane B. Alavi, M.D., Institute for Human Gene Therapy, University of Pennsylvania School of Medicine, 409 Stellar-Chance Laboratories, 422 Curie Boulevard, Philadelphia, Pennsylvania 19104

W. French Anderson, M.D., Gene Therapy Laboratories, University of Southern California, Norris Cancer Center, 1441 Eastlake Ave., Room 614, Los Angeles, California 90033

Carlos L. Arteaga, M.D., Department of Cell Biology, Vanderbilt University, 21st Avenue and Garland St., Nashville, Tennessee 37232

Colleen Baker, Ph.D., Institute of Human Gene Therapy, University of Pennsylvania School of Medicine, 262 Wistar Institute, 3601 Spruce St., Philadelphia, Pennsylvania 19104

David Y. Bouffard, Ph.D., Berlex Biosciences, 15049 San Pablo Ave., Richmond, California 94804

Paul Clarke, M.D., McElwain Laboratories, Institute of Cancer Research, 15 Cotswold Road, Sutton, Surrey SM2 5NE, UK

O. Michael Colvin, M.D., Comprehensive Cancer Center, Duke University, Durham, North Carolina 27710

Kenneth Cornetta, M.D., Indiana Cancer Research Institute, 1044 W. Walnut St., Room 202, Indianapolis, Indiana 46202

Finbarr E. Cotter, M.D., Molecular Haematology Unit, Institute of Child Health, University College London, 30 Guilford St., London WC1N 1EH, UK

Stanley T. Crooke, M.D., Isis Pharmaceuticals, Inc., 2280 Faraday Avenue, Carlsbad, California 92008

David Cunningham, M.D., Lymphoma Unit, Royal Marsden Hospital, Downs Road, Sutton, Surrey SM2 5PT, UK

Alan R. Davis, Ph.D., Institute of Human Gene Therapy, University of Pennsylvania School of Medicine, 262 Wistar Institute, 3601 Spruce St., Philadelphia, Pennsylvania 19104

Joseph J. DeGeorge, Ph.D., Oncology Drug Products, HFD 150, Center for Drug Evaluation and Research, Food and Drug Administration, 5600 Fishers Lane, Rockville, Maryland 20857

Victor J. Dzau, M.D., Research Institute and Department of Medicine, Harvard Medical School and Brigham and Women's Hospital, 75 Francis Street, Boston, Massachusetts 02115

Stephen L. Eck, M.D., Ph.D., Institute for Human Gene Therapy, University of Pennsylvania School of Medicine, 409 Stellar-Chance Laboratories, 422 Curie Boulevard, Philadelphia, Pennsylvania 19104-6100

Richard G. Einig, Ph.D., Hybridon, Inc., 155 Fortune Blvd., Milford, Massachusetts 01757

Laurence C. Eisenlohr, V.M.D, Ph.D., Department of Microbiology and Immunology, Thomas Jefferson University, 233 S. 10th St., Room 726, Philadelphia, Pennsylvania 19107-5541

Suzanne L. Epstein, Ph.D., Center for Biologic Evaluations and Research, OTRR, DCGT, Food and Drug Administration, 1401 Rockville Pike, HFM-521, Rockville, Maryland 20852

Alan M. Gewirtz, M.D., Department of Pathology and Laboratory Medicine, CRB 813, University of Pennsylvania School of Medicine, 422 Curie St., Philadelphia, Pennsylvania 19104

Jose E. Gonzalez, Ph.D., Hybridon, Inc., 155 Fortune Blvd., Milford, Massachusetts 01757

Erlinda M. Gordon, M.D., Gene Therapy Laboratories, University of Southern California, Norris Cancer Center, 1441 Eastlake Ave., Room 614, Los Angeles, California 90033

James W. Hawkins, Ph.D., *Antisense and Nucleic Acid Drug Development*, 24510 Burnt Hill Road, Clarksburg, Maryland 20871

Evan M. Hersh, M.D., Arizona Cancer Center, 1501 N. Campbell Ave., Tucson, Arizona 85724

Jeffrey T. Holt, M.D., Department of Cell Biology, Vanderbilt University, 21st Avenue and Garland St., Nashville, Tennessee 37232

Patrick L. Iversen, Ph.D., AntiVirals, Inc., 4575 South West Research Way, Suite 200, Corvallis, Oregon 97333

Mohammed Kashani-Sabet, M.D., Mt. Zion Cancer Center, University of California at San Francisco, 2356 Sutter St., Box 1706, San Francisco, California 94143

Paul E. Kennedy, Ph.D., Hybridon, Inc., 620 Memorial Drive, Cambridge, Massachusetts 02139

Edmund C. Lattime, Ph.D., Division of Neoplastic Diseases, Thomas Jefferson University, 1025 Walnut St., Room 1014, Philadelphia, Pennsylvania 19107

Michael J. Mann, M.D., Department of Medicine, Harvard Medical School, and Thom 12 – Cardiovascular Research, Brigham and Women's Hospital, 75 Francis Street, Boston, Massachusetts 02115

John D. Mannion, M.D., Division of Cardiology, Thomas Jefferson University, 1025 Walnut St., Suite 410, Philadelphia, Pennsylvania 19107

Michael J. Mastrangelo, M.D., Division of Neoplastic Diseases, Thomas Jefferson University, 1025 Walnut St., Suite 1024, Philadelphia, Pennsylvania 19107

Timothy P. Noonan, Ph.D., Hybridon, Inc., 155 Fortune Blvd., Milford, Massachusetts 01757

Patrice S. Obermiller, M.D., Department of Cell Biology, Vanderbilt University, 21st Avenue and Garland St., Nashville, Tennessee 37232

Anne M. Pilaro, Ph.D., Center for Biologic Evaluations and Research, Food and Drug Administration, 1401 Rockville Pike, HFM-579, Rockville, Maryland 20852

Patricia Puma, Ph.D., Hybridon, Inc., 155 Fortune Blvd., Milford, Massachusetts 01757

Fernando Roqué, Departamento Cardiovascular, Clinica Olivos, Buenos Aires 1636, Argentina

Jack Roth, M.D., Department of Thoracic and Cardiovascular Surgery, Box 109, University of Texas, M. D. Anderson Cancer Center, 1515 Holcombe Boulevard, Houston, Texas 77030

Kevin J. Scanlon, M.D., Berlex Biosciences, 15049 San Pablo Ave., Richmond, California 94804

Yi Shi, Ph.D., Division of Cardiology, Thomas Jefferson University, 1025 Walnut St., Suite 410, Philadelphia, Pennsylvania 19107

Eric J. Small, M.D., Department of Medicine and Mt. Zion Cancer Center, University of California, San Francisco, California 94143

Jason G. Smith, B.S., Institute for Human Gene Therapy, University of Pennsylvania School of Medicine, 409 Stellar-Chance Laboratories, 422 Curie Boulevard, Philadelphia, Pennsylvania 19104

Deborah Lee Sokol, Ph.D., Department of Pathology and Laboratory Medicine, University of Pennsylvania School of Medicine, 515 Stellar-Chance Laboratories, 422 Curie Boulevard, Philadelphia, Pennsylvania 19104

Alison T. Stopeck, M.D., Arizona Cancer Center, P.O. Box 245024, 1515 N. Campbell Ave., Tucson, Arizona 85724-5024

Bruce G. Sturgeon, B.S., Hybridon, Inc., 155 Fortune Blvd., Milford, Massachusetts 01757

Jin-yan Tang, Ph.D., Hybridon, Inc., 155 Fortune Blvd., Milford, Massachusetts 01757

Heiko E. von der Leyen, M.D., Abteilung Kardiologie, Medizinische Hochschule, 30625 Hannover, Germany

Bing H. Wang, Ph.D., Hybridon, Inc., 155 Fortune Blvd., Milford, Massachusetts 01757

Andrew Webb, M.D., Lymphoma Unit, Royal Marsden Hospital, Downs Road, Sutton, Surrey SM2 5PT, UK

Eric Wickstrom, Ph.D., Department of Microbiology and Immunology, Thomas Jefferson University, 1025 Walnut St., Suite 420, Philadelphia, Pennsylvania 19107

Andrew Zalewski, M.D., Division of Cardiology, Thomas Jefferson University, 1025 Walnut St., Suite 410, Philadelphia, Pennsylvania 19107

1

A Brief History of Genetic Therapy: Gene Therapy, Antisense Technology, and Genomics

James W. Hawkins

Antisense and Nucleic Acid Drug Development,
Clarksburg, Maryland

I. INTRODUCTION

At first glance, the biology of life appears to offer a complexity beyond comprehension. Hundreds of thousands of different organisms present themselves in diverse physical forms and pursue countless strategies for survival and reproduction. Since its inception, experimental science has sought to devise models that unify our understanding of biology. In turn, modern medicine and therapeutics have increasingly relied on such unifying systems in order to develop technologies to effectively address human disease.

In this century, our appreciation of biology has progressively shifted from one based upon biological structure to one focused on underlying biological function. Anatomy and histology have given way to physiology and cell biology which have spawned biochemistry, leading to molecular biology and, finally, molecular genetics. Although the reader may take pause with such a sweeping conceptualization of the recent history of biology, the direction is unmistakable--the focus of biological thought has made the transition from phenotype to genotype. We now stand in an era where molecular genetics forms the dominant model for biology.

In response to this paradigm shift, our efforts to develop new therapeutics have started to change from those associated with traditional drugs, which address the symptoms of disease, to new genetically based drugs which address the causes of disease. As we proceed into the 21st century, the march toward a comprehensive functional model--a genetic model--of biology will only pick up pace. Likewise, the drive to develop new therapies--genetic therapies--must be expected to remain in step.

The first two branches of genetic therapy to emerge within the new paradigm of biology are gene therapy and antisense technology. This article seeks to briefly introduce these two disciplines to the level necessary to make them understandable and give them a historical context. It further invites the reader to put these technologies into a new integrated frame of genetic reference--the frame of reference of the emerging field of genomics.

In addressing the task of offering a historical treatment of genetic therapy, or perhaps more broadly, genetically targeted research and development, the author has not attempted to set down a rigid historical sequence of events or to attribute the birth of ideas or materials to individuals or groups. Rather, events, concepts, key papers and technologies which have converged to identify a distinct approach to a branch of genetic therapy have been noted. Generally, genetic therapy technologies are treated in detail up to historical events around 1990, a point at which most disciplines had addressed proof of principle with cellular and *in vivo* experimentation. The discussion of the field after 1990 is presented with more emphasis on emerging clinical development, and the reader is pointed to review articles for the intimate details and progress of basic research in each discipline. In short, this chapter is designed more to offer the beginnings of historical threads to entice the inquisitive reader into any or all of the new fields contributing to genetic therapy rather than to offer historical closure to ideas or career contributions.

This article also seeks to describe the potential interrelationships among gene therapy, antisense technology, and the evolving foundation of molecular genetics on which the future progress and proposed unification of these fields is ultimately based.

II. CLASSICAL GENETICS AND THE HISTORICAL FOUNDATION OF GENETIC THERAPY

II.A. Classical Genetics: Genotype from Phenotype

Many people believe that genetic therapy--the notion of controlling biology or treating disease through genetic mechanisms--is a modern invention. It is not. Early societies recognized genetic forces at work in populations and moved to promulgate social forces against incest; these prohibitions implicitly sought to control deleterious genetic traits resulting from human crosses which could result in nonadaptive homozygous recessive offspring.

Other early efforts at genetic manipulation were empirical, employing phenotype, the physical manifestation of genetic forces, to predict the outcome of genetics. Although crude by today's standards, an understanding of genetics through phenotype proved simple and powerful. The selective breeding of animals and crops, which has taken place worldwide over centuries, is a testament to the potency of the phenotypic approach to genetics.

With the rise of modern science, the practice of genetics moved from an empirical to an experimental discipline. The scientific study of phenotype led to Mendel's laws of heredity which were published in 1866 (Mendel, 1866). However, the field of science did not place emphasis on Mendel's work at the time. While perhaps bewildering to us, the events of this period demonstrate an intellectual struggle to rationalize other established and competing perspectives on biology, none of which were satisfactory, to form a unifying explanation of genetic phenomena (Mayer, 1982).

Beginning in the 1870s, however, technical advances including improved microscopes and new aniline dyes allowed the detailed observation of the cell nucleus; in 1879 Flemming described heavily staining cellular figures, *chromatin*, which were subsequently termed *chromosomes* to describe the same structures we know today. In 1883, Weismann concluded that nuclear material was the substance of heredity; in 1889 (Weismann, 1889) he wrote that his theory was,

> founded upon the idea that heredity is brought about by the transmission from one generation to another of a substance with a definite chemical and, above all, molecular constitution.

The moment was ripe. The abstract Mendelian notion of an underlying pattern to genetic forces and the concept that the discrete physical material on chromosomes could mediate these forces were united. By 1904, the Sutton-Boveri chromosomal theory of inheritance provided the proof that some chromosomes are individually recognizable during mitosis and meiosis and that chromosomes with the same characteristics occur again and again at each cell division. With the introduction of new genetic experimental systems such as *Drosophila*, in 1909, Morgan and associates provided further convincing proof for the association of Mendelian laws and physical properties of chromosomes. Moreover, within the next few years, all the major aspects of transmission genetics, such as the independent assortment of genes, had been elucidated and the new field of cytogenetics was under way.

The challenge of bringing a direct and detailed physical interpretation to classical genetic phenomena occupied scientists throughout the first half of the twentieth century. The term *genetics* was first proposed by Bateson (1906). Before the existence of a physical definition of genetic material, Johannsen (1909) originated the term *gene* [a shortened version of the term *pangen* employed by de Vries (1889)] to refer to an elemental unit of heredity.

From the earliest points in the development of classical genetics, physical traits conforming to the laws of genetic inheritance were observed in association with specific disease states. The inference was made that the principles of genetics may be linked in a physical manner to disease. For example, Garrod (1902) studied the incidence of the condition alkaptonuria, and later discussed this disease as an "inborn error of metabolism" (Garrod, 1909), thus linking a genetic inheritance pattern with the occurrence of a disease state. With this new method of disease classification, a list of genetically associated diseases began to grow. Indeed, through the extension of classical genetic studies and an expansion of associated technologies, 6,678 entries describing human genetic disorders were made by 1994 (McKusick, 1994). We refer to these diseases collectively as *hereditary* diseases. Indeed, hereditary diseases, such as cystic fibrosis and hemophilia, were among the first proposed targets for gene therapy.

By 1940, Beadle and Tatum had demonstrated that some genes were linked directly with the production of a unique protein product, and the concept of "one gene-one protein" emerged. The unification of classical genetics and modern biochemical sciences took place when Avery *et al.* (1944) demonstrated genetic material to be composed of DNA. The stage was set for the birth of modern molecular genetics.

Notwithstanding the formal study and successful manipulation of genetics in nature, the actual genetic structure responsible for dictating the phenotype of organisms remained unknown. In other words, classical genetics was--and is--an indirect study of genetic forces.

II.B. Molecular Genetics: Genotype From Genes

Three major events which have taken place in the latter half of the 20th century have led to a new understanding of genetics. The first event was the discovery of the structure of DNA; the second was the advent of molecular biology; and the third is the recent emergence of a new branch of molecular genetics termed, *genomics*. Collectively, these three events have moved our perspective of genetics from that of an indirect science to a direct science--from the study of phenotype to the direct study of genotype. These events have also made possible a direct approach to genetic therapy.

The first event initiating a major change in our perspective of genetics took place in 1952, with the discovery of the basic structure of DNA (Watson and Crick, 1953); this event precipitated the elucidation of the genetic code during the 1950s and 1960s (Judson, 1979). For the first time it became possible to conceive of a gene in physical terms. A simple four-letter code, used by all living cells and viruses could give rise to the production of every protein found in nature; the basic genetics of all organisms were unified.

However, despite the conceptual leap relating to the universal chemical makeup of DNA and its direct relationship to protein production, as well as other tantalizing work in microbial genetics that pointed the way to gene identity and the organization of genes in simple organisms, it was still not possible to isolate and characterize any gene at the DNA level. Further, the sheer size of most eukaryotic genomes, orders of magnitude larger than simple microbial genomes, seemed to place the promise of this new genetics (the promise of a direct physical study of genes) at an unreachable distance from experimental science. In this context, genetic therapy, too, seemed only theoretical.

The direct study of genetics through genes remained a remote hope until the early 1970s when a profusion of technologies, collectively termed *molecular biology,* made the manipulation of any gene for which there was a characterized protein a relatively straightforward exercise. Genes could be isolated, mapped and their exact DNA sequences determined (Maniatis *et al.,* 1976). This new approach to the direct study of genetics came into sharp focus around individual genes as hundreds of scientists became consumed with structural analysis and the list of new genes grew exponentially from less than 100 in the early 1980s to thousands by the 1990s.

II.C. Genomics: From Genotype to Genome

In the late 1980s the proposal was advanced to sequence all human genes. This effort, termed the Human Genome Project (U.S. Department of Health and Human Services and Department of Energy, 1990), represented a quantum leap in the size and scope of gene discovery. In addition, for the first time, gene discovery began to be contemplated in the context of an entire organism. Concomitantly, the perception of gene discovery began to move away from the early notion of developing a simple catalog of gene sequences--each defining a solitary function-- toward the concept of adding gene sequence information to enrich a vast complement of genes.

This new holistic attitude toward gene discovery encouraged further thinking about the relationship of gene structure to gene function. Clearer links could be conceptualized between the action of multiple genes and the maintenance of basic biological function at cellular and tissue levels; a new approach to examining dynamic biological functions such as growth, cellular differentiation, and organ development could also be envisioned. Involved pathologies such as cancer could be interpreted through the dysfunction of a set of interrelated gene sequences. Even new routes to therapeutic development could be suggested. In short, when one contemplated the DNA sequences of all the genes of an organism together, the whole was now greater than the sum of the parts.

In the face of these new attitudes and perceptions, scientists groped for a term to fuse the descriptive elements of primary gene sequence information with the complex dynamics of the biological expression of this information. The term genomics emerged to express the integrated study of the DNA structure of a genome and the multi-tiered, gene- directed functions of a genome.

III. THE TECHNOLOGIES OF GENETIC THERAPY

There are three basic technologies which form the foundation of modern genetic therapy: gene therapy, antisense technology, and genomics. Each of these technologies has evolved from a different branch of science, giving each technology a unique path toward therapeutic development. However, since the inception of each, there has been a steadily increasing interrelationship. The key historic events which have taken place to form and connect each of these technologies is considered below.

III.A. Gene Therapy

The impetus, if not the format, for modern gene therapy is found in molecular biology with the isolation of the first eukaryotic genes in the 1970s. The dramatic entrance of eukaryotic genes onto the experimental stage prompted many scientists to supply new details to older conceptual experimental designs for the transfer of "healthy" genes into bodies of individuals with "defective" genes to effect the cures of a host of well-characterized hereditary genetic diseases.

Despite an awareness of the possibility of gene therapy, however, several obstacles had to be overcome in inventive or new ways before the first clinical trial in gene therapy could actually be performed in 1990. The major obstacles confronting gene therapy included: 1) the need to develop a method to transfer and stably express human genes *in vivo*, 2) the need to provide an institutional structure for public acceptance and professional review of emerging clinical protocols, and 3) lack of financial support to accelerate the growth of research and clinical development once a technical and institutional platform for gene therapy had been established. Historical treatments of the development of gene therapy may be found in several reviews (Friedman, 1992; Friedman, 1993; Wolff *et al.,* 1994).

III.A.1. Transfer of Genetic Material: A Phenomenon of Nature

The capacity for viruses to transmit genes was first demonstrated in *Salmonella* (Zinder *et al.,* 1952). The concept that viral genomes could become integrated into cellular genomes followed from elegant work employing bacteria and bacteriophages (Lederberg, 1956). During the mid- to late-1960s a number of investigators began to work with DNA tumor viruses and observed that the DNA of these viruses could infect a cell and become stably integrated into the cellular DNA and subsequently be expressed (Sambrook *et al.,* 1968; Temin, 1976). These observations led to speculation that viruses might be used as "Trojan horses" to transport therapeutic genes into diseased cells to alleviate disease symptoms. Today's approach to gene therapy, which uses viruses or plasmid vectors as shuttles for the introduction of therapeutic genes into host cells, derives from these early studies with DNA viruses.

Unfortunately, there were no methods available at that time to test these concepts. Indeed, the actual structure of human genes was not understood, and isolation of a gene had never been attempted.

III.A.2. Recombinant DNA Technology and the Facile Manipulation of Human Genes

As mentioned, the 1970s ushered in a molecular biology. This new discipline rapidly contributed a set of interrelated protocols and reagents which together allowed for the isolation and characterization of any gene for which a protein counterpart could be identified. Once isolated, these genes could be recombined with DNA from any other biological source. The profound utility of this new recombinant DNA approach was first demonstrated in 1972 when Berg and colleagues introduced recombined bacterial genes into mammalian cells using a DNA virus vector (Jackson *et al.*, 1972). In short order, genes were cloned which had relevance to human disease. In 1981, an experiment was designed to introduce a gene for hypoxanthine guanine phosphoribosyl transferase (HPRT) into cell cultures derived from patients with Lesch-Nyhan syndrome, a hereditary disease characterized by a deficiency of HPRT caused by a defective HPRT gene. The introduction of the recombinant normal gene and the subsequent expression of HPRT corrected the original metabolic defect of the cell (Mulligan *et al.*, 1981).

Despite the proof of principle demonstrated in cellular systems, technical obstacles prevented the development of techniques for testing gene transfer in a clinical setting. Most important, DNA-based vectors were very inefficient in transferring genetic information into cellular targets.

III.A.3. The Cline Experiment

By the late 1970s, with the widening application of molecular biology, a growing number of human genes were becoming available. Techniques for cloning these genes into vectors were becoming refined and, though inefficient, techniques for transferring genetic constructs into cells had become reproducible. There was a widespread feeling that it would only be a matter of time until the first genes were transferred to humans to achieve therapeutic effects.

Seizing the opportunity of the moment, Cline, a hematologist at the University of California, Los Angeles, attempted to transfer a simple vector containing the human β-globin gene into the isolated bone marrow cells of two patients with thalassemia, a blood-cell disease characterized by defective globin genes. These treated cells were introduced into the patients with the hope that the cells with the healthy globin gene would colonize the patients' bodies, allowing on-going production of enough non-defective globin to permanently cure the disease. This result was not obtained and the experiment came under intense criticism from technical, administrative and ethical perspectives (Wade, 1980).

In retrospect, it is understandable that the Cline experiment did not yield a therapeutic result and that it was severely criticized. From a scientific perspective, the methods for transfer of genetic material were very inefficient and themselves could have accounted for lack of success. From an administrative and ethical standpoint, it is apparent that institutional resources had not been organized in advance to provide a suitable framework for human experimentation.

The Cline experiment did, however, contribute indirectly to the development of the field of gene therapy. This event sparked a debate which led to the first critical discussions of the scientific and public-policy aspects of gene therapy. As a result of these debates, a gene therapy subcommittee was added to the newly established Recombinant DNA Advisory Committee (RAC) of the National Institutes of Health (NIH). From the late 1980s, the RAC has played an instrumental role in nurturing the clinical science and promulgating the public policy associated with almost all gene therapy protocols.

III.A.4. The First Useful Clinical Tools and the Institutionalization of Gene Therapy

During the early 1980s as new molecular biology tools became available, the potential for the recruitment and application to gene therapy of related but independent technologies became apparent. For example, it was demonstrated in the early 1980s that retroviruses could be engineered to carry foreign genetic information and transfer this information with high efficiency into mammalian cells (Wei *et al.*, 1981). By the mid-1980s, techniques became available to produce large quantities of retroviral vectors (Mann *et al.*, 1983).

Also by the mid-1980s, success had been achieved in applying improved cell transfer technology in model disease systems including the HPRT (Miller *et al.*, 1983) and adenosine deaminase (ADA)-deficiency severe combined immunodeficiency disease (SCID) models (Kantoff *et al.*, 1986). This experimental foundation served as a final proof of principle for many scientists interested in pursuing gene therapy as a specialized field.

In 1990, a "Points to Consider" document was prepared by the RAC, in close cooperation with the Food and Drug Administration (FDA) Center for Biologics Evaluation and Research (Food and Drug Administration, 1990). This document provided guidelines and an institutional oversight structure for those interested in conducting gene therapy in humans.

With this guidance, *ex vivo* treatment of ADA-deficiency SCID was then advanced as a clinical protocol. This disease is caused by a defective gene for ADA. Through the introduction of a retrovirus containing the ADA gene, researchers were able to restore ADA expression in cultured human blood cells and abolish, by cell therapy, the cell toxicity caused by ADA deficiency. Using an experimental design carefully tested in animals and approved by the RAC and the FDA, the first gene therapy protocol was carried out in a human at the NIH in Bethesda, Maryland, by Anderson and coworkers (Blaese *et al.*, 1990, and this volume, Chapter 6). In this first protocol, the ADA gene was introduced into a patient with SCID; subsequent follow-up indicated that a lasting therapeutic effect had been achieved with this treatment. The era of gene therapy had begun.

Table 1. Annual Number of Gene Therapy Literature Citations

Year	1985	86	87	88	89	90	91	92	93	94	95
Citations	0	0	4	12	43	118	106	172	285	448	572

Number of literature citations per year, generated by a computer search of the controlled vocabulary term "gene therapy" using the Grateful Med™, Medline® scientific literature database.

III.A.5. Growth of Gene Therapy Research

Progress of the field of gene therapy through the mid-1990s can be measured indirectly by the increase in publication in the area since the mid-1980s, illustrated in Table 1. Moreover, a scientific literature devoted exclusively to gene therapy has emerged with the publication of the new journals *Human Gene Therapy* (1989), *Cancer Gene Therapy* (1994) and *Gene Therapy* (1994). Research in the field has been reviewed periodically (Anderson, 1984; Friedman, 1989; Miller, 1992; Mulligan, 1993) for the interested reader.

III.B. Antisense Technology

The second basic technology of modern genetic therapy is antisense technology. This article will adopt the term *antisense* in a catholic application to refer collectively to four distinct disciplines. These include: classical antisense or anticode; ribozyme or catalytic RNA; triplex, triple helix or antigene; and aptamer approaches. It should be noted that a unified terminology has not been adopted to describe any or all of these technologies.

Antisense technologies are related by the fact that the active biomolecule for each is a polymeric nucleic acid. Except in the case of some aptamer applications, each of these technologies also focuses on a specific hybridization interaction with another unique nucleic acid target to achieve activity and selectivity in biological action. Sequence specificity is also at the heart of the therapeutic promise of antisense compounds; antisense addresses itself to the genotype of an organism and can target a specific gene or gene transcription product to achieve selective therapeutic effects not possible through drugs which target protein phenotype.

It is not the purpose of this article to review the technical subtleties of each antisense technology. In this regard, the reader is directed to several basic review articles of varying levels of sophistication which highlight principles and fundamental research aspects of classical antisense (Weintraub, 1990; Zon, 1995; Gibson, 1994; Putnam, 1996;

Table 2. A Simplified Comparison of Different Antisense Technologies Indicating Their Most Common Form of Application

	Classical Antisense	Ribozyme or Catalytic RNA	Triplex or Antigene	Aptamer
Molecular Target	RNA	RNA	DNA	protein
Molecule of action	DNA or RNA	RNA	DNA	DNA or RNA
Site of action	cytoplasm	cytoplasm	nucleus	cytoplasm and/or nucleus or extracellular
Mechanism of action	duplex formation: translation arrest and/or accompanied destruction of RNA by RNase H	duplex formation: translation arrest and/or direct destruction of RNA	triplex formation: block of transcription	protein binding: interference with protein function

Adapted from Stull and Szoka (1995).

Whitton, 1994; Scanlon *et al.*, 1995), ribozyme (Gibson, 1994; Putnam, 1996; Stull *et al.*, 1995; Whitton, 1994; Scanlon *et al.*, 1995; Bartolomé *et al.*, 1995; Castanotto *et al.*, 1994), triplex (Weintraub, 1990; Gibson, 1994; Putnam, 1996; Stull *et al.*, 1995; Scanlon *et al.*, 1995; Maher III, 1996), and aptamer (Stull *et al.*, 1995) technologies; reviews which expand on these topics and emphasize the therapeutic development of antisense technologies are provided below. A succession of books which have been published since the late 1980s provide a much more detailed coverage of all aspects of the antisense field (Melton, 1988; Cohen, 1989b; Wickstrom, 1991; Mol *et al.*, 1991; Baserga *et al.*, 1992; Murray, 1992; Erickson *et al.*, 1992; Crooke, *et al.*, 1993; Meyer, 1995).

A brief historical sketch of the development of each antisense technology is presented below. Reference will be made to basic features of each technology, progress leading to cellular applications, and progress toward proof of principle. Table 2 presents a simplified but useful chart comparing these different technologies and their basic modes of action.

III.B.1. Synthetic DNA: A Preface to Advances in All Antisense Technologies

As it may be argued that molecular biologists provided the impetus for modern gene therapy, so it can be argued that organic chemists have enabled modern antisense technology. Although the initial observations which led to the conceptualization of classical antisense, ribozyme and triplex technologies have their origins in biological studies, the differentiation of these areas into distinct fields has depended significantly on one element--synthetic DNA. It is appropriate to digress for a moment to consider the advent of synthetic DNA and its application to antisense technology.

The total synthesis of an oligonucleotide ("oligo" from the Greek for "a few") posed a bewildering array of problems to the first chemists who addressed the possibility. In large part, the strategy for the synthesis of DNA--selective protection and deprotection of functional groups--is similar to that for other natural products. However, the magnitude of the task of synthesizing a macromolecule was very great in scope for its day.

A number of reviews have been written on the development of DNA synthesis techniques (Khorana, 1979; Crockett, 1983; Itakura *et al.*, 1984). The steady refinement of appropriate chemistries and improved methods to assemble bases into oligonucleotides eventually culminated in the full automation of DNA synthesis (Letsinger *et al.*, 1976; Caruthers, 1985). With automation, DNA synthesis moved from an arcane science practiced by a few skilled chemists to a technology widely available to molecular biologists. Moreover, by the mid-1980s, DNA synthesizers were widely available which could routinely produce single-stranded oligonucleotides up to approximately 50 residues, technically sufficient for any small scale antisense experiment.

In addition to DNA synthesis itself, it should also be noted that chemists became interested at an early date in the functional interactions between chemically modified polymeric nucleic acids. For example, Belikova *et al.* (1967) suggested:

> In the case of nudeic acids it seems promising for the purpose of obtaining oligonucleotide residue capable of specific base pairing with the compounds, containing a reactive group bound to [an] complementary nudeotide sequence.

III.B.2. Classical Antisense Technology: Phase I

Before the advent of synthetic oligonucleotides, the phenomenon which underpins most classical antisense activity--the disruption of translation through the hybridization of a complementary oligonucleotide to a specific mRNA molecule--had been experimentally observed. Spiegelman *et al.* (1972), employing a phage lambda system, noted that a gene which could be bidirectionally transcribed could give rise to different gene expression patterns depending on the pattern of transcription. In the paper he noted:

To measure antisense RNA, an assay based on the formation of nuclease-resistant, double-stranded RNA, specific to the *cro* region, was developed. These results raise the possibility that bidirectional transcription of *cro* has a regulatory function in phage lambda.

With the introduction of synthetic oligonucleotides, the first experiments were initiated to directly block translation. Taniguchi and Weissmann (1978) used an oligonucleotide specific for *Escherichia coli* ribosomal RNA to block translation. In the same year, Stephenson and Zamecnik (1978) employed a synthetic oligonucleotide to disrupt protein synthesis in Rous sarcoma virus 35S RNA in a cell-free system, and to prevent viral replication in mammalian cells (Zamecnik and Stephenson, 1978). Zamecnik went on to speculate on the therapeutic use of antisense oligonucleotides:

It might be possible to inhibit the translation of specific cell proteins, if that were desirable. The inhibition of globin synthesis might, for example, be useful in polycythemia vera.

Another technical approach to antisense appeared with new cloning technologies. This technique permitted the construction of vectors which could express antisense transcript sequences in cell systems. Inouye *et al.* (Coleman *et al.*, 1984), and Izant and Weintraub (1984), demonstrated the utility of these systems to suppress the expression of genes in prokaryotic and eukaryotic systems respectively.

III.B.3. Classical Antisense Technology: Phase II

With the wider availability of high quality synthetic DNA in the mid-1980s, it was possible to conduct an increasing number of experiments to determine if oligonucleotides could be used to selectively block the translation of a homologous mRNA transcript in cell systems. These efforts were met with mixed results, however.

In some systems consistent antisense effects could be achieved with conventional DNA oligonucleotides. For example, Wickstrom *et al.* (1988) were able to inhibit c-*myc* proto-oncogene expression in human promyelocytic HL-60 cells. Alternatively, in other systems many early studies which demonstrated an antisense effect by a particular oligonucleotide in an initial experiment were not later reproducible in similar subsequent experiments. This phenomenon was gradually understood to be due to differing experimental systems and particularly the sensitivity of regular phosphodiester oligonucleotides to nuclease digestion in biological systems (Wickstrom, 1986) in general. Due to reproducibly irreproducible results such as these, antisense technology was held in suspicion in some quarters and the publication of antisense experiments was somewhat problematic during the mid to late 1980s. This situation was resolved, however, by the second major contribution of chemists to antisense technology--the development of modified oligonucleotides.

One of the most important events in the growth of antisense technology was the development of new chemical formulations of synthetic DNA, specifically backbone modifications (Miller *et al.*, 1979; Stec *et al.*, 1984; Eckstein, 1986; Zon, 1987; Agrawal *et al.*, 1987; Iyer *et al.*, 1990). Several laboratories discovered that by substituting various molecules for one or more of the nonbridging oxygen atoms present at the phosphorus atoms in the internucleotide linkages of an oligonucleotide, the new compound would become more resistant to degradation in cell systems and *in vivo*. These modifications also preserved important hybridization characteristics of oligonucleotides. The most commonly used oligonucleotide modification today is the backbone substitution of a single sulfur atom for a nonbridging oxygen atom; this compound is termed a phosphorothioate oligonucleotide.

Success in the modification of oligonucleotide backbone structure led to efforts to develop a host of new chemical formulations of oligonucleotides--efforts which continue today. These modifications range from different extents of backbone modification to complete redesign of internucleotide linkages, sugar structure, sidechain chemistry and to oligonucleotides conjugated with other bioactive molecules (Hélène, 1989; Letsinger *et al.*, 1989; Perbost *et al.*, 1989; Rayner *et al.*, 1989; Goodchild, 1990). In addition to new developments in DNA synthesis, techniques were also developed to perform automated RNA synthesis (Pfister *et al.*, 1989).

Notwithstanding a certain level of unexplained phenomena in biological experiments, as new oligonucleotide chemistries and vector systems came to bear on the field, by 1990, antisense effects were documented in a number of diverse systems including animal cells (Holt *et al.*, 1986; Gewirtz *et al.*, 1988; Mercola *et al.*, 1988), virally infected cells (Matsukura *et al.*, 1987; Krieg *et al.*, 1989), and plants (Smith *et al.*, 1988; Hiatt *et al.*, 1989; Van der Krol *et al.*, 1989).

In connection with classical antisense technology, it is important to note that a significant literature exists which documents the existence of endogenous systems for antisense RNA control of gene expression in bacteria, bacteriophages, and plasmids (Wagner *et al.*, 1994; Eguchi *et al.*, 1991; Simons, 1993) as well as in eukaryotes (Simons, 1993; Austin *et al.*, 1994; Nellen *et al.*, 1993).

III. B. 4. Catalytic RNAs: Ribozymes

During the early 1980s, independent efforts were under way to understand a question which had arisen due to the discovery of intervening sequences (IVS), non-coding sequences present within a protein coding region of primary RNA transcripts in eukaryotic genes. Specifically, scientists wanted to understand how intervening sequences or *introns*, were removed to produce a mature mRNA lacking these extra RNA sequences. The answer turned out to be quite apocalyptic--the sequences were removed through an RNA-mediated catalytic event.

Catalytic RNAs are RNA structures capable of cleaving covalent bonds in an RNA molecule. Catalytic RNAs were first described by Cech's laboratory (Zaug *et al.*, 1980; Cech *et al.*, 1981; Kruger *et al.*, 1982) and Altman *et al.* (Guerrier-Takada *et al.*, 1983). Cech described the catalytic removal of introns in the pre-rRNA of the ciliated protozoan, *Tetrahymena thermophila*. The term "ribozyme" describes RNA (ribonucleic acid) molecules with catalytic properties analogous to enzymes, as explained by Kruger *et al.* (1982),

> Because the IVS RNA is not an enzyme but has some enzyme-like characteristics, we call it a ribozyme, an RNA molecule that has the intrinsic ability to break and form covalent bonds.

Altman made similar and independent observations employing the self-cleaving pre-tRNA of RNase P. Since first discovered, ribozymes have been identified in a variety of organisms including plant viroids and virusoids (Forster and Symons, 1987a,b); bacteria (Xu *et al.*, 1990); the animal virus, hepatitis delta virus (Wu *et al.*, 1989); and newt (Epstein *et al.*, 1987).

There are an assortment of different natural ribozyme structures which are capable of intramolecular, or *cis*, cleavage. Many of these RNA structures can be modified to act upon a second RNA molecule or act "in *trans*" (Uhlenbeck, 1987). The most common trans-acting ribozymes studied are the "hammerhead" and "hairpin" ribozymes, so named for the two dimensional shapes into which they can be drawn.

Haseloff and Gerlach (1987) published general design parameters for a trans-acting hammerhead ribozyme and suggested the use of ribozymes as potential therapeutic agents. These *in vitro* studies were followed in the late 1980s by the earliest tests of ribozymes in cell systems (Cameron *et al.*, 1989).

A great deal of effort has been directed at understanding and modifying the stability, specificity and catalytic activity of trans-acting ribozymes by altering their nucleotide sequence, length and chemistry

(Cech, 1992; Symons, 1992). The best characterized ribozyme is the hammerhead ribozyme; it is composed of two separate functional regions: a conserved *catalytic core* which cleaves the target RNA, and *flanking regions*, which are complementary to the target sequence and position the catalytic core for target cleavage. It is important to note that certain target sequences are not susceptible to ribozyme targeting because of sequence incompatibility; for hammerhead ribozymes, cleavage sites in most RNAs occur with a random frequency of 3/16 (Scanlon *et al.*, 1995).

As work has continued, ribozymes have been synthesized in a variety of forms including molecules with mixed DNA and RNA structures. Due to the fact that ribozymes are larger and more complex in structure than oligonucleotides used in classical antisense experiments, they present unique cellular delivery challenges. For this reason, considerable efforts have been made to produce ribozymes intracellularly through the use of plasmid and retroviral vectors.

III.B.5. Triplex or Antigene Technologies

Classical antisense and ribozyme technologies both affect gene expression primarily at the level of RNA translation. A third technology, triplex, triple helix, or antigene technology, targets gene expression at the level of transcription. Triplex technology employs oligonucleotides to bind to a specific sequence within the major groove of duplex DNA; the resulting triplex nucleic acid structure is presumed to interfere with the transcription of the particular gene locus to which it has been specifically targeted.

Interest in the nucleic acids following the initial description of DNA duplex structure in the 1950s led to the study of a variety of RNA, DNA and RNA/DNA structures. A description of the first triple helix structure was provided by Felsenfeld and Rich (1957). In the 1960s, before the isolation of genes and the discovery of sequence-specific binding proteins, proposals were advanced that triple helix formation by sequence specific RNA molecules might provide a mechanism for gene regulation (Miller *et al.*, 1966; Britten *et al.*, 1969). Indeed, evidence for inhibition of *in vitro* translation by triplex formation was first provided by Morgan (1968).

With the advent of synthetic oligonucleotides, triple helix structures could be more readily studied. Oligonucleotide-directed triple helix formation was reported independently by Dervan's group (Moser *et al.*, 1987) and Hélène's group (Le Doan *et al.*, 1987). These and subsequent studies have proposed various motifs or recognition patterns for the hybridization of different base combinations in triplex structures (Cooney *et al.*, 1988).

There has been considerable experimental evidence in cell-free systems that oligonucleotides bound to duplex DNA can inhibit protein-DNA interactions; the earliest of these studies inhibited the binding of restriction endonucleases (François *et al.*, 1989) and transcription factors (Maher III *et al.*, 1989). Biological responses have also been suggested in some cell systems, the earliest of which targeted viral replication (Birg *et al.*, 1990) and eukaryotic gene expression (Postel *et al.*, 1991).

A great deal of rigorous work is taking place in the development of triplex technology. However, a number of basic scientific issues must be addressed before any type of clinical effort can get under way (Maher III, 1996). For example, due to the limited number of base sequence matches for triplex formation currently available, it is necessary to have a nearly uninterrupted homopurine-homopyrimidine sequence in duplex DNA to carry out studies; such sequences of at least 10 base pairs may occur once in every 250 bases in a eukaryotic genome but this frequency is too low for the reliable targeting of most gene promoters. In addition to target-sequence restrictions, few studies have addressed the question of the accessibility of oligonucleotides to the nucleoprotein complexes within which target sequences exist. Moreover, a host of more technical issues which relate to the stability and specificity of triplex formation must be addressed before the challenge of drug development can be realistically embarked upon. A number of reviews relating to triplex technology have been written (Hélène, 1991; Maher III *et al.*, 1991; Gee *et al.*, 1992).

III.B.6. Aptamer Technology

The use of oligonucleotides for experimentation has uncovered some interesting findings related to the interaction of short nucleic acid sequences with other cellular constituents involved in gene regulation or other types of biological function. Ellington *et al.* (1990), while describing an *in vitro* RNA selection technique for binding specific ligands, noted "We have termed these individual RNA sequences 'aptamers,' from the Latin 'aptus', to fit."

One model for aptamer use envisions the oligonucleotide as a sink for protein factors which mediate the expression of cellular genes; by controlling the relative availability of these proteins, it should be possible to modulate gene expression. One of the first demonstrations of an aptamer effect involved the modulation of IL-2 gene expression in B-cell and T-cell lines with double stranded DNA molecules (Bielinska *et al.*, 1990). Another aptamer approach involves the intracellular production of RNA transcripts which act as "decoy" molecules. This method has been used to interrupt HIV viral replication (Sullenger *et al.*, 1990). Aptamers have also been employed to bind proteins with functions unrelated to gene expression. For example, thrombin-catalyzed coagulation can be blocked *in vitro*, and this anticoagulation effect has been demonstrated *in vivo* by the use of a 15-base DNA aptamer (Griffin *et al.*, 1993).

III.B.7. Growth of Antisense Research

On the strength of experimentation in the late 1980s which addressed the proof of principle for antisense technology, a period of steady growth in the field has ensued (*Antisense Research and Development*, 1991) as demonstrated by Table 3. Moreover, a new journal devoted exclusively to antisense technology, *Antisense Research and Development* (now *Antisense & Nucleic Acid Drug Development*), was initiated in 1991.

IV. THERAPEUTIC DEVELOPMENT OF GENETIC THERAPY

IV.A. Therapeutic Development of Gene Therapy

There has been a steady increase in clinical development of gene therapy since 1988. This growth is reflected in the number of clinical protocols, as shown in Table 4, which have been approved by the RAC or registered with the RAC for sole FDA review, except in 1996 (at which time the RAC underwent a review and reorganization). Protocols involving vectors are reviewed by the Center for Biologics Evaluation and Research (CBER) within the FDA.

Table 3. Annual Antisense Literature Citations

Year	1985	86	87	88	89	90	91	92	93	94	95
Citations: Medline®	0	0	0	0	2	47	166	239	243	326	368
Citations: Antisense	28	53	80	133	209	115	-	-	-	-	-

Literature citations were generated by a computer search of the controlled vocabulary terms "oligonucleotides, antisense" and "RNA, antisense" using the Grateful Med™, Medline® scientific literature database. Literature citations were also compiled from a comprehensive bibliography search published in *Antisense Research and Development* (1991). The second search is more sensitive since it was generated through retrieval of original papers and cross-referencing of these papers as well as the inclusion of multidisciplinary antisense topics such as oligonucleotide chemistry. Discrepancies are due to the different natures of the searches and to the fact that the term "antisense" was not in widespread use in the literature before the early 1990s.

Table 4. Annual Approved Gene Therapy Clinical Protocols in the US

	1988	89	90	91	92	93	94	95	96
Protocols	1	0	2	9	23	31	32	44	25

Number of clinical protocols approved by the RAC or registered with the RAC for sole FDA review for each year since the beginning of clinical activity in the field. [Source: Office of Recombinant DNA Activities (ORDA) (http://www.nih.gov/od/orda).]

As of December 1996, there are 167 clinical gene transfer protocols in the USA which have achieved various levels of NIH or FDA review or approval (http://www.nih.gov/od/orda). The application of current clinical efforts in therapeutic gene transfer to specific diseases is summarized in Table 5. TMC Development of Paris, France, has estimated that there are now 232 gene therapy protocols worldwide; of these, 175 are reported to have been initiated and include an estimated 1,537 patients (TMC Development Report, 1996; Anderson, 1996).

IV.B. Therapeutic Development of Antisense Technology

The transition of antisense research to therapeutic development has been one which superficially resembles the traditional route of drug development which has attended prospective small molecule drugs for decades. While similar in its general process, the profound difference between conventional drug development and antisense therapeutic development was strikingly articulated (Crooke, 1995):

The technology focuses on chemicals called oligonucleotides which have not been studied as potential drugs before and uses them to intervene in processes that have not been studied as sites at which drugs might act.

As a result of the excitement generated by this new approach to drug development, many investigators have speculated upon or initiated studies to determine the potential of antisense therapeutics. In this regard, the reader is directed to several review articles of varying levels of sophistication which highlight principles and pharmacological considerations involved in of therapeutic development of classical antisense (Miller *et al.*, 1987; Loose-Mitchell, 1988; Miller *et al.*, 1988; Paoletti, 1988; Cohen, 1989a; Le Doan *et al.*, 1989; Tikhonenko, 1989; Zon, 1989; Agrawal, 1992; Crooke, 1992; Neckers *et al.*, 1992; Wickstrom, 1992; Crooke, 1993; Nagel *et al.*, 1993; Whitton, 1994; Agrawal *et al.*, 1995; Crooke, 1995; Sharma *et al.*, 1995; Srinivasan *et al.*, 1995; Zon,

1995; Askari *et al.*, 1996; Tonkinson *et al.*, 1996; Agrawal, 1996), ribozyme (Whitton, 1994; Castanotto *et al.*, 1994; Rossi *et al.*, 1990; Sarver *et al.*, 1990; Christoffersen *et al.*, 1995; Kiehntopf *et al.*, 1995a; Kiehntopf *et al.*, 1995b; Sullenger, 1995), and triplex (Maher III, 1996) therapeutics.

On the merit of early results, the field has progressed rapidly into the clinic. In fact, progress of antisense drug development has been rapid considering the relative novelty of the antisense approach and the fact that even a conventional drug takes an average of 11.8 years to reach market (Hawkins, 1995a).

Table 5. Disease Targets for Therapeutic Gene Transfer Protocols

Category	Disease/Disorder	No. of Protocols	Total Protocols
Cancer	Various Types		87
Viruses	Human Immunodeficiency Virus (HIV)		17
Monogenetic	Alpha-1-Antitrypsin Deficiency	1	
Diseases	Chronic Granulomatous Disease	1	
	Cystic Fibrosis	13	
	Familial Hypercholesterolemia	1	
	Fanconi Anemia	1	
	Gaucher Disease	3	
	Hunter Syndrome	1	
	Ornithine Transcarbamylase	1	
	Purine Nucleoside		
	Phosphorylase Deficiency	1	
	Severe Immunodeficiency Disease		
	(SCID)-Adenosine Deaminase (ADA)	1	
	(SCID)-X-linked	1	
	Total		25
Other	Arterial Restenosis	1	
Diseases	Peripheral Artery Disease	1	
	Rheumatoid Arthritis	1	
	Total		3
	Grand Total		139

The protocols above were approved by the RAC or registered with the RAC for sole FDA review through 1996. In addition to the number of therapeutic gene transfer protocols, there are 28 gene marking protocols making a total of 167 gene transfer protocols. [Source: ORDA (http://www.nih.gov/od/orda).]

Table 6. Clinical Trials Currently Underway for Antisense Oligonucleotides in Association with Various Commercial Entities

Disease Target(s)	Molecular Target	Drug	Developer	Status
Cytomegalovirus (CMV)-induced retinitis in AIDS	CMV viral sequences	ISIS 2922 (Fomivirsen)	Isis	Phase III
Renal transplant rejection, rheumatoid arthritis, psoriasis, Crohn's disease, ulcerative colitis	Intercellular adhesion molecule-1 (ICAM-1)	ISIS 2302	Isis	Phase II
Acquired Immunodeficiency Syndrome (AIDS)	Human Immunodeficiency Virus (HIV) *gag* gene	GEM®91	Hybridon	Phase lb/II
	HIV viral integrase protein	AR177	Aronex	Phase I
Follicular lymphoma	*bcl*-2 gene	G3139	Genta	Phase I/II: U.K.
Cancer	Protein kinase C-alpha (PKC-a)	ISIS 3521	Isis	Phase I
	C-*raf* kinase	ISIS 5132	Isis	Phase I
Systemic CMV and CMV retinitis	CMV viral sequences	GEM®132	Hybridon	Phase I: Europe

As of this writing, there are eight antisense compounds that are currently in clinical development, Table 6. In addition, there have been six compounds which have entered clinical trials but whose development has been discontinued during phase I or phase II trials. Oligonucleotide-based drugs are reviewed by the Center for Drug Evaluation and Research (CDER) of the FDA. All of these drug candidates represent a classical oligonucleotide approach.

V. RESEARCH AND COMMERCIAL DEVELOPMENT OF GENETIC THERAPY

V.A. Research and Commercial Development of Gene Therapy

There are now approximately 30 companies worldwide whose major focus is gene therapy or who have substantial commitments in the area, able 7 (Hawkins, 1995b). These companies were formed in three cohorts.

One group of companies, such as Genzyme Corp., originated in the early 1980s and have applied gene therapy to complement areas of developed expertise. A second cohort of companies, such as Genetic Therapy Inc., were formed in the mid to late 1980s and originated with a focus on an *ex vivo*, viral vector-mediated gene therapy approach. A third group of companies, such as GeneMedicine, Inc. were formed in the 1990s and placed an emphasis on *in vivo*, nonviral vector-mediated gene therapy technologies. It has been estimated that over $2 billion (Anderson, 1996) has been invested by the biotechnology and pharmaceutical industries in the research and development of gene therapy over the past 3 years. In addition, the NIH has made a significant commitment to fund gene therapy; the U.S. National Institutes of Health Budget Office reported that the fiscal 1996 budget for gene therapy was $200.7 million from a total NIH budget of $11.9 billion.

V.B. Research and Commercial Development of Antisense Technology

There are now approximately 15 companies worldwide that focus on antisense technologies, Table 8 (Hawkins, 1995b). In general, each company tends to focus on one of the antisense technologies. For example: Isis Pharmaceuticals Inc., focuses on classical oligonucleotide antisense technology; Ribozyme Pharmaceuticals, Inc. specializes in ribozymes; and Aronex Pharmaceuticals Inc. has a major effort in triplex technology.

Approximately $500 million has been invested in the largest antisense companies, to date, to support the development of antisense technology (Mack, 1996). To put this investment in perspective, it is interesting to note that it takes a total investment of $359 million dollars to bring a conventional drug to market (Hawkins, 1995a). With regard to basic research funding, there were a total of 61 active research grants being funded through the NIH, in fiscal year 1996, whose primary designation of interest was in antisense technology; by comparison, there were 406 primary research grants in gene therapy.

VI. FUTURE PROSPECTS FOR GENETIC THERAPY

A science caught in the rush forward is often preoccupied with concerns of immediate application and less with matters related to its long-term development. Moreover, a near term perspective is likely to lead to a cycle of over-enthusiasm and unrealistic expectation followed by disenchantment and negative bias. This is a cycle familiar to many who

Table 7. Companies With Major Activities in Gene Therapy

Company	Location	Public/Private
Applied Immune Sciences, Inc.	Santa Clara, CA	Subsidiary
ARIAD Pharmaceuticals, Inc.	Cambridge, MA	Public
Avigen, Inc.	Alameda, CA	Public
Bavarian Nordic Research Institute A/S	Denmark	Private
Canji, Inc.	San Diego, CA	Subsidiary
Cell Genesys, Inc.	Foster City, CA	Public
Codon Pharmaceuticals, Inc.	Gaithersburg, MD	Private
GeneMedicine, Inc.	The Woodlands, TX	Public
Genetic Therapy, Inc.	Gaithersburg, MD	Subsidiary
Genetix Pharmaceuticals, Inc.	Rye, NY	Private
Genovo, Inc.	Philadelphia, PA	Private
GenVec, Inc.	Rockville, MD	Private
Genzyme Corp.	Cambridge, MA	Public
Immune Response Corp.	Carlsbad, CA	Public
Immusol, Inc.	San Diego, CA	Private
Ingenex, Inc.	Menlo Park, CA	Private
Introgen Therapeutics, Inc.	Houston, TX	Private
IntroGene BV	The Netherlands	Private
Megabios Corp.	Burlingame, CA	Private
Onyx Pharmaceuticals, Inc.	Richmond, CA	Public
Progenitor, Inc.	Columbus, OH	Private
RGene Therapeutics, Inc.	The Woodlands, TX	Private
Somatix Therapy Corp.	Alameda, CA	Public
SyStemix, Inc.	Palo Alto, CA	Public
Targeted Genetics, Corp.	Seattle, WA	Public
Therexsys Ltd.	England	Private
Transgene S.A.	France	Private
Transkaryotic Therapies, Inc.	Cambridge, MA	Private
Chiron Viagene	San Diego, CA	Subsidiary
Vical, Inc.	San Diego, CA	Public

Adapted from Hawkins (1995b).

Table 8. Companies with Major Activities in Antisense Therapy

Company	Location	Public/Private
Aronex Pharmaceuticals, Inc.	The Woodlands, TX	Public
Antivirals, Inc.	Portland, OR	Private
Chugai Biopharmaceuticals, Inc.	San Diego, CA	Private
Dyad Pharmaceutical Corp.	Columbia, MD	Private
Epoch Pharmaceuticals, Inc.	Bothell, WA	Public
Gene Shears PTY, Ltd.	Australia	Private
Genset S.A.	France	Public
Genta, Inc.	San Diego, CA	Public
Gilead Sciences, Inc.	Foster City, CA	Public
Hybridon, Inc.	Worcester, MA	Public
Innovir Laboratories, Inc.	New York, NY	Public
Isis Pharmaceuticals, Inc.	Carlsbad, CA	Public
Lynx Therapeutics, Inc.	Hayward, CA	Public
NeXstar Pharmaceuticals, Inc.	Boulder, CO	Public
Ribozyme Pharmaceuticals, Inc.	Boulder, CO	Public

Adapted from Hawkins (1995b).

have followed different technologies within the biotechnology industry--
and genetic therapy has not been spared these trials in its coming of age.
In this regard, it is perhaps useful to conclude this historical overview of
genetic therapy with some reflection on the longer-term challenges which
face the field, and the different tools with which the field might overcome
these challenges.

Based on a historical perspective, there are three major ways in
which a scientific field can progress: through improved experimental
design and technique; through the convergence of established and related
fields; and through the adoption of a new scientific context for the field.
All three of these avenues are open to genetic therapy and a few
examples of present initiatives and possible future directions of the field
are discussed below.

VI.A. Refinement of Experimental Design

Classical antisense technology, since its inception, has profited a
great deal from a refinement of experimental design. With the wide
application of oligonucleotides for the first time in a variety of biological

systems, a range of effects have been observed. From these studies it has become clear that oligonucleotides, like all other biomolecules, interact nonspecifically as well as specifically with a variety of different cellular constituents. Thoughtful recommendations for the specific design of basic experiments in antisense technology have been proposed (Wagner, 1994; Stein and Krieg, 1994) which will improve the interpretation of an array of basic research questions in the field.

The diverse observations which have been made in a basic research vein have raised issues in some quarters with respect to the rate or manner in which an antisense approach to therapeutic development should be pursued. Some of these issues arise from fundamentally different perspectives regarding the application of antisense technology. From a basic research perspective, there is a tremendous desire to understand many of the parameters surrounding antisense activity. These include cellular uptake, site and mechanism of oligonucleotide action. There is also a desire on the part of basic researchers to fully interpret every phenomenon which is found in association with antisense activity-- to understand the whole story before advancing the application of the technology. In this regard, some have sounded a cautionary note with regard to antisense therapeutic development (Stein and Cheng, 1993).

On the other hand, from an applied therapeutic development perspective, issues focus more on the use of existing information to produce workable prototypes from which therapeutic development can proceed incrementally. In assessing the progress of antisense drugs, the applied scientist is more likely to be guided by broader measures of a compound's biological activity which relate more to physiology than cell biology. These parameters include toxicity, pharmacokinetics, tissue distribution, metabolism, and efficacy at an organismal level. Other, incidental phenomena draw less attention. In this regard, progress has also been made in the refinement of techniques needed to explore antisense oligonucleotide as prospective therapeutic agents and place them on a solid footing of pharmacological science (Crooke, 1994,1996).

Going forward, antisense technology offers a new forum for the association of basic and applied researcher beyond that achieved between biologist and pharmacologist in traditional drug development. In this regard the prospect for further experimental refinement of the field seems great. Moreover, the premise of antisense technology--that drugs can be designed in a rational fashion to address the fundamental mechanisms of disease--offers a new point of departure for dialogue which both addresses the curiosity of the basic researcher and offers a practical advantage for the applied scientist.

VI.B. Convergence of Gene Therapy and Antisense Technology

A second force which is under way in genetic therapy is the convergence of some gene therapy and antisense technologies. For example, vector technologies are now being applied widely to both gene therapy and antisense; there is an increasing propensity to employ vectors to express antisense transcripts or ribozymes as well as to recruit antisense strategies for gene therapy applications. In fact, within the last two years, four gene therapy protocols employing antisense strategies--three involving antisense transcripts and one using a ribozyme approach--have been approved by the RAC for clinical applications of gene therapy (http://www.nih.gov/od/orda).

The convergence of antisense and gene therapy vector technologies may spur others to further explore these areas of joint interest. For example, a major problem facing gene therapy and the vector applications of antisense is the ability to quantitatively control vector expression. Yet it is interesting to note that endogenous antisense vector systems have achieved sensitive, dynamic transcriptional control through the production of antisense transcripts. Greater understanding of these endogenous antisense systems may suggest approaches to simple quantitative vector control and new gene therapy applications. Moreover, if progress such as this takes place, it is not hard to conceive of the application of antisense oligonucleotides as potential modulators of these same vector expression systems.

VII. TOWARD GENOMIC THERAPY

In the long term, there are two major obstacles which appear to limit the widescale application of genetic therapy; these are 1) the inability for gene therapy technology to effectively deliver therapeutics employing today's vector systems, and 2) the inability of all genetic therapy technologies to effectively address a complete spectrum of diseases.

While significant advances may be made in addressing these issues through the refinement of experimental design and the convergence of technologies, these approaches may ultimately be insufficient to deliver the full promise of genetic therapy. Instead, it may be necessary to formulate an entirely new context in which to pursue the development of comprehensive genetic therapeutics.

With regard to the first issue of effectively delivering gene therapies, let us examine the context in which we currently view delivery. In short, our perspective derives from viruses--small genetic elements that have evolved a specialized but separate function from the endogenous genetic constituents of cells i.e. genomic DNA and mitochondria. In fact, when it is considered that even the most efficient delivery system we possess, the retroviral vector, recombines with human DNA in an almost random manner with mechanics which resemble no other process of human genetic recombination--what are the long-term prospects for extending this technology to be compatible with human biology for comprehensive treatment of disease? It is also worth underscoring that the current vector model for addressing even the simplest hereditary diseases requires adding a new gene to the genome--and the defective gene cannot even be removed. In other words, the scope of gene therapy viewed in the context of vector application may severely limit the context in which we think about delivering gene therapies as well as our overall success in the future.

Would it not be useful to expand the context of our experimental genetic transfer systems to embrace the processes by which human cells regularly and efficiently transmit and express large stretches of DNA? Examples of this include the process of recombination during meiosis where the cell itself deftly removes and inserts large segments of genomic DNA. Or might it be useful to adopt the biological context of the mechanism by which specialized endogenous genetic material--chromosomes and mitochondrial genomes--coexists with the greater genome of the cells?

With regard to the second issue of addressing disease comprehensively, let us examine the context in which we currently view a drug which treats disease and the process of drug development. Historically, modern drug discovery has been pursued with the objective of obtaining one defined molecule to address one disease indication. In fact, the recognition of the active ingredient as the only essential component within traditional medicinal mixtures represents an earlier shift in scientific context which gave rise to modern pharmacology. Moreover, the process of searching for a single compound was sufficiently simple and independent of any knowledge of biological effect or mechanism of action as to become very successful in its day. We may simply examine our prejudices against compound medicines of older cultures to understand how deeply rooted our belief in single-ingredient drugs has remained.

As noted earlier, there is a fundamental difference between traditional drugs and genetic therapies, however; the former targets phenotype, the later targets genotype. Through a knowledge of the genetic etiology of a disease, multiple genotypic sites of action for a genetic therapy can be identified. Further, several different nucleic acid drugs can be rationally designed to act in concert at these different sites to bring a fuller spectrum of action to a disease therapeutic - an action superior to any of the individual constituent compounds. In short, the best context in which to pursue genetic therapeutics may be - heretically - that of compound medicines. Stated another way, if genetic therapies are pursued in the context of single compounds, as they are now, they may be underestimated or limited from delivering their full potential as therapeutics.

There may be several reasons why the pursuit of genetic therapies has not been pursued in the context of compound medicines to date. As mentioned, prejudice may be one. The absence of any drug development model for compound medicines may be another. Probably the most compelling reason is the fact that a full genetic understanding of most diseases has been lacking--and this type of understanding is prerequisite to compound genetic drug development.

This situation has changed dramatically, however. It is now possible to address the genetic definition of the most discrete diseases--those caused by cellular pathogens through the determination of their genomic sequence. In fact, the complete genomic sequence and the identification of a significant number of organismal genes is now available for a few microorganisms, some of which are pathogenic to man (Table 9).

With the identity of a pathogen's genetic makeup, it may be possible to design compound genetic therapeutics which target multiple sites of action, i.e. a number of different enzymes in an essential metabolic pathway peculiar to the organism.

Table 9. Four Organisms for Which the Entire Genomic DNA Sequence Is Available and Public

Organism	Size of Genome (base pairs)	Total Protein-coding Genes	% of Genes Identified
Saccharomyces cerevisiae	13,105,020	6613	66
Haemophilus influenzae	1,830,137	1743	58
Methanococcus jannaschii	1,739,933	1738	38
Mycoplasma genetalium	580,070	470	79

Genome size and identification of protein-coding genes, by clear sequence homology to genes of known function, is indicated. Source: *Saccharomyces cerevisiae* (http://genome-www.stanford.edu/Saccharomyces/sequence_done. html), *Haemophilus influenzae* (Fleischmann *et al.,* 1995), *Methanococcus jannaschii* (Bult *et al.,* 1996), *Mycoplasma genetalium* (Fraser *et al.,* 1995).

Individual members of a compound therapeutic such as these might require significantly lower efficacies than conventional drugs since their accumulated effect could become exponential as opposed to simply additive. In addition, the ability to flexibly target multiple genetic sites of action might offer a therapeutic strategy much more difficult for a pathogen to adapt to through mutation or other mechanism which frustrates current drug applications or single-compound genetic strategies. Moreover, through developing compound genetic therapeutics against simple pathogenic diseases, it may become clearer how to develop these therapeutics for complex human diseases such as cancer.

The two new perspectives for genetic therapy discussed above--endogenous systems for genetic transfer and compound medicines--represent ideas outside the context of our current practice of genetic therapy development. However, as the field of genomics continues to expand, these ideas and others will present themselves as intuitive and obvious. Further, if genomics is embraced as a new context for genetic therapy--much as Mendel's laws of inheritance were embraced at the turn of the last century--it may propel our appreciation of the field into a conceptual realm for which we have little current awareness. Stated another way, our genetic therapy of today may be a prelude to a new genomic therapy of tomorrow.

VIII. SUMMARY

There is little doubt that genetic therapy has advanced remarkably in only the last ten years. Indeed it has been transformed from a tantalizing idea just beyond the technical reach of early investigators who were drawn by its potential, to a precocious reality subject to the conflicting influences of scientific scrutiny, commercial momentum and public wonder. From a historical perspective, the future success of genetic therapy will depend on three factors: the rate at which experimental design can be refined, the extent to which related and new disciplines can be incorporated, and the boldness with which we are able to conceptualize biology through our new genetic paradigm.

IX. REFERENCES

Agrawal, S. (1992) Antisense oligonucleotides as antiviral agents. *Trends in Biotechnology* **10**:152-157.

Agrawal, S. (1996) Antisense oligonucleotides: Toward clinical trials. *Trends in Biotechnology* **14**:376-387.

Agrawal, S., and Goodchild, J. (1987) Oligodeoxynucleoside methyl-phosphonates: Synthesis and enzymic degradation. *Tet. Lett.* **28**:3539-3542.

Agrawal, S., Temsamani, J., Galbraith, W., and Tang, J. (1995) Pharmaco-kinetics of antisense oligonucleotides. *Clin. Pharmacokinet* **28**:7-16.

Anderson, W.F. (1984) Prospects for human gene therapy. *Science* **226**:401-409.

Antisense Research and Development. (1991) An indexed bibliography of antisense literature, 1978-1990. *Antisense Res. Dev.* **1**:65-113.

Anderson, W. F. (1996) End-of-the-year potpourri--1996. *Hum. Gene Ther.* **7**:2201-2202.

Askari, F.K., and McDonnell, W.M. (1996) Molecular medicine: Antisense-oligonucleotide therapy. *New Engl. J. Med.* **334**:316-318.

Austin, J. and Kenyon, C. (1994) Making time with antisense. *Curr. Biol.* **4**:366-369.

Avery, O.T., Macleod, C.M., and McCarty, M. (1944) Studies on the chemical nature of the substance inducing transformation of pneumococcal types. *J. Exp. Med.* **79**:137-158.

Bartolomé, J.B., Madejón, A., and Carreño, V. (1995) Ribozymes: Structure, characteristics and use as potential antiviral agents. *J. Hepatol.* **22**:57-64.

Baserga, R., and Denhardt, D.T. (eds.) (1992) *Antisense Strategies.* The New York Academy of Sciences, New York.

Belikova, A.M., Zarytova, V.F., and Grineva, N.I. (1967) Synthesis of ribonucleosides and diribonucleoside phosphates containing 2-chloroethylamine and nitrogen mustard residues. *Tet Lett.* **37**:3557-3562.

Bielinska, A., Shivdasani, R.A., Zhang, L., and Nabel, G.J. (1990) Regulation of gene expression with double-stranded phosphorothioate oligonucleotides. *Science* **250**:997-1000.

Birg, F., Praseuth, D., Zerial, A., Thuong, N.T., Asseline, U., Le Doan, T., Hélène, C. (1990) Inhibition of simian virus 40 DNA replication in CV-1 cells by an oligodeoxynucleotide covalently linked to an intercalating agent. *Nucleic Acids Res.* **18**:2901-2907.

Blaese, R.M., Anderson, W.F., Culver, K.W., and Rosenberg, S.A. (1990) Treatment of severe combined immune deficiency disease (SCID) due to adenosine deaminase (ADA) deficiency with autologous lymphocytes transduced with a human ADA gene. *Hum. Gene Ther.* **1**:327-362.

Britten, R.J., and Davidson, E.H. (1969) Gene regulation in higher cells: A theory. *Science* **165**:349-357.

Bult, C. J., White, O., Olsen, G.J., Zhou, L., Fleischmann, R.D., Sutton, G.G., Blake, J.A., FitzGerald, L.M., Clayton, R.A., Gocayne, J.D., Kerlavage, A.R., Dougherty, B.A., Tomb, J.F., Adams, M.D., Reich, C.I., Overbeek, R., Kirkness, E.F., Weinstock, K.G., Merrick, J.M., Glodek, A., Scott, J.L., Geoghagen, N.S.M., Weidman, J.F., Fuhrmann, J.L., Venter, J.C. *et al.* (1996) Complete genome sequence of the methanogenic archaeon, *Methanococcus jannaschii. Science* **273**:1058-1073.

Cameron, F.H., and Jennings, P.A. (1989) Specific gene suppression by engineering ribozymes in monkey cells. *Proc. Natl. Acad. Sci. USA* **86**:9139-9143.

Caruthers, M.H. (1985) Gene synthesis machines: DNA chemistry and its uses. *Science* **230**:281-285.

Castanotto, D., Rossi, J.J., and Sarver, N. (1994) Antisense catalytic RNAs as therapeutic agents. *Adv. Pharmacol.* **25**:289-317.

Cech, T.R. (1992) Ribozyme engineering. *Curr. Opinion Struc. Biol.* **2**:605-609.

Cech, T.R., Zaug, A.J., and Grabowski, P.J. (1981) *In vitro* splicing of the ribosomal RNA precursor of *Tetrahymena*: Involvement of a guanosine nucleotide in the excision of the intervening sequence. *Cell* **29**:487-496.

Christoffersen, R.E., and Marr, J.J. (1995) Ribozymes as human therapeutic agents. *J. Med. Chem.* **38**:2023-2037.

Cohen, J.S. (1989a) Designing antisense oligonucleotides as pharmaceutical agents. *Trends in Pharm. Sci.* **10**:435-437.

Cohen, J.S. (ed.) (1989b) *Oligodeoxynucleotides. Antisense Inhibitors of Gene Expression*. CRC Press, Boca Raton, FL.

Cohen, J.S., and Hogan, M.E. (1994) The new genetic medicines. *Sci. Am.* **271**(6):76-82.

Coleman, J., Green, P.J., and Inouye, M. (1984) The use of RNAs comple-mentary to specific mRNAs to regulate the expression of individual bacterial genes. *Cell* **37**:429-436.

Cooncy, M., Czernuszewicz, G., Postel, E.H., Flint, S.J., and Hogan, M.E. (1988) Site-specific oligonucleotide binding represses transcription of the human *c-myc* gene *in vitro*. *Science* **241**:456-459.

Crockett, G.C. (1983) The chemical synthesis of DNA. *Aldrichimica Acta* **16**:47-54.

Crooke, S.T. (1992) Therapeutic applications of oligonucleotides. *Annu. Rev. Pharmacol. Toxicol.* **32**:329-376.

Crooke, S.T. (1993) Progress toward oligonucleotide therapeutics: Pharma-codynamic properties. *FASEB J.* **7**:533-539.

Crooke, S.T. (1994) Progress in evaluation of the potential of antisense tech-nology. *Antisense Res. Dev.* **3**:145-146.

Crooke, S.T. (1995) Progress in Antisense Therapeutics. *Hematol. Pathol.* **9**:59-72.

Crooke, S.T. (1996) Proof of mechanism of antisense drugs. *Antisense Res. Dev.* **6**:145-147.

Crooke, S.T., and Lebleu, B. (eds.) (1993) *Antisense Research and Applications*. CRC Press, Boca Raton, FL.

De Vries, H (1889) *Intracellulare Pangenesis*. Gustav Fisher, Jena. (English translation, 1910, Open Court Publishing, Chicago.)

Eckstein, F. (1986) Nucleoside phosphorothioates. *Ann. Rev. Biochem.* **54**:367-402.

Eguchi, Y., Itoh, T., and Tomizawa, J. (1991) Antisense RNA. *Annu. Rev. Biochem.* **60**:631-652.

Ellington, A.D., and Szostak, J.W. (1990) *In vitro* selection of RNA molecules that bind specific ligands. *Nature* **346**:818-822.

Epstein, L.M., and Gall, J.G. (1987) Self-splicing transcripts of satellite DNA from the newt. *Cell* **48**:535-543.

Erickson, R.P., and Izant, J.G. (eds.) (1992) *Gene Regulation: Biology of Antisense*

RNA and DNA. Raven Press, New York.

Felsenfeld, G., Davies, D.R., and Rich, A. (1957) Formation of a three-stranded polynucleotide molecule. *J. Am. Chem. Soc.* **79**:2023-2024.

Fleischmann, R.D., Adams, M.D., White, O., Clayton, R.A., Kirkness, E.F., Kerlavage, A.R., Bult, C.J., Tomb, J.F., Dougherty, B.A., Merrick, J.M. *et al.* (1995) Whole-genome random sequencing and assembly of *Haemophilus influenzae* Rd. *Science* **269**:496-512.

Food and Drug Administration (1990) Points to consider in the design and submission of protocols for the transfer of recombinant DNA into the genome of human subjects. *Federal Register* **55**:7443-7447.

Forster, A.C. and Symons, R.H. (1987a) Self-cleavage of plus and minus RNAs of a virusoid and a structural model for the active site. *Cell* **49**:211-220.

Forster, A.C. and Symons, R.H. (1987b) Self-cleavage of virusoid RNA is performed by the proposed 55-nucleotide active site. *Cell* **50**:9-16.

François, J.C., Saison-Behmoaras, T., Thuong, N.T., and Hélène, C. (1989) Inhibition of restriction endonuclease cleavage via triple helix formation by homopyrimidine oligonucleotides. *Biochemistry* **28**:9617-9619.

Fraser, C. M., Gocayne, J. D., White, O., Adams, M. D., Clayton, R. A., Fleischmann, R. D., Bult, C. J., Kerlavage, A. R., Sutton, G., Kelley, J. M., Fritchman, J. L., Weidman, J. F., Small, K. V., Sandusky, M., Fuhrmann, J., Nguyen, D., Utterback, T. R., Saudek, D. M., Phillips, C. A., Merrick, J. M., Tomb, J.-F., Dougherty, B. A., Bott, K. F., Hu, P.-C., Lucier, T. S., Peterson, S. N., Smith, H. O., Hutchison III, C. A., and Venter, J. C. (1995) Life with 6000 genes. *Science* **270**:397-403.

Friedman, T. (1989) Progress toward human gene therapy. *Science* **244**:1275-1281.

Friedman, T. (1992) A brief history of gene therapy. *Nature Genet.* **2**:93-98.

Friedman, T. (1993) Milestones and events in the early development of human gene therapy. *Molec. Genet. Med.* **3**:1-32.

Garrod, A. E. (1902) The incidence of alkaptonuria: A study in chemical individuality. *Lancet* **2**:1616-1620.

Garrod, A. E. (1909) Inborn errors of metabolism. Frowde & Hodder, London.

Gee, J.E., and Miller, D.M. (1992) Structure and applications of intermolecular DNA triplexes. *Am. J. Med. Sci.* **304**:366-372.

Gewirtz, A.M., and Calabretta, B. (1988) A *c-myb* antisense oligodeoxynucleotide inhibits normal human hematopoiesis *in vitro*. *Science* **242**:1303-1306.

Gibson, I. (1994) Antisense DNA and RNA strategies: New approaches to therapy. *J. Royal Col. Physicians London* **28**:507-511.

Goodchild, J. (1990) Conjugates of oligonucleotides and modified oligonucleotides: A review of their synthesis and properties. *Bioconj. Chem.* **1**:165-187.

Griffin, L.C., Tidmarsh, G.F., Bock, L.C., Toole, J.J., and Leung, L.L.K. (1993) *In vivo* anticoagulant properties of a novel nucleotide-based thrombin inhibitor and demonstration of regional anticoagulation in extracorporeal circuits. *Blood* **12**:3271-3276.

Guerrier-Takada, C., Gardiner, K. Marsh, T., Pace, N., and Altman, S. (1983) The

RNA moiety of ribosomal Ribonuclease P is the catalytic subunit of the enzyme. *Cell* **35**:849-857.

Haseloff, J. and Gerlach, W.L. (1988) Simple RNA enzymes with new and highly specific endoribonuclease activity. *Nature* **334**:585-591.

Hawkins, J.W. (1995a) Oligonucleotide therapeutics: Coming 'round the clubhouse turn. *Antisense Res. Dev.* **5**:1.

Hawkins, J.W. (1995b) Human Gene Therapy. *Drug & Market Development*, Boston.

Hélène, C. (1989) Artificial control of gene expression by oligonucleotides covalently linked to intercalating agents. *Br. J. Cancer* **60**:157-160.

Hélène, C. (1991) The anti-gene strategy: Control of gene expression by triplex-forming-oligonucleotides. *Anti-Cancer Drug Design* **6**:569-584.

Hiatt, W.R., Kramer, M., and Sheehy, R.E. (1989) The application of antisense RNA technology to plants - *Agrobacterium tumefaciens*-mediated transformation of tomato cotyledon culture with plasmid pCGN1416. *Gen. Eng.* **11**:49.

Holt, J.T., Gopal, T.V., Moulton, A.D., and Nienhuis, A.W. (1986) Inducible production of c-*fos* antisense RNA inhibits 3T3 cell proliferation. *Proc Natl. Acad. Sci. USA* **83**:4794-4798.

Itakura, K., Rossi, J.J., and Wallace, R.B. (1984) Synthesis and use of synthetic oligonucleotides. *Ann. Rev. Biochem.* **53**:323-356.

Iyer, R.P., Egan, W., Regan, J.B., and Beaucage, S.L. (1990) The automated synthesis of sulfur-containing oligodeoxyribonucleotides using 3H-1, 2-benzodithiol-3-one 1,1-dioxide as a sulfur-transfer reagent. *J. Org. Chem.* **55**:4693-4699.

Izant, J.G., and Weintraub, H. (1984) Inhibition of thymidine kinase gene expression by anti-sense RNA: A molecular approach to genetic analysis. *Cell* **36**:1007-1015.

Jackson, D.A., Symons, R.H., and Berg, P. (1972) Biochemical method for inserting new genetic information into DNA of simian virus 40: Circular SV40 DNA molecules containing lambda phage genes and the galactose operon of *Escherichia coli*. *Proc. Natl. Acad. Sci. USA* **69**:2904-2909.

Johannsen, W. (1909) *Elemente der Exakten Erblichkeitslehre*. Gustav Fischer, Jena.

Judson, H.F. (ed.) (1979) *The Eighth Day of Creation: The Makers of the Revolution in Biology*. Simon and Schuster, New York.

Kantoff, P.W., Kohn, D.B., Mitsuya, H., *et al.* (1986) Correction of adenosine deaminase deficiency in cultured human T and B cells by retrovirus-mediated gene transfer. *Proc. Natl. Acad. Sci. USA* **83**:6563-6567.

Khorana, H.G. (1979) Total synthesis of a gene. *Science* **203**:614-625.

Kiehntopf, M., Esquivel, E.L., Brach, M.A., and Herrmann, F. (1995a) Clinical applications of ribozymes. *Lancet* **345**:1027-1031.

Kiehntopf, M., Esquivel, E.L., Brach, M.A., and Herrmann, F. (1995b) Ribozymes: Biology, biochemistry, and implications for clinical medicine. *J. Mol. Med.* **73**:65-71.

Krieg, A.M., Gause, W.C., Gourley, M.F., and Steinberg, A.D. (1989) A role for endogenous retroviral sequences in the regulation of lymphocyte activation. *J. Immunol.* **143**:2448-2451.

Kruger, K., Grabowski, P.J., Zaug, A. J., Sands, J., Gottschling, D.F., and Cech, T. R. (1982) Self-splicing RNA: Autoexcision and autocyclization of the ribosomal RNA intervening sequence of *Tetrahymena*. *Cell* **31**:147-157.

Lederberg, J. (1956) Genetic transduction. *Am. Sci.* **44**:264-280.

Le Doan, T., Chavany, C., and Hélène, C. (1989) Antisense oligonucleotides as potential antiviral and anticancer agents. *Bull. Cancer* **76**:849-852.

Le Doan, T., Perrouault, L., Praseuth, D., Habhoub, N., Decout, J., Thuong, N.T., L'homme, J., and Hélène, C. (1987) Sequence-specific recognition, photocrosslinking and cleavage of the DNA double helix by an oligo-[alpha]-thymidylate covalently linked to azidoproflavine derivative. *Nucleic Acids Res.* **15**:7749-7760.

Letsinger, R.L., and Lunsford, W.B. (1976) Synthesis of thymidine oligonucleotides by phosphite triester intermediates. *J. Am. Chem. Soc.* **98**:3655-3661.

Letsinger, R.L., Zhang, G.R., Sun, D.K., Ikeuchi, T., and Sarin, P.S. (1989) Cholesteryl-conjugated oligonucleotides: Synthesis, properties, and activity as inhibitors of replication of immunodeficiency virus in cell culture. *Proc. Natl. Acad. Sci. USA* **86**:6553-6556.

Loose-Mitchell, D.S. (1988) Antisense nucleic acids as a potential class of pharmaceutical agents. *Trends Pharm. Sci.* **9**:45-47.

Mack, A. (1996) Companies create new antisense drugs as clinical trials pro-gress. *The Scientist* **10**:8.

Maher III, L.J. (1996) Prospects for the therapeutic use of antigene oligo-nucleotides. *Cancer Invest.* **14**:66-82.

Maher III, L. J., Dervan, P.B., and Wold, B. (1991) Inhibition of DNA/protein interactions by oligonucleotide-directed DNA triple helix formation: Progress and prospects. In Wickstrom, E., ed., *Prospects for Antisense Nucleic Acid Therapy of Cancer and AIDS*. Wiley-Liss, New York, 227-242.

Maher III, L.J., Wold, B., and Dervan, P.B. (1989) Inhibition of DNA binding proteins by oligonucleotide-directed triple helix formation. *Science* **245**:725-730.

Maniatis, T., Jim, G.K., Efstradiadis, A., and Kafatos, F. (1976) Amplification and characterization of beta-globin synthesized *in vitro*. *Cell* **8**:163-182.

Mann R., Mulligan, R.C., and Baltimore, D. (1983) Construction of a retrovirus packaging mutant and its use to produce helper-free defective retrovirus. *Cell* **33**:153-159.

Matsukura, M., Shinozuka, K., Zon, G., Mitsuya, H., Reitz, M., Cohen, J.S., and Broder, S. (1987) Phosphorothioate analogs of oligodeoxynucleotides: Inhibitors of replication and cytopathic effects of human immuno-deficiency virus. *Proc Natl. Acad. Sci. USA* **84**:7706-7710.

Mayer, E. (1982) *The Growth of Biological Thought*. Harvard University Press, Cambridge, Massachusetts.

McKusick, V.A. (ed.) (1994) *Mendelian Inheritance in Man: A Catalog of Human Genes and Genetic Disorders*. The Johns Hopkins University Press, Baltimore.

Melton, D.A. (ed.) (1988) *Antisense RNA and DNA*. Cold Spring Harbor, New

York.

Mendel, J. (Gregor) (1866) Versuche über pflanzen-hybriden. *Verh. Natur. Veriens Brünn.* **4**:3-57.

Mercola, D., Westwick, J., Rundell, A.Y., Adamson, E.D., and Edwards, S.A. (1988) Analysis of a transformed cell line using antisense c-*fos* RNA. *Gene* **72**:253-265.

Meyer, P. (ed.) (1995) *Gene Silencing in Higher Plants and Related Phenomena in Other Eukaryotes.* Springer-Verlag, New York.

Miller, A.D., Jolly, D.J., Friedman, T., and Verma, I.M. (1983) A transmissible retrovirus expressing human hypoxanthine phosphoribosyltransferase (HPRT): Gene transfer into cells obtained from humans deficient in HPRT. *Proc. Natl. Acad. Sci. USA* **80**:4709-4713.

Miller, A.D. (1992) Human gene therapy comes of age. *Nature* **357**:455-460.

Miller, J.H., and Sobell, H.M. (1966) A molecular model for gene repression. *Proc. Natl. Acad. Sci. USA* **55**:1201-1205.

Miller, P.S., and Ts'o, P.O.P. (1987) A new approach to chemotherapy based on molecular biology and nucleic acid chemistry: Matagen (masking tape for gene expression). *Anti-Cancer Drug Design* **2**.117-128.

Miller, P.S., and Ts'o, P.O.P. (1988) Oligonucleotide inhibitors of gene expression in living cells: New opportunities in drug design. *Ann. Reports Med. Chem.* **23**:295-304.

Miller, P.S., Yano, J., Yano, E., Carroll, C., Jayaraman, K., and Ts'o, P.O.P (1979) Nonionic nucleic acid analogues. Synthesis and characterization of dideoxyribonucleoside methylphosphonates. *Biochemistry* **18**:5134-5143.

Mol, J.N.M., van der Krol, A.R. (eds.) (1991) *Antisense Nucleic Acids and Proteins.* Marcel Dekker, New York.

Morgan, A.R., and Wells, R.D. (1968) Specificity of the three-stranded complex formation between double-stranded DNA and single-stranded RNA containing repeated nucleotide sequences. *J. Mol. Biol.* **37**:63-80.

Moser, H.E., and Dervan, P.B. (1987) Sequence-specific cleavage of double helical DNA by triple helix formation. *Science* **238**:645-650.

Mulligan, R.C. (1993) The basic science of gene therapy. *Science* **260**:926-932.

Mulligan, R.C., and Berg, P. (1981) Selection for animal cells that express the *Escherichia coli* gene for xanthine-guanine phosphoribosyltransferase. *Proc. Natl. Acad. Sci. USA* **78**:2072-2076.

Murray, J.A.H. (ed.) (1992) *Antisense RNA and DNA.* Wiley-Liss, New York.

Nagel, K.M., Holstad, S.G., and Isenberh, K.E. (1993) Oligonucleotide pharmacotherapy: An antigene strategy. *Pharmacotherapy* **13**:177-188.

Neckers, L., Whitesell, L., Rosolen, A., and Geselowitz, D.A. (1992) Antisense inhibition of oncogene expression. *Crit. Rev. Oncogenesis* **3**:175-231.

Nellen, W. and Lichtenstein, C. (1993) What makes an mRNA anti-sense-itive? *Trends Biomed. Sci.* **18**:419-423.

Paoletti, C. (1988) Antisense oligonucleotides as potential antitumor agents: Prospective views and preliminary results. *Anti-Cancer Drug Design* **2**:325-331.

Perbost, M., Lucas, M., Chavis, C., Pompon, A., Baumgartner, H., Rayner, B.,

Griengl, H., and Imbach, J.-L. (1989) Sugar modified oligonucleotides. I. Carbo-oligonucleotides as potential antisense agents. *Biochem. Biophys. Res. Comm.* **165**:742-747.

Pfister, M., Pfleiderer, W. (1989) New results in oligoribonucleotide synthesis. *Nucleosides Nucleotides* **8**:1001-1006.

Postel, E.H., Flint, S.J., Kessler, D.J., and Hogan, M.E. (1991) Evidence that a triplex-forming oligodeoxyribonucleotide binds to the c-*myc* promoter in HeLa cells, thereby reducing c-*myc* mRNA levels. *Proc. Natl. Acad. Sci. USA* **88**:8227-8231.

Putnam, D.A. (1996) Antisense strategies and therapeutic applications. *Am. J. Health-Syst. Pharm.* **53**:151-160.

Rayner, B., Malvy, C., Paoletti, J., Lebleu, B., Paoletti, C., and Imbach, J.-L. (1989) Alpha-oligonucleotide analogs. In Cohen, J.S., ed., *Oligonucleotides: Antisense Inhibitors of Gene Expression.* CRC Press, Boca Raton, FL, 119-136.

Rossi, J.J., and Sarver, N. (1990) RNA enzymes (ribozymes) as antiviral therapeutic agents. *Trends Biotechnol.* **8**:179-183.

Sambrook, J., Westphal, H., Srinivasan, P.R., and Dulbecco, R. (1968) The integrated state of viral DNA in SV40-transformed cells. *Proc. Natl. Acad. Sci. USA* **60**:1288-1295.

Sarver, N., Cantin, E.M., Chang, P.S., Zaia, J.A., Ladna, P.A., Stephens, D.A. and Rossi, J.J. (1990) Ribozymes as potential anti-HIV-1 therapeutic agents. *Science* **247**:1222-1225.

Scanlon, K.J., Ohta, Y., Ishida, H., Kijima, H., Ohkawa, T., Kaminski, A., Tsai, J., Horng, G., and Kashani-Sabet, M. (1995) Oligonucleotide-mediated modulation of mammalian gene expression. *FASEB J.* **9**:1288-1296.

Sharma, H.W., and Narayanan, R. (1995) The therapeutic potential of antisense oligonucleotides. *BioEssays* **17**:1055-1063.

Simons, R.W. (1993) The control of prokaryote and eukaryote gene expression by naturally occurring antisense RNA. In Crooke, S.T., and Lebleu, B., eds., *Antisense Research and Applications.* CRC Press, Boca Raton, FL, 97-124.

Smith, C.J.S., Watson, C.F., Ray, J., Bird, C.R., Morris, P.C., Schuch, W., and Grierson, D. (1988) Antisense RNA inhibition of polygalacturonase gene expression in transgenic tomatoes. *Nature* **334**:724-726.

Spiegelman, W.G., Reichardt, L.F., Yaniv, M., Heinemann, S.F., Kaiser, A.D., and Eisen, H. (1972) *Proc. Natl. Acad. Sci. USA* **69**:3156-3160.

Srinivasan, S.K., and Iversen, P. (1995) Review of *in vivo* pharmacokinetics and toxicology of phosphorothioate oligonucleotides. *J. Clin. Lab. Anal.* **9**:129-137.

Stec, W.J., Zon G., Egan, W., and Stec, B. (1984) Automated solid-phase synthesis, separation and stereochemistry of phosphorothioate analogues of oligodeoxynucleotides. *J. Am. Chem. Soc.* **106**:6077-6080.

Stein, C.A., and Cheng, Y.-C. (1993) Antisense oligonucleotides as therapeutic agents--is the bullet really magic? *Science* **261**:1004-1012.

Stein, C.A., and Krieg, A.M. (1994) Problems in interpretation of data derived from *in vitro* and *in vivo* use of antisense oligonucleotides. *Antisense Res. Dev.* **4**:67-69.

Stephenson, M.L., and Zamecnik, P.C. (1978) Inhibition of Rous sarcoma virus RNA translation by a specific oligodeoxyribonucleotide. *Proc. Natl. Acad. Sci. USA* **75**:285-288.

Stull, R.A., and Szoka, F.C., Jr., (1995) Antigene, ribozyme and aptamer nucleic acid drugs: Progress and prospects. *Pharm. Res.* **12**:465-483.

Sullenger, B.A. (1995) Alternative approaches for the application of ribozymes as gene therapies for retroviral infections. *Adv. Pharmacol.* **33**:143-178.

Sullenger, B.A., Gallardo, H.F., Ungers, G.E., and Gilboa, E. (1990) Over-expression of TAR sequences renders cells resistant to human immunodeficiency virus replication. *Cell* **63**:601-608.

Symons, R.H. (1992) Small catalytic RNAs. *Annu. Rev. Biochem.* **61**:641-671.

Taniguchi, T. and Weissmann, C. (1978) Inhibition of Qβ RNA 70S ribosome initiation complex formation by an oligonucleotide complementary to the 3' terminal region of *E. coli* 16S ribosomal RNA. *Nature* **275**:770-772.

Temin, H. (1976) The DNA provirus hypothesis. *Science* **192**:1075-1080.

Tikhonenko, T.I. (1989) Antisense polynucleotides and prospects for their use in fighting viruses. *Molekulyarnaya Biologiya* **23**:629-638.

TMC Development Report. (1996) *Hum. Gene Ther.* **7**:2025-2046.

Tonkinson, J.L., and Stein, C.A. (1996) Antisense oligodeoxynucleotides as clinical therapeutic agents. *Cancer Invest.* **14**:54-65.

U.S. Department of Health and Human Services and Department of Energy (1990) *Understanding our genetic inheritance. The U.S. Human Genome Project: The first five years.*

Uhlenbeck, O.C. (1987) A small catalytic oligonucleotide. *Nature* **328**:596-600.

Van der Krol, A.R., Lenting, P.E., Veenstra, J., van der Meer, I.M., Koes, R.E., Gerats, A.G.M., Mol, J.N.M., and Stuitje, A.R. (1989) An anti-sense chalcone synthase gene in transgenic plants inhibits flower pigmentation. *Nature* **333**:866-869.

Wade, N. (1980) UCLA gene therapy racked by friendly fire. *Science* **210**:509-511.

Wagner, E.G.H., and Simons, R.W. (1994) Antisense RNA control in bacteria, phages, and plasmids. *Annu. Rev. Microbiol.* **48**:713-742.

Wagner, R.W. (1994) Gene inhibition using antisense oligodeoxynucleotides. *Nature* **372**:333-335.

Watson, J.D., and Crick, F.H.C. (1953) A structure for desoxyribose nucleic acid. *Nature* **171**:737-738.

Wei, C., Gibson, M., Spear, P.G., and Scolnick, E.M. (1981) Construction and isolation of a transmissible retrovirus containing the *src* gene from Harvey Murine Sarcoma virus and the thymidine kinase gene from herpes simplex virus 1. *J. Virol.* **39**:935-944.

Weintraub, H.M. (1990) Antisense RNA and DNA. *Sci. Am.* **262**:40-46.

Weismann, A. (1889) *Essays upon Heredity.* (2nd edition, 1892) Clarendon Press, Oxford, UK.

Whitton, J.L. (1994) Antisense treatment of viral infection. *Adv. Virus Res.* **44**:267-303.

Wickstrom, E. (1986) Oligodeoxynucleotide stability in subcellular extracts and culture media. *J. Biochem. Biophys. Methods* **13**:97-102.

Wickstrom, E. (ed.) (1991) *Prospects for Antisense Nucleic Acid Therapy of Cancer and AIDS.* Wiley-Liss, New York.

Wickstrom, E. (1992) Strategies for administering targeted therapeutic oligo-deoxynucleotides. *Trends in Biotechnology* **10**:281-287.

Wickstrom, E. L., Bacon, T. A., Gonzalez, A., Freeman, D. L., Lyman, G. H., and Wickstrom, E. (1988) Human promyelocytic leukemia HL-60 cell proliferation and c-*myc* protein expression are inhibited by an antisense pentadecadeoxynucleotide targeted against c-*myc* mRNA. *Proc. Natl. Acad. Sci. USA* **85**:1028-1032.

Wolff, J.A., and Lederberg, J. (1994) An early history of gene transfer and therapy. *Hum. Gene Ther.* **5**:469-480.

Wu, H.N., Lin, Y.J., Lin, F.P., Makino, S., Chang, M.F., and Lai, M.M. (1989) Human delta hepatitis virus RNA subfragments contain an autocleavage activity. *Proc. Natl. Acad. Sci. USA* **86**:1831-1835.

Xu, M., Kathe, S.D., Goodrich-Blair, H., Nierzwicki, S.A., and Shub, D.A. (1990) Bacterial origin of a chloroplast intron: Conserved self-splicing group I introns in cyanobacteria. *Science* **250**:1566-1570.

Zamecnik, P. C., and Stephenson, M. L. (1978) Inhibition of Rous sarcoma virus replication and cell transformation by a specific oligodeoxy-nucleotide. *Proc. Natl. Acad. Sci. USA* **75**:280-284.

Zaug, A.J., and Cech, T.R. (1980) *In vitro* splicing of the ribosomal RNA precursor in nuclei of Tetrahymena. *Cell* **19**:331-338.

Zinder, N.D., and Lederberg, J. (1952) Genetic exchange in *Salmonella. J. Bacteriol.* **64**:679-699.

Zon, G. (1987) Synthesis of backbone-modified DNA for biological appli-cations. *J. Prot. Chem.* **6**:131-145.

Zon, G. (1989) Oligonucleotide analogues as potential chemotherapeutic agents, *Pharm. Res.* **5**:539-549.

Zon, G. (1995) Brief overview of control of genetic expression by antisense oligonucleotides and *in vivo* applications. *Molec. Neurobiol.* **10**:219-229.

2

Preclinical Development of Antisense Oligonucleotide Therapeutics for Cancer: Regulatory Aspects

Chang-Ho Ahn and Joseph J. DeGeorge
Food and Drug Administration, Rockville, Maryland

I. INTRODUCTION

Activation of oncogenes and deletion or mutation of tumor suppressor genes have been identified as key factors for malignant transformation of normal cells. This has suggested possibilities of exploitation of those oncogenes and altered tumor suppressor genes as targets for selective anticancer chemotherapy. It is thought that inhibition of expression or functions of oncogenes may trigger cancer cells to revert back to a normal phenotype or undergo apoptosis and may make normal cells less susceptible to transformation. Antisense oligonucleotide (AS ODN) technology is one method for intervening in the expression of transforming genes at the level of mRNA, or at the level of DNA by triplex formation. Unlike typical pharmaceuticals which regulate functions of important proteins (e.g., receptors, ion channels, enzymes and components of cytoskeleton) or many standard oncolytics that directly modify DNA, AS ODNs directly bind RNA or DNA without modification and control expression of the targeted gene and subsequent expression of its protein product. The therapeutic utilities of AS ODN are not limited to selective anticancer therapy, but potentially range from genetic diseases to infectious diseases, including cardiovascular disease and AIDS (Bennett and Schwartz, 1995; Carter and Lemoine, 1993; Hélène, 1991; Neckers *et al.*, 1992). Since late 1991, significant and growing numbers of Investigational New Drug Applications (INDs) have been received for clinical studies of AS ODNs. The majority of these investigations are at phase I-II level, but at least one has progressed to phase III. Primarily AS ODNs with phosphorothioate modified backbones have been used in these INDs. Though still limited, significant amounts of both clinical and preclinical information are available to us from sources such as the scientific literature and the proprietary data from the INDs. In an effort to expedite the entry of new and potentially effective AS ODN products into clinical testing and to facilitate the marketing approval of AS ODNs that are found to be safe and effective in clinical trials, this chapter discusses the design and timing of preclinical studies considered important for the development of anticancer AS ODN drugs based on our current experience. The issues discussed are similar to those previously outlined in our earlier publications that provide general guidance for preclinical development of AS ODNs (Black *et al.*, 1993,1994). This discussion focuses on those issues in the context of oncolytic drug development.

II. PRECLINICAL STUDIES FOR INITIAL CLINICAL TRIALS

In the early development of AS ODN drugs, the expectation was that the specificity and selectivity for target mRNAs would produce few unintended adverse effects. Unlike the initial expectation, however, AS ODNs have shown many toxicities similar to typical pharmaceuticals and numerous non-antisense effects including interactions with many functional proteins rather than interaction with RNAs alone (Krieg and Stein, 1995; Sarmiento *et al.*, 1994; Schick *et al.*, 1995; Sharma *et al.* 1996). As it is now recognized that AS ODNs are capable of producing toxicities including lethality in animals in a manner similar to other classes of therapeutics, i.e., toxicities as a consequence of chemical structure, pharmacological activities and impurities, it is clear that preclinical evaluation programs for anticancer AS ODN drugs should follow a preclinical development plan similar to that for standard chemotherapeutics.

The life threatening nature of cancer can justify acceptance of increased risks of serious treatment toxicities, and raises different issues in the preclinical development of AS ODNs as oncologic drugs compared to development of AS ODNs for non-oncologic indications. For example, genotoxicity tests are considered important for AS ODN drugs prior to phase I clinical studies when healthy normal volunteers will be entered in the study, an uncommon clinical development plan for anticancer therapies (Food and Drug Administration, 1996a). The design and timing of the preclinical studies are determined by the nature and the phase of the clinical studies proposed and the proposed cancer patient population. The most appropriate test species, a range of doses and a dosing schedule of the AS ODN, however, may be determined based on the targeted sequence of an AS ODN and initial pharmacology and pharmacokinetic studies. The duration of initial preclinical studies should be based on the duration of the drug treatment period of the proposed clinical studies. However, preclinical study durations longer than twenty-eight days are uncommon in support of treatment for advanced malignancies. A similar strategy is applicable to develop AS ODNs to treat advanced malignancies.

II.A. Preclinical pharmacology and pharmacokinetic studies

Preclinical pharmacology and pharmacokinetic studies are not a regulatory requirement for IND studies as described in CFR 312.22 and 312.42. Information on pharmacodynamics and pharmacokinetics of AS ODN drugs, however, are important for the efficient development of AS ODNs. These preclinical studies provide a basis for optimizing clinical trials and assist in determination of the route, schedule, dose escalation scheme, and effective and safe dosage regimens that are most likely to yield benefit (Newell, 1990). Pharmacokinetic (or toxico-kinetic) data can be obtained from a separate, single or repeated dose pharmacokinetic studies or as a part of toxicity studies (DeGeorge, 1995).

Pharmacology studies of anticancer AS ODN drugs are conducted both at the cellular and molecular level and in animal models. Evaluation of the specificity and selectivity of anticancer AS ODN drugs is desirable. The unique features of AS ODNs (c.g., specificity on genes and species) sometimes make it difficult to find appropriate *in vivo* pharmacological or toxicological models. For AS ODNs targeting genes which are found only in humans (or which are present in the test species, but the target sequences differ in humans and the test species), use of established human cell lines *in vitro* that are expressing the target gene has been common. Testing analog ODNs complementary to the animal sequence or using animals with human transgenes may be useful for *in vivo* studies as well.

AS ODNs may cause toxicity, thus compromising their safety, by unintended interaction of AS ODNs with functional proteins or nontargeted genes, or by interference of AS ODNs or their metabolites with normal nucleic acid synthesis and metabolism. Phosphorothioate ODNs can interact, in a sequence-independent manner, with many functional proteins (e.g., bFGF, CD4, HIV-1 reverse transcriptase, gp120, p210$^{bcr-abl}$, H-Ras, thrombin, p53, and protein kinase C) (Barton and Lemoine, 1995; Bergan *et al.*, 1994; Guvakova *et al.*, 1995; Jansen *et al.*, 1995; Majumdar *et al.*, 1989; Stein and Cheng, 1993; Stein *et al.*, 1993; Smetsers *et al.*, 1995; Tonkinson *et al.*, 1994; Yakubov *et al.*, 1993). Some AS ODNs containing the -GGGG- sequence (G-quartet) possess antiproliferative effects (Ho, *et al.*, 1991; Burgess *et al.*, 1995, Sharma *et al.*, 1995, Yaswen *et al.*, 1993), whereas ODNs containing unmethylated 5'-PuPuCpGPyPy-3' nucleotides activate B lymphocytes in a T cell-independent manner (Krieg *et al.*, 1995; Kuramoto *et al.*, 1992). Both unmodified and modified ODNs are eventually degraded to monomers. The monomers potentially change intracellular pools of free nucleotides, and can result in altered DNA synthesis (Agrawal *et al.*, 1988; Heikkila *et al.*, 1987; Matson and Krieg, 1992). These unintended effects of AS ODNs can be detected by carefully designed pharmacological studies. For AS ODN drugs targeting specific genes by a sequence-dependent mechanism, it is useful to demonstrate dose-dependent inhibition of expression of targeted genes at the RNA and protein level and to show no inhibition of expression of internal control genes, which are selectively inhibited by different ODN sequences. The sequence-dependent effects of a specific AS ODN drug may be further supported by demonstrating that sense and scrambled sequences and/or a mismatched sequence exert no effects on expression of the targeted genes or protein products. Additional controls may include demonstrating a lack of AS ODN effect on cell lines that do not have the target sequence (e.g., cells with a mutant gene or deleted gene). For AS ODNs containing G-quartet or 5'-PuPuCpGPyPy-3' sequences, the control sense and scrambled sequences and/or mismatched sequence containing G-quartet or 5'-PuPuCpGPyPy-3' may be especially useful to correctly establish the mechanism of action.

II.B. Toxicity studies

It is important that a safety profile of an AS ODN be thoroughly established as a part of preclinical and clinical drug development. Much of this can be appraised through preclinical pharmacodynamic, pharmacokinetic (and toxicokinetic), and toxicity studies. These studies are used to guide determination of a starting dose and dose escalation scheme of an AS ODN drug for clinical trials, and to identify potential end-organ toxicities. The animal toxicity studies of AS ODNs can be most effectively performed using routes, durations and schedules comparable to the proposed clinical studies. These toxicity studies are expected to be conducted in accordance with current Good Laboratory Practices (cGLP)

(Food and Drug Administration, 1994). When studies are not conducted in accord with GLP, the deviations should be described and the potential consequences of the deviation on the study outcome should be evaluated and described. Animal toxicity studies using two animal species are usually conducted to support initial phase I clinical studies: a study in rodent species that identifies life-threatening and non-life threatening toxic doses and a second study in a non-rodent species to confirm non-life threatening doses to be used in the clinical trial. For AS ODN drugs, non-human primates have often been used as appropriate non-rodent species in part due to the high sequence homology between non-human primates and human genes. Examinations of clinical pathology and histopathology should be performed in at least one of the two toxicity studies. When the intended use of an AS ODN drug is for *ex vivo* therapies (e.g., bone marrow purging in CML with subsequent marrow transplantation), a multiple level, single dose toxicity study in one species that is deemed most appropriate is generally sufficient for the initiation of phase I clinical trials. In such a case, *in vitro* toxicity studies are conducted to assess adverse effects of the AS ODN drug on human bone marrow and blood cells. For AS ODN drugs complementary to a gene expressed only in humans, or for AS ODNs for which the target gene is present in the test species but the sequence differs from humans, the preclinical toxicity studies with the AS ODN will only produce target sequence-independent toxic effects. In such cases, target sequence-dependent toxic effects of the AS ODN drug may be evaluated in a test species with an AS ODN complementary to the sequence of the target gene in the test species. The target sequence-dependent toxic effects of an AS ODN drug may be further characterized by adding sense and scrambled ODN groups to the toxicity studies, but this is not considered essential to the assessment of safety. Alternative strategies addressing these concerns for sequence-specific effects can also be considered.

In general, interspecies extrapolation from a dose determined in test species to a recommended initial human dose for anticancer drugs is normalized by estimated body surface area rather than body mass (Freireich *et al.*, 1966). Regression analysis of cytotoxic anticancer drugs has shown that there exists a generally linear relationship between the mouse equivalent LD_{10} dose ($MELD_{10}$) in rodents and maximum tolerated dose (MTD) in non-rodent species on a body surface area basis (Freireich *et al.*, 1966; Homan, 1972; Lowe and Davis, 1987). Sensitive non-rodent species have also been used to help determine a starting dose of anticancer drugs (Penta *et al.*, 1979). While there is insufficient experience to date to determine if this approach is essential to assure the safety of AS ODN starting doses in initial human studies, there have been no adverse reactions reported for initial doses derived by using this approach. As currently practiced with cytotoxic anticancer drugs, a safe starting dose of an AS ODN for the initial clinical study is estimated as one tenth the $MELD_{10}$ (or severely toxic dose, if no lethality is observed) or one sixth of the highest non-lethal dose tested in non-rodent species, on a body surface area basis (mg/m^2) (DeGeorge *et al.*,1996). For dose escalation, the standard or modified Fibonacci procedure has usually been used (Goldsmith *et al.*, 1975). Besides this relatively fixed system for dose escalation, alternative methods such as a pharmacokinetically guided escalation, the modified continual reassessment, or escalation based on dose doubling until biological activity is observed have been employed for escalation of cytotoxic drug doses (Collins *et al.*, 1986,1990; Graham and Workman, 1992; Peck *et al.*, 1992; Faries, 1994), but to date have not been used for AS ODNs.

III. SPECIAL PRECLINICAL STUDIES

Considering the nature of the toxicities that have been observed with AS ODNs in animals and in humans, special toxicity studies for assessments of cardiotoxicity, hepatotoxicity, and/or immunotoxicity may be useful in support of early studies and ultimate marketing applications (Cornish *et al.*, 1993; Galbraith *et al.*, 1994; Pisetsky, 1995; Yamamoto *et al.*, 1992a,b). However, it is recognized that when studying any agent for treatment of cancer, encountering significant toxicity in the clinical trials is not uncommon. Our database and the scientific literature suggest that a specific class of AS ODNs, i.e., AS ODNs with phosphorothioate backbones, are able to induce cardiovascular changes in a sequence-independent manner (Cornish *et al.*, 1993; Galbraith *et al.*, 1994). These cardiovascular effects, observed in animal test systems and during clinical trials, have included hypotension, arrhythmias, and changes in coagulation parameters and cardiac output as well as heart block. The underlying pathogenic mechanism(s) have not been identified. Considering the potentially significant safety risks involved with these cardiovascular changes, any new phosphorothioate AS ODN drug should be evaluated for its potential toxic effects on cardiovascular system as a part of early safety testing programs (Black *et al.*, 1994). Although non-human primates are a recommended test species at the present time, use of other non-rodent species may also be considered if the AS ODN can be shown to elicit these effects in the test species. As recommended in the previous guidance for development of AS ODNs, preclinical studies of the cardiotoxic potential of an AS ODN should evaluate centrally monitored mean arterial pressure, electrocardiography, coagulation parameters and clinical pathology, and should include measurement of plasma levels of total and parent ODNs in the test species. Clinical signs and cardiovascular responses should be monitored for the entire duration of the treatment.

AS ODNs also appear to possess a variety of sequence-dependent or independent immunological activities (Branda *et al.*, 1993; Krieg *et al.*, 1995; Kuramoto *et al.*, 1992; Pisetsky and Reich, 1994; Yamamoto *et al.*, 1992a; Zhao *et al.*, 1996). The antigenic and immunogenic potential of AS ODNs may depend on a number of factors, including backbone structure, charge, configuration, molecular size, and metabolic susceptibility (Kuramoto *et al.*, 1992; Messina *et al.*, 1993). Since the polyanionic nature of phosphorothioate ODNs enables sequence-independent binding to certain proteins, the antigenic and immunogenic potential of this class of ODNs has been shown to be high (Branda *et al.*, 1993; Krieg *et al.*, 1995; Kuramoto *et al.*, 1992; McKintyre *et al.*, 1993; Pisetsky and Reich, 1994; Zhao *et al.*, 1996). Although no antibodies specific to AS ODNs have been identified in the blood of test species, due in part to lack of efforts and limitation of detection, strong suggestive evidence of antigenic potential of AS ODNs exists in the literature and in our database: IgG and IgM production (which also may be due to immune stimulation), massive histiocytic and lymphoid infiltrations in most organs, and inflammations and necrosis in the kidneys, liver and heart, etc. These observations are more severe in rodents than in primates, as AS ODNs as a class are more toxic in rodents than in primates on mg/m^2 basis. A reason for this disparity could be related to differences in antigenicity and hypersensitivity between these species. When antigenic and/or immunogenic effects occur, the safety assessment or therapeutic profile of an AS ODN drug in some cases may be compromised. For example, anti-drug antibody responses can result in hypersensitivity responses, neutralization of drug activity and/or pharmacokinetic alteration (e.g., changes in clearance, tissue distribution or plasma half-life) of the AS ODN. If this occurs in the preclinical test species, the determination of safe starting doses for the initial clinical studies could be inaccurate, and/or might incorrectly be ascribed to pathological findings as direct target organ toxicity. It is important to monitor immune responses in order to correctly interpret findings in nonclinical studies with this class of compounds. Clearly, demonstration of antigenicity in an animal model is not necessarily predictive of antigenic potential in humans.

Phosphorothioate AS ODN-induced immunostimulation has been observed from both *in vitro* and *in vivo* models and is frequently manifested as activation and proliferation of B lymphocytes and increased production of M and G type immunoglobulins. These immunostimulative effects of AS ODNs appear to be B cell-specific (and T cell-independent) and are generally target sequence-independent (Branda *et al.*, 1993; Krieg *et al.*, 1995; McKintyre *et al.*, 1993; Pisetsky and Reich, 1994; Zhao *et al.*, 1996). In addition to sequence-independent immunostimulative responses produced by phosphorothioate ODNs, there is evidence for non-antisense sequence-dependent, class-nonspecific immunostimulation by some ODNs. Phosphorothioate or phosphodiester ODNs containing 5'-PuPuCpGPyPy-3' sequences have shown strong immunostimulative activity by stimulating natural killer cell activity, by directly inducing B-cell activation, and by increasing immunoglobulin production (Krieg *et al.*, 1995; Kuramoto *et al.*, 1992).

Based on information from INDs and the scientific literature, ODNs appear to activate complement pathways, in a sequence-independent manner. Altered levels of C5a and/or C3a as well as Bb, an alternative pathway product, were observed in monkeys immediately after intravenous infusion of phosphorothioate ODNs (Galbraith *et al.*, 1994). These activations appear to be associated with peak plasma concentrations (40 µg/ml) regardless of rates of infusion. Frequently, the activation preceded cardiovascular effects associated with AS ODNs and was accompanied by biphasic effects on neutrophil counts. Since activation of complement pathways is often followed by release of endogenous vasoactive peptides, the cardiovascular effects induced by AS ODNs as noted above may be related to complement activation (Edmunds, 1993; Mugge *et al.*, 1993).

The types of AS ODN-induced immunostimulation, i.e., production of IgG and IgM, activation of complement pathways and proliferation of B cells, suggest that these ODNs may induce Type II and III hypersensitivity reactions, which are characterized as antibody-dependent cell cytotoxicity, complement activation, and immune complex formation and deposition. Signs consistent with these types of immunopathies observed in repeat-dose toxicology studies of ODNs are leukopenia, thrombocytopenia, and lesions in the kidneys. In most cases, tissue distribution studies demonstrate that ODNs deposit at high levels in the liver and kidneys, which may be associated with the observed toxicities in these organs (Bayever *et al.*, 1993; Iversen *et al.*, 1994; Zhang *et al.*, 1995). Although most of the observed toxicities are usually attributed to direct drug toxicity, indirect immune mechanisms have not been thoroughly investigated. Some methods for determining involvement of immune reactions in this toxicity would be demonstration of antibody and complement deposition in the affected tissue by immunohistochemical staining, measurement of IgG and IgM production, and complement and complement activation. We are unaware of such methods having been employed.

IV. PRECLINICAL STUDIES TO SUPPORT NEW DRUG APPLICATION (NDA) FILING

To support a NDA for an AS ODN drug in treatment of cancer, the non-clinical studies needed are determined by multiple factors (e.g., the proposed patient population, duration of clinical treatment, and the nature of toxicities observed in animals and in humans). Most AS ODNs used to treat advanced malignancies would not need studies of durations longer than twenty-eight days. One setting in oncology when longer studies would be valuable is in adjuvant therapy. In this case, the maximum duration of long-term toxicity studies is six months in a rodent species and one year in non-rodent species. Reproductive toxicology studies covering ICH Stage C-D (i.e., segment II) (Food and Drug Administration, 1994) are expected for drugs used in the treatment of cancer, but may be deferred until late in development unless patient populations of special concern are to be included in clinical trials. Carcinogenicity studies may be important if the AS ODN drugs are intended for chronic administration in adjuvant therapy (Food and Drug Administration, 1996b). The current standard for carcinogenicity studies is the two-year rodent bioassay (U.S. Interagency Staff Group on Carcinogens, 1986), although alternatives may be equally acceptable provided they are scientifically justified.

V. CONCLUSIONS

AS ODNs interfere with expression of specific genes at the transcription and translation level in a sequence-specific manner, and have potential as promising new therapies for cancer. Though still limited, a significant and growing amount of information on the clinical safety and toxicity of AS ODNs is available. Despite of their highly touted selectivity and specificity to target gene sequences, these ODNs also produce toxicities in animals in a manner similar to other chemotherapeutics. Therefore, it is necessary that preclinical evaluation programs for anticancer AS ODN drugs follow a preclinical development plan similar to that for standard chemotherapeutics. Furthermore, some classes of AS ODNs have been found to exhibit unique toxicologic effects. Thus, it is important to have discussion with FDA staff on the preclinical aspects of the drug development plan, both early in the planning for initial clinical studies and at or before the end of Phase 2. Such interactions can lead to a more focused and efficient preclinical development plan.

ACKNOWLEDGMENTS

The authors appreciate Drs. Paul Andrews and Robert DeLap for reading and useful suggestions on this manuscript.

VI. REFERENCES

Agrawal, S., Goodchild, J. Civeira, M.P., Thornton, A.H., Sarin, P.S., and Zamecnik, P.C. (1988) Oligodeoxynucleotide phosphoramidites and phosphorothioates as inhibitors of human immunodeficiency virus. *Proc. Natl. Sci. Acad. USA* **85**:7079-7083.

Barton, C.M., and Lemoine, N.R. (1995) Antisense oligonucleotides directed against p53 have antiproliferative effects unrelated to effects on p53 expression. *Br. J. Cancer* **71**:429-437.

Bayever, E., Iversen, P.L., Bishop, M.R., Sharp, J. G., Tewary, H.K., Arneson, M.A., Pirruccello, S.J., Ruddon, R.W., Kessinger, A., Zon, G., and Armitage, J.O. (1993) Systemic administration of a phosphorothioate oligonucleotide with a sequence complementary to p53 for acute myelogenous leukemia and myelodysplastic syndrome: initial results of a phase I trial. *Antisense Res. Dev.* **3**:383-390.

Bennett, M.R., and Schwartz, S.M. (1995) Antisense therapy for angioplasty restenosis: some critical considerations. *Circulation* **92**:1981-1993.

Bergan, R.C., Connell, Y., Fahmy, B., Kyle, E., and Neckers, L. (1994) Aptameric inhibition of $p210^{bcr-abl}$ tyrosine kinase autophosphorylation by oligodeoxynucleotides of defined sequence and backbone structure. *Nucleic Acids Res.* **22**:2150-2154.

Black, L.E., DeGeorge, J.J., Cavagnaro, J.A., Jordan, A., and C-H. Ahn, (1993) Regulatory considerations for evaluating the pharmacology and toxicology of antisense drugs. *Antisense Res. Dev.* **3**:399-404.

Black, L.E., Farrelly, J.G., Cavagnaro, J.A., C.-H. Ahn, DeGeorge, J.J., Taylor, A.S., DeFelice, A.F., and Jordan, A. (1994) Regulatory considerations for oligonucleotide drugs: updated recommendations for pharmacology and toxicology studies. *Antisense Res. Dev.* **4**:299-301.

Branda, R.F., Moore, A.L., Mathews, L., McCormack, J.J., and Zon, G. (1993) Immune stimulation by an antisense oligomer complementary to the *rev* gene of HIV-1. *Biochem. Pharmacol.* **45**:2037-2043.

Burgess, T.L., Fisher, E.F., Ross, S.L., Bready, J.V., Qian, Y.X., Bayewitch, L.A., Cohen, A.M., Herrera, C.J., Hu, S.S., Kramer, T.B., Lott, R.D., Martin, F.H., Pierce, G.F., Simonet, L., and Farrell, C.L. (1995) The antiproliferative activity of c-*myb* and c-*myc* antisense oligonucleotides in smooth muscle cells is caused by a non-antisense mechanism. *Proc. Natl. Acad. Sci. USA* **92**:4051-4055.

Carter, G., and Lemoine, N. R. (1993) Antisense technology for cancer therapy: does it make sense? *Br. J. Cancer*, **67**:869-876.

Collins, J.M., Zaharko, D.S. Dedrick, R.L., and Chabner, B.A. (1986) Potential roles for preclinical pharmacology in phase I clinical trials. *Cancer Treat. Rep.* **70**:73-80.

Collins, J.M., Grieshaber, C.K., and Chabner, B.A. (1990) Pharmacologically guided phase I clinical trials based upon preclinical drug development. *J. Natl. Cancer Inst.* **82**:1321-1326.

Cornish, K.G., Iversen, P., Smith, L., Arneson, M., and Bayever, E. (1993) Cardiovascular effects of a phosphorothioate oligonucleotide with sequence antisense to p53 in the conscious rhesus monkey. *Pharmacol. Commun.* **3**:239-247.

Dawson, K.H., and Bell, D.A. (1991) Production and pathogenic effects of anti-DNA antibodies: relevance to antisense research. *Antisense Res. Dev.* **1**:351-360.

DeGeorge, J.J. (1995) Food and Drug Administration viewpoints on toxicokinetics: the view from review. *Toxicol. Pathol.* **23**.220 225.

DeGeorge, J.J. Ahn, C.-H., Andrews, P.A., Brower, M.E., Giorgio, D., Goheer, M.A., Lee-Ham, D.Y., McGuinn, W.D., Schmidt, W., Sun, C.J., and Tripathi, S.C. (1997) Regulatory considerations for preclinical development of anticancer drugs. *Cancer Chemother. Pharmacol.*, in press.

Edmunds, L.H. (1993) Blood-surface interactions during cardiopulmonary bypass. *J. Card. Surg.* **8**:404-410.

Faries, D. (1994) Practical modifications of the continual reassessment method for phase I cancer clinical trials. *J. Biopharm. Statistics* **4**:147-164.

Food and Drug Administration (1994) International Conference on Harmonization: guideline for detection of toxicity to reproduction for medicinal products. *Fed. Register* **59**:48746-48752.

Food and Drug Administration (1995) Good laboratory practices for nonclinical laboratory studies. 21 *Code Federal Regulations* **58**:265-278.

Food and Drug Administration (1996a) International Conference on Harmonization: guidance on specific aspects of regulatory genotoxicity tests. *Fed. Register* **61**:18198-18202.

Food and Drug Administration. (1996b) International Conference on Harmonization: final guideline on the need for long-term rodent carcinogenicity studies of pharmaceuticals. *Fed. Register* **61**:8154-8156.

Freireich, E.J., Gehan, E.A., Rall, D.P., Schmidt, L.H., and Skipper, H.E. (1966) Quantitative comparison of toxicity of anticancer agents in mouse, rat, hamster, dog, monkey, and man. *Cancer Chemother. Rep.* **50**:219-244.

Galbraith, W.M., Hobson, W.C., Giclas, P.C., Schechter, P.J., and Agrawal, S. (1994) Complement activation and hemodynamic changes following intravenous administration of phosphorothioate oligonucleotides in the monkey. *Antisense Res. Dev.* **4**:201-206.

Goldsmith, M.A., Slavik, M., and Carter, S.K. (1975) Quantitative prediction of drug toxicity in humans from toxicology in small and large animals. *Cancer Res.* **35**:1354-1364.

Graham, M.A., and Workman, P. (1992) The impact of pharmacokinetically guided dose escalation strategies in phase 1 clinical trials: Critical evalu-ation and recommendation for future studies. *Ann. Oncology* **3**:339-347.

Grieshaber, C.K., and Marsoni, S. (1986) Relation of preclinical toxicology to findings in early clinical trials. *Cancer Treat. Rep.* **70**:65-72.

Guvakova, M.A., Yakubov, L.A., Vlodavsky, I., Tonkinson, J. L., and Stein, C.A. (1995) Phosphorothioate oligodeoxynucleotides bind to basic fibroblast growth factor, inhibit its binding to cell surface receptors and remove it from low affinity binding sites on extracellular matrix. *J. Biol. Chem.* **270**:2620-2627.

Hélène, C. (1991) The anti-gene strategy: control of gene expression by triplex-forming oligonucleotides. *Anticancer Drug Design,* **6**:569-584.

Heikkila, R., Schwab, G., Wickstrom, E., Loke, S.L., Pluznik, D.H., Watt, R., and Neckers, L.M. (1987) A c-*myc* antisense oligonucleotide inhibits entry into S phase but not progress from G0 to G1. *Nature* **328**:445-449.

Ho, P. T. C., Ishiguro, K., Wickstrom, E., and Sartorelli, A. C. (1991) Non-sequence-specific inhibition of transferrin receptor expression in HL-60 leukemia cells by phosphorothioate antisense oligonucleotides. *Antisense Res. Dev.* **1**:329-342.

Homan, E.R. (1972) Quantitative relationships between toxic doses of antitumor chemotherapeutic agents in animals and man. *Cancer Chemother. Rep.* **3**:13-19.

Iversen, P.L., Mata, J., Tracewell, W.G., and Zon, G. (1994) Pharmacokinetics of an antisense phosphorothioate oligodeoxynucleotide against *rev* from human immunodeficiency virus type I in the adult male rat following single injections and continuous infusion. *Antisense Res. Dev.* **4**:43-52.

Jansen, B., Wadl, H., Inoue, S.A., Trulzsch, B., Selzer, E., Duchene, M., Eichler, H-G., Wolff, K., and Pehamberger, H. (1995) Phosphorothioate oligonucleotides reduce melanoma growth in a SCID-hu mouse model by a nonantisense mechanism. *Antisense Res. Dev.* **5**:271-277.

Krieg, A., and Stein, C.A. (1995) Phosphorothioate oligodeoxynucleotides: Antisense or anti-protein? *Antisense Res. Dev.* **5**:241.

Krieg, A.M., Yi, A-K., Matson, S., Waldschmidt, T.J., Bishop, G.A., Teasdale, R., Koretzky, G.A., and Klinman, D.M. (1995) CpG motifs in bacterial DNA trigger direct B-cell activation. *Nature* **374**:546-549.

Kuramoto, E., Yano, O., Kimura, Y., Baba, M., Makino, T., Yamamoto, S., Yamamoto, T., Kataoka, T., and Tokunaga, T. (1992) Oligonucleotide sequences required for natural killer cell activation. *Jpn. J. Cancer Res.* **83**:1128-1131.

Lowe, M.C., and Davis, R.D. (1987). The current toxicology protocol of the National Cancer Institute. In Hellmann, K., and Carter, S.K., eds., *Fundamentals of Cancer Chemotherapy,* McGraw-Hill, New York, 228-235.

Majumdar, C., Stein, C.A., Cohen, J.S., Broder, S., and Wilson, S. (1989) HIV reverse transcriptase stepwise mechanism: phosphorothioate oligodeoxynucleotides as primer. *Biochemistry* **28**:1340-1346.

Matson, S., and Krieg, A. (1992) Non-specific suppression of ^{3}H-thymidine incorporation by 'control' oligonucleotides. *Antisense Res. Dev.* **2**:325-336.

McKintyre, K., Lombardo-Gillooly, K., Perez, J., Kunch, C., Sarmiento, U., Larigan, D., Landreth, D., and Narayanan, R. (1993) A sense phosphorothioate oligonucleotide directed to the initiation codon of transcription factor NF-κB p65 cause sequence-specific immune stimu-lation. *Antisense Res. Dev.* **3**:309-322.

Messina, J.P., Gilkeson, G.A., and Pisetsky, D.S. (1993) The influence of DNA structure on the *in vitro* stimulation of murine lymphocytes by natural and synthetic polynucleotide antigens. *Cell Immunol.* **147**:148-157.

Mugge, A., Lopez, J.A., Heistad, D.D., and Lichtlen, P.R. (1993) Vasocon-striction in response to activated leukocytes: implications for vasospasm. *Eur. Heart J.* **14** (suppl. 1):87-92.

Neckers, L., Whitesell, L., Rosolen, A., and Geselowitz, D.A. (1992) Antisense inhibition of oncogene expression. *Crit. Rev. Oncogenesis* **3**:175-231.

Newell, D.R. (1990) Phase 1 clinical studies with cytotoxic drugs: pharma-cokinetic and pharmacodynamic considerations. *Br. J. Cancer* **61**:189-191.

Peck, C.C., Barr, W.H., Benet, L.Z., Collins, J., Desjardins, R.E., Furst, D.E., Harter, J.G., Levy, G., Ludden, T., Rodman, J.H., Sanathanan, L., Schentag, J.J., Shah, V.P., Sheiner, L.B., Skelly, J.P., Stanski, D.R., Temple, R.J., Viswanathan, C.T., Weissinger, J., and Yacobi, A. (1992) Opportunities for integration of pharmacokinetics, pharmacodynamics, and toxicokinetics in rational drug development. *Clin. Pharmacol. Therap.* **51**:465-473.

Penta, J.S., Rozencweig, M., Guarino, A.M., and Muggia, R.M. (1979) Mouse and large animal toxicology studies of twelve antitumor agents: relevance to starting dose for phase 1 clinical trials. *Cancer Chemother. Pharmacol.* **3**:97-101.

Pisetsky, D.S., and Reich, C.F. (1994) Stimulation of murine lymphocyte proli-feration by a phosphorothioate oligonucleotide with antisense activity for herpes simplex virus. *Life Sci.* **54**:101-107.

Pisetsky, D.S. (1995) Immunologic consequences of nucleic acid therapy. *Antisense Res. Dev.* **5**:219-225.

Sarmiento, U.M., Perez, J.R., Becker, J.M., and Narayanan, R. (1994) *In vivo* toxicological effects of rel A antisense phosphothioate in CD-1 mice. *Antisense Res. Dev.* **4**:99-107.

Schick, B.P., Eras, J.L., and Mintz, P.S. (1995) Phosphothioate oligonucleotides cause degradation of secretory but not intracellular serglycin proteoglycan core protein in a sequence-independent manner in human megakaryocytic tumor cells. *Antisense Res. Dev.* **5**:59-65.

Sharma, H.W., Hsiao, R., and Narayanan, R. (1996) Telomerase as a potential molecular target to study G-quartet phosphorothioates. *Antisense Nucleic Acid Drug Dev.* **6**:3-7.

Smetsers, T.F.C.M., van de Locht, L.T.F., Pennings, A.H.M., Wessels, H.M.C., de Witte, T.M., and Mensink, E.J.B.M. (1995) Phosphorothioate *bcr-abl* antisense oligonucleotides induce cell death, but fail to reduce cellular *bcr-abl* protein levels. *Leukemia* **9**:118-130.

Stein, C.A., Cleary, A., Yakubov, L., and Lederman, S. (1993) Phosphorothioate oligodeoxynucleotides bind to the third variable loop domain (V3) of HIV-1 gp120. *Antisense Res. Dev.* **3**:19-31.

Stein, C.A., and Cheng, Y-C. (1993) Antisense oligonucleotides as therapeutic agents - is the bullet really magical? *Science* **261**:1004-1012.

Tonkinson, J.L., Guvakova, M., Khaled, Z., Lee, J., Yakubov, L., Marshall, W.S., Caruthers, M.H., and Stein, C.A. (1994) Cellular pharmacology and protein binding of phosphoromonothioate and phosphorodithioate oligodeoxynucleotides: a comparative study. *Antisense Res. Dev.* **4**:269-278.

U.S. Interagency Staff Group on Carcinogens. (1986) Chemical carcinogens: a review of the science and its associated principles. *Environ. Health Perspect.* **67**:201-282.

Yakubov, L., Khaled, Z. Zhang, L.M., Truneh, A., Vlassov, V., and Stein, C.A. (1993) Mode of interaction of oligodeoxynucleotides with recombinant sCD4. *J. Biol. Chem.* **268**:18818-18823.

Yamamoto, S., Yamamoto, T., Kataoka, T., and Tokunaka, T. (1992a) Unique palindromic sequences in synthetic oligonucleotides are required to induced INF and augment INF-mediated natural killer activity. *J. Immunol.* **148**:4072-4076.

Yamamoto, S., Yamamoto, T., Shimada, S., Kuramoto, E., Yano, O., Kataoka, T., and Tokunaka, T. (1992b) DNA from bacteria, but not from vertebrates, induces interferons, activates natural killer cells and inhibits tumor growth. *Microbiol. Immunol.* **36**:983-997.

Yaswen, P., Stampfer, M.R., Ghosh, K., and Cohen, J.S. (1993) Effects of sequence of thioated oligonucleotides on cultured human mammary epithelial cells. *Antisense Res. Dev.* **3**:67-77.

Zhang, R., Diasio, R.B., Lu, Z., Liu, T., Jiang, Z., Galbraith, W.M., and Agrawal, S. (1995) Pharmacokinetics and tissue distribution in rats of an oligodeoxynucleotide phosphorothioate (GEM®91) developed as a therapeutic agent for human immunodeficiency virus type-I. *Biochem. Pharmacol.* **49**:929-939.

Zhao, Q, Temsamani, J., Iadarola, P.L., Jiang, Z., and Agrawal, S. (1996) Effect of different chemically modified oligonucleotides on immune stimulation. *Biochem. Pharmacol.* **51**:173-182.

3

Commercial Scale Manufacturing of Oligonucleotides Under Good Manufacturing Practices

Jose E. Gonzalez, Richard G. Einig, Patricia Puma, Timothy P. Noonan, Paul E. Kennedy, Bruce G. Sturgeon, Bing H. Wang, and Jin-yan Tang
Hybridon, Inc., Milford, Massachusetts

I. INTRODUCTION

The ability to inhibit gene expression by synthetic oligonucleotides has attracted intense interest to antisense DNA as a novel class of chemotherapeutic agents (Agrawal, 1996; Zamecnik, 1991). As the number of oligonucleotides entering clinical trials increases, so does the demand for increasing amounts of highly pure synthetic oligonucleotides. In the recent past, one apparent obstacle to the successful application of oligonucleotides for therapeutic use had been the inability to produce them at large scale (Seliger, 1993; Sinha, 1993). Another perceived hurdle to the commercialization of synthetic oligonucleotides was the inability to produce them at sufficient purity under current Good Manufacturing Practices (cGMPs). As recently as 1991, it was speculated that 100 grams of a 21-mer phosphorothioate (PS) would cost $1.2 million (Zon and Geiser, 1991).

By virtue of technological breakthroughs, implementation of quality systems, and economies of scale, we now have the ability to produce synthetic oligonucleotides at the scale, purity, and cost to support the commercialization of oligonucleotides as chemotherapeutic agents. Today,

Hybridon can regularly produce oligonucleotides at the rate of several hundred grams per day. Ultimately, lot sizes will be measured in kilograms, with annual capacities reaching several metric tons. Currently, our recently completed active pharmaceutical ingredient (API) plant has an annual throughput capacity of over one metric ton of purified oligonucleotides.

II. LARGE SCALE PRODUCTION OF cGMP ACTIVE PHARMACEUTICAL INGREDIENT OLIGONUCLEOTIDES

II.A. Synthesis

Presently, the chemistry of choice for large scale synthesis of oligonucleotides is solid phase phosphoramidite chemistry (Beaucage, 1993; Beaucage and Iyer, 1992; Caruthers, 1989). In this technology, oligonucleotides are synthesized in linear sequential fashion, usually by adding one nucleotide per cycle on solid supports initially functionalized with a protected nucleoside via a 3'-succinyl linker. Both polymer-based supports (e.g., polystyrene divinyl benzene) and inorganic supports (e.g., controlled pore glass) are used. Commonly, the nucleoside sugar 5' hydroxyl is protected with a dimethoxytrityl (DMT) group.

Similarly, the reactive groups in the nucleoside bases (adenine, guanine, and cytosine) are also protected with reagents specifically designed for this purpose. The synthesis follows a series of programmed cycles, each of which contains four reaction steps: deblock, coupling, oxidation, and capping, alternating with four wash steps (Figure 1). Washing the solid support with acetonitrile reduces interference between the reagents used in the sequential reaction steps.

After packing the reactor column with the DMT-protected nucleoside solid support, the initial reaction step in the chain elongation cycle is the deblock step. The deblock step specifically removes the acid labile 5'-DMT group with a solution of dichloroacetic acid in dichloromethane, while leaving the base-sensitive protecting group of the nucleosides intact. The effluent from the deblocking step provides a useful monitor of the efficiency of synthesis. Analysis at smaller scales is often done off-line using a colorimetric trityl assay as a measure of coupling efficiency. To accomplish this at the larger scales, the synthesizers monitor the trityl peak on-line by measuring optical absorbance or conductivity.

After DMT removal, the resultant free 5'-hydroxyl group of the sugar moiety reacts in the coupling step with the protected nucleoside phosphoramidite, which is next in the oligonucleotide sequence. To obtain

A. Synthesis:

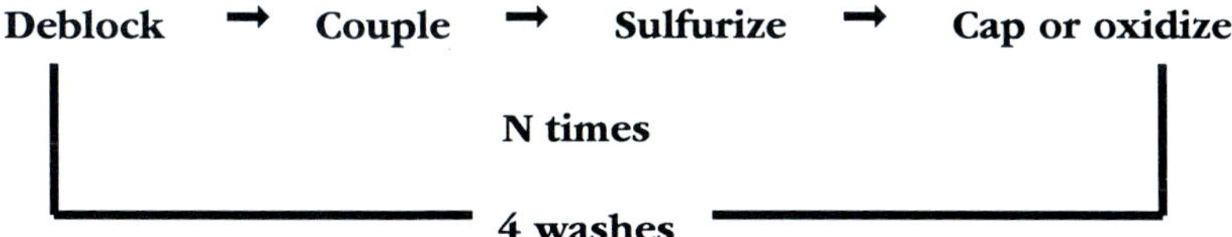

N = number of cycles to obtain the desired length of oligonucleotide

B. Post-synthesis:

$$NH_3 \, (aq)$$

Oligonucleotide on solid support → Crude API solution

Figure 1. Synthetic Process. **A:** Each synthetic cycle contains four reaction steps: deblock, coupling, oxidation, and capping. These steps alternate with four wash steps. **B:** After synthesis is completed, the protected oligonucleotide, still attached to the synthesis support, is incubated in concentrated ammonium hydroxide to cleave the oligonucleotide from the supports (e.g., controlled pore glass) are used. Commonly, the nucleoside sugar 5′ hydroxyl is protected with a dimethoxytrityl (DMT) group support and to remove protecting groups. The solid support is removed from the ammoniacal solution, and the solution, containing the crude API, is ready for purification.

effective coupling, the amidite solution is activated with tetrazole before contacting the solid support. The molar stoichiometric excess ("molar equivalents") of phosphoramidite in this step is also used to drive the reaction towards completion. Efficient delivery of reagents and optimized design of process equipment permits use of only 1.5 molar equivalents of phosphoramidites, compared with 20 or more in a typical laboratory scale synthesizer.

The resultant phosphite intermediate is oxidized to the more stable phosphate triester during the oxidation step, or thiolation step, when a sulfur transfer reagent is used during synthesis of phosphorothioates. Unreacted 5′-hydroxyl groups are capped during the capping step by contacting the anchored oligonucleotide with a solution of acetic anhydride activated by N-methylimidazole.

Once synthesis is complete, concentrated ammonium hydroxide is used to cleave the oligonucleotide from the synthesis support and remove protecting groups. This step can be done directly on the synthesis column. The cleaved, deprotected oligonucleotide, or crude API, becomes the starting material for purification.

Large scale synthesis is currently performed at up to the 100 mmol scale on automated solid phase DNA synthesis machines. A synthesis file on a PC workstation is downloaded to the synthesizer and executed in a sequential manner until all programmed steps are completed. The large scale machines used currently are the OligoProcess Synthesizer (Pharmacia Biotech) and the Hybridon 601 Large Scale Synthesizer, which was designed and built in-house. These machines have a maximum capacity of 200 mmol.

The Hybridon 601 Large-Scale Synthesizer (Figure 2) was designed and built in-house. It contains eight amidite ports and incorporates the ability to recycle reagents. Recycling has proven to be of great use in the coupling steps used in synthesis of second generation oligonucleotides using 2'-O-alkyl amidites (see II.E. Second Generation Oligonucleotides).

II.B. Purification

Early purification schemes relied on a relatively direct scale-up of the reverse phase liquid chromatography (RPLC) methodology used routinely for analysis and preparation of small amounts of oligonucleotides (Beaucage, 1993). RPLC removed the DMT-off capped oligonucleotide failure sequences. The eluted DMT-on oligonucleotide was subsequently detritylated with concentrated acetic acid and precipitated in ethanol. The ammonium salt was converted to a sodium salt by use of ion-exchange on Geigon® cation exchange resins charged with sodium, or by chromatography on an anion exchange support. The sodium form of the oligonucleotide was then concentrated by evaporation, and desalted using gel filtration chromatography. While RPLC purification remains an option for use with certain chemistries, the approach using aqueous buffers (Puma, 1996) is superior for large scale manufacturing due to its scalability, adaptability, and final product purity. We routinely use this aqueous purification process at large scale (Figure 3). It incorporates hydrophobic interaction chromatography (HIC) to remove failure sequences lacking the DMT group. The desired product, still bearing a DMT group, is eluted in low salt buffer or water and subsequently detritylated.

The total time required for removal of the DMT group varies with the sequence. The oligonucleotide product is then purified using anion exchange chromatography. With phosphodiester (PO) oligonucleotides, full-length product is adequately separated from shorter impurties on anion exchange columns. Moreover, anion exchange chromatography is effective in separating PO containing oligonucleotide impurities from phosphorothioate (PS) oligonucleotides, n + x impurities, and n − 2 and smaller oligonucleotides.

Figure 2. Hybridon 601 Large Scale Oligonucleotide Synthesizer. This synthesizer was designed and built in house. Besides synthesizing phosphodiester (PO), and phosphorothioate (PS) oligonucleotides, it was specifically designed to support efficiently the synthesis of modified oligonucleotides. This synthesizer has a batch capacity of 200 mmol of oligonucleotide and a cycle time per batch of less than 12 hours.

However, anion exchange chromatography is largely ineffective in separating n − 1 PS from n, full length PS. In spite of this limitation, excellent purities and recoveries are obtained for phosphorothioates by designing the anion exchange method to set up a displacement train in which full-length PS displaces oxidized and shorter chain length species.

Ultrafiltration is used to desalt, diafilter, and concentrate the purified oligonucleotide. Ultrafiltration is also the method of choice to reduce endotoxin levels, and to produce final concentrations of approximately 50 mg/mL. For a more concentrated final product, thin film evaporation is used. Since aqueous solutions of most purified oligonucleotides are very stable, freeze drying is optional.

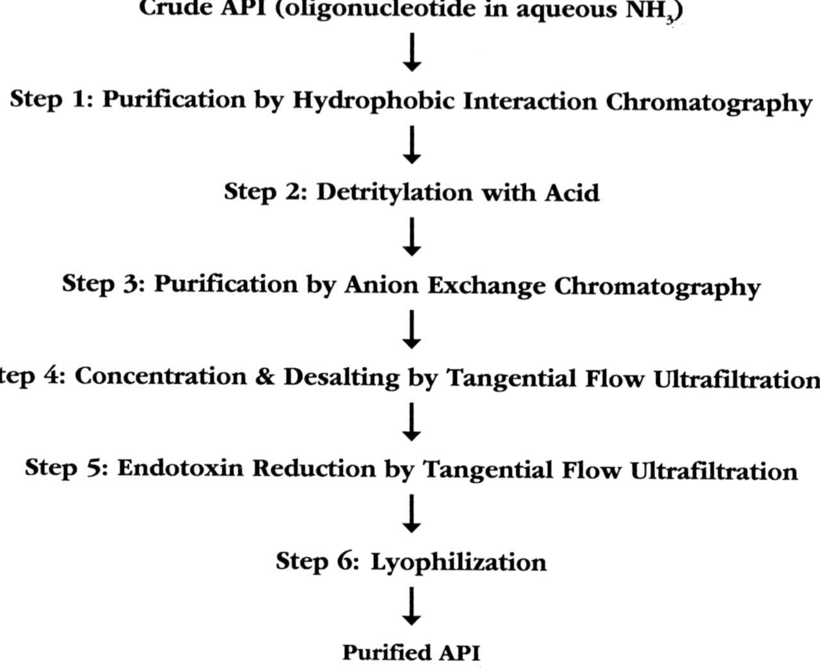

Figure 3. Purification Process. An aqueous purification process is routinely used at large scale. The crude API is passed over a hydrophobic interaction chromatography column. Trityl-off sequences (i.e., failure sequences) do not bind. Trityl-on sequences are eluted from the column and subsequently detritylated in acid. The detritylated oligonucleotide is further purified by anion exchange chromatography. The product, which elutes in high salt, is further processed using ultrafiltration. The first ultrafiltration step desalts, exchanges the buffer, and concentrates the product. The second ultrafiltration step reduces endotoxin contaminants. For some products a final lyophilization step is performed.

II.C. In-Process Controls

II.C.1. Synthesis

Large-scale processing, while presenting challenges to oligonucleotide production, also offers distinct advantages relating to process control and documentation during both synthesis and purification. This is due to the extensive automation required to carry out large scale production.

By necessity, the preparation and delivery of reagents to a large-scale synthesis is more automated than at smaller scales. In addition to the obvious operational advantages of decreased cost and increased throughput, automation also results in higher quality and reproducibility. The reagents prepared for large scale syntheses are mixed in a dedicated solution preparation area in large volume, reducing the variability that can occur when a number of small lots are prepared separately. Since moisture can have an adverse effect on synthesis yield, the stainless steel tanks used as reagent reservoirs at large scale provide a more effective barrier to moisture, and the use of stainless steel couplings enables moisture-free transfer of reagents both in preparing solutions and in charging the synthesizers.

The large-scale synthesizers have validatable software programs for more reliable operation. While smaller-scale synthesizers can monitor catastrophic failures (e.g., a failed coupling), this information is of limited utility and is primarily useful after synthesis in analyzing process performance. In large scale synthesizers the software monitors peaks, pressures, and flow rates in real time. Real-time observation of the flows, absorbances, and pressures during delivery of all reagents enables the operator to evaluate performance of the synthesizer and to recognize any developing problem before it interferes with synthesis.

In addition, the software can trigger alarms if certain critical parameters are not achieved. An example of this is monitoring of the absorbance of the detritylation peak in each cycle. A decreased peak height in any cycle triggers an alarm that pauses the synthesis and must be acknowledged for the synthesis to continue. As a routine practice, the operator is required to evaluate the situation and take one of several appropriate actions. Another example is a "dry line" alarm indicating that a reagent delivery line is dry and a "level sensor" alarm indicating that the level in a reservoir is low. High pressure alarms indicate potential problems either in reagent delivery or in the column bed before these problems interfere with synthesis.

II.C.2 Purification and analysis

The large-scale chromatographic purification platforms, or skids, use validatable software, sanitary fittings, and components, which enable more efficient cleaning. As is the case with synthesis, the software is used to control the process and prevent problems from developing.

For example, parameters such as ultraviolet absorbance, pH, and conductivity are used to monitor delivery of reagents and elution of product, and alarms are programmed to indicate if required setpoints have not been reached. Impurities may be encountered from reaction by-products, reactants, nucleotide or base deletions from the chain ($n - 1$, $n - 2$, etc.) and salt.

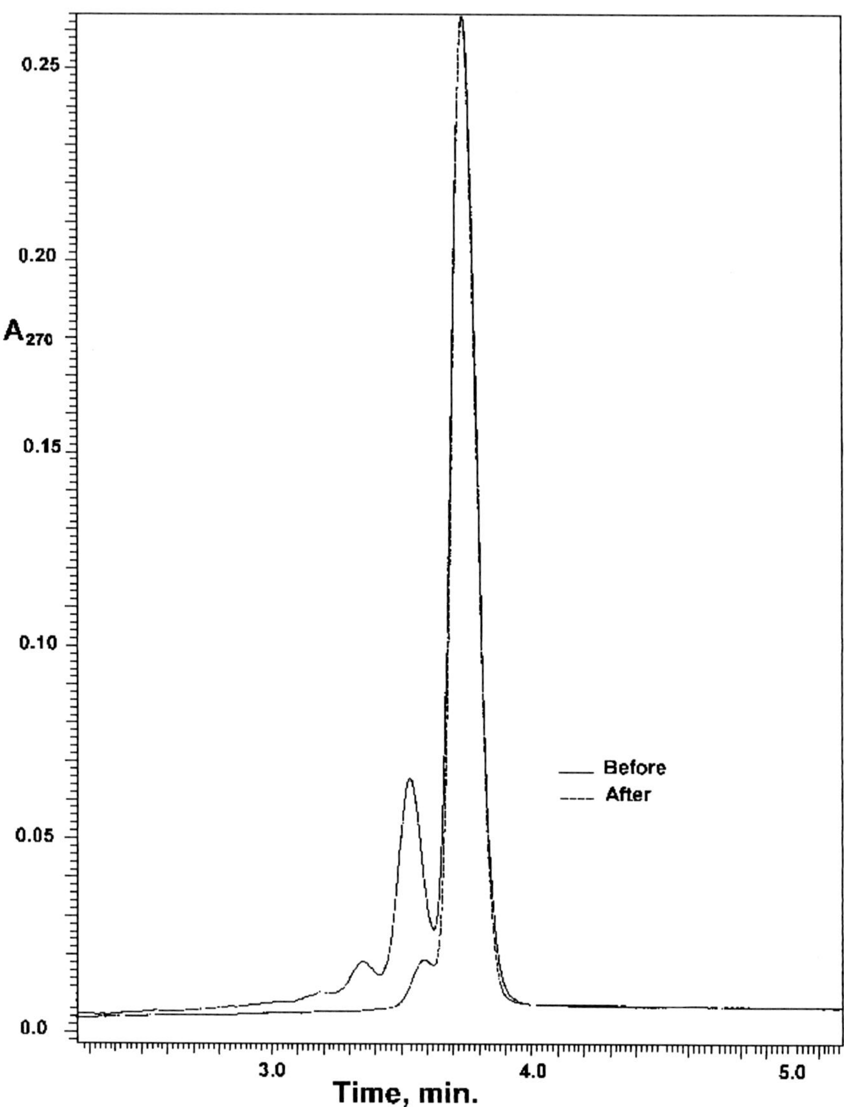

Figure 4. Ion-exchange HPLC analysis of GEM®91, a 25-mer phosphorothioate (PS), 5'-dCTCTCGCACCCATCTCTCTCCTTCT, before purification (—), and after purification (---). GEM®91 was analyzed on a Dionex NucleoPac SP3302 column, maintained at 65°C, eluted at 2 mL/min with a 3 min. gradient from 0.45 – 2.25 M NH₄Cl in 1.0 mM EDTA, 20 mM Tris-HCl, pH 8.0, 10% CH₃CN. Oligonucleotides were detected by absorbance at 270 nm. Subsequent mass spectral analysis revealed the impurities to include oxygenated species.

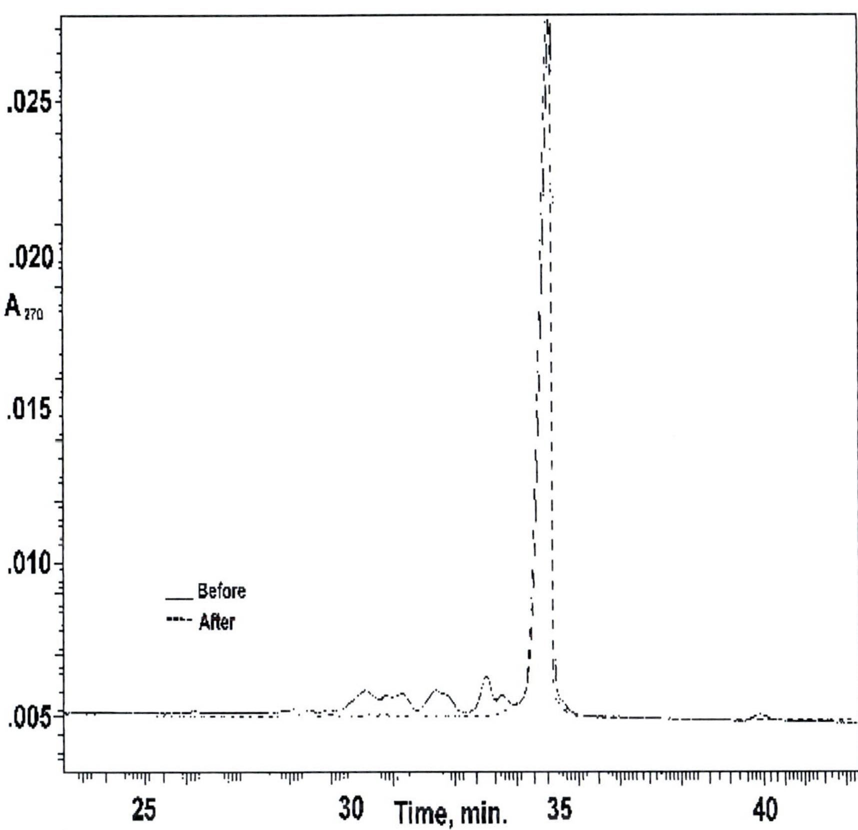

Figure 5. CGE analysis of an 18-mer 2'-O-methyl hybrid oligonucleotide with a phosphorothioate backbone, before purification (—), and after purification (---). An aliquot of a 0.1 g/L solution was injected onto a 66 cm denaturing polyacrylamide gel-filled capillary and electrophoresed on a modular system with a Glassman High Voltage power supply, at 400 V/cm, at ambient temperature. Oligonucleotides were detected by absorbance at 270 nm with a Thermo Separation Spectra 100 variable wavelength detector.

Analysis of the purity of the crude API, in terms of ion-exchange (IEX) high performance liquid chromatography (HPLC) (Figure 4) and capillary gel electrophoresis (CGE) (Figure 5), is used as a control point in purification. For example, if the percent phosphodiester (PO) in a phosphorothioate (PS) is above a certain level, the purification can be adjusted by increasing the column wash volumes to ensure that PO contaminants are washed away before the PS product elutes.

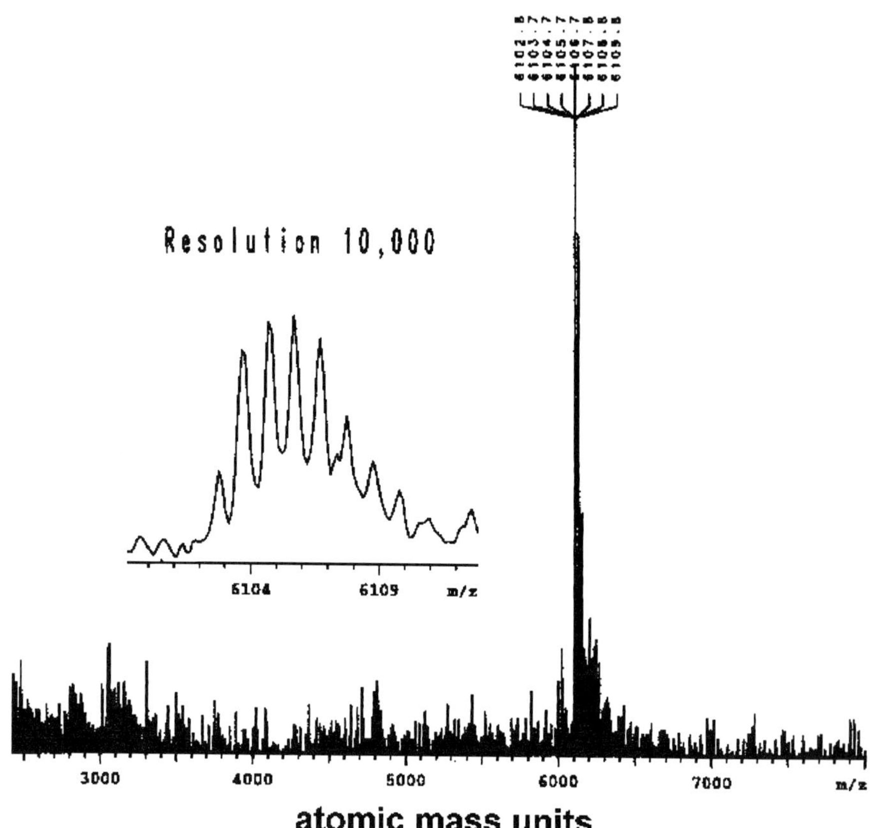

Figure 6. MALDI-TOF MS analysis of pdT$_{20}$ phosphodiester utilizing pulsed delayed extraction. The icosamer, 50 μM, was dissolved in a matrix of 65 g/L 3-hydroxypicolinic acid, 16 g/L N-(3-indolylacetyl)-l-leucine, 50% CH$_3$CN, 50% water. A nL aliquot was deposited on the target, and analyzed on a Bruker Reflex II MALDI-TOF mass spectrometer equipped with a pulsed delay extraction ion source and an ion reflector, operated in the positive ion mode. Pulsed delay voltage was 8 kV with total acceleration voltage 25 kV. The spectrum was externally calibrated with a standard dT$_n$ mixture. The central peak yielded an experimental mass of 6106.7 amu.

Sophisticated analytical techniques such as matrix assisted laser desorption time of flight mass spectrometry (MALDI-TOF MS) (Figure 6), and ^{31}P nuclear magnetic resonance (NMR) spectroscopy (Figure 7) are used to confirm the specific identity of these large molecules.

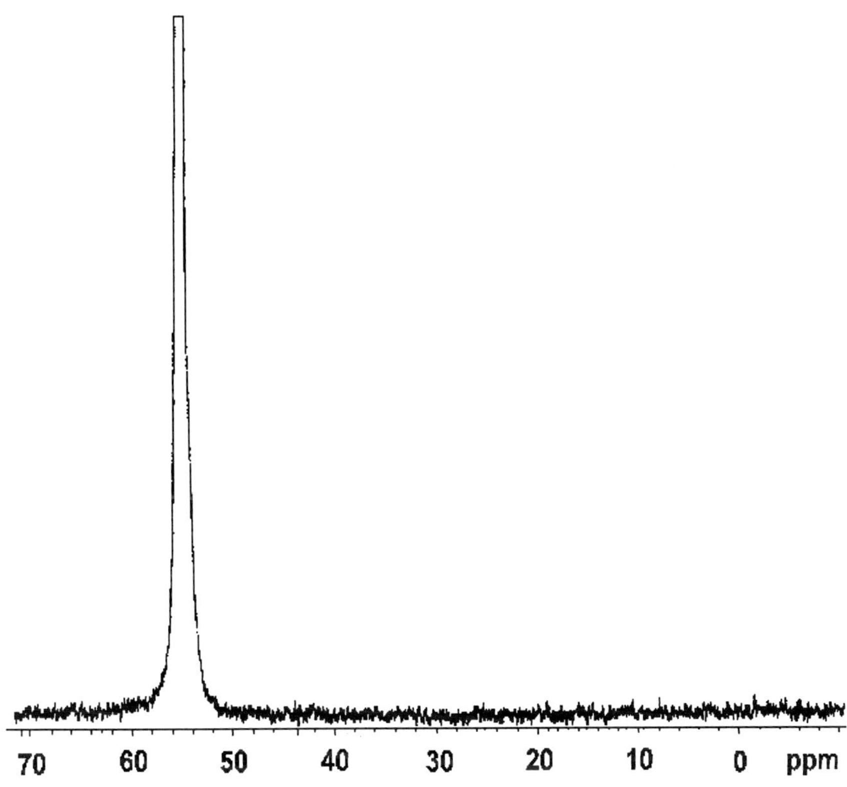

Figure 7. ^{31}P-NMR spectrum of 13 g/L GEM®91, a 25-mer phosphorothioate (PS) oligonucleotide, in D_2O on a Varian Unity 300 MHz NMR with a ^{31}P probe at 121 MHz. 27696 repetitions were acquired in 12 hours. The main P=S peak appeared at 55 ppm, but no detectable P=O peak at 2 ppm. The limit of quantitation was determined from a standard curve prepared by adding a solution of a phosphodiester (PO) 20-mer to the phosphorothioate solution at concentrations of 6.7, 13.4 and 20.1 •g/mL equivalent to 0.05, 0.10 and 0.15% (w/w) PO in a PS backbone.

Advances in MALDI-TOF MS technology such as pulsed delayed extraction have improved on this specificity by enhancing the ion resolution. Figure 6 illustrates the resolution of isotopes achieved with the delayed extraction technique for pdT_{20}.

In general for both synthesis and purification, in-process trends can be identified before they have a negative impact on product quality. Trend charts are also generated retrospectively and used to evaluate performance and to identify problem areas.

II.D. Validation

Validation of new processes defines the conditions under which the manufacturing process is reproducible, and provides assurance to the manufacturer that the process is under control (Puma, 1996).

cGMP compliance requires that drug manufacturers perform prospective validation of a process. In this approach, approved validation protocols are executed in consecutive stages, with the first stage consisting of execution of protocols for the installation and operation qualification (referred to as IQ and OQ) of process equipment and facilities. When these are complete, the next stage is to qualify the process itself (process qualification or PQ).

Validation protocols are supported by approved calibration programs and validated analytical assays. Formal IQ, OQ, and PQ reports are written and circulated for review and approval by production, technical, engineering, and quality departments. At a minimum, three consecutive process qualification runs demonstrating adherence to critical process control parameters are required to demonstrate control. Criteria used to demonstrate control of a synthesis process includes consistency in yield and purity of the crude API. Shorter oligonucleotide failure sequences, referred to as "n − x" impurities, are a complex mixture of impurities, which vary from product to product. Except for n − 1 species in PS oligonucleotides, anion exchange chromatography (Figure 4) is highly effective in the separation of all other n − x species. Control of the presence of n − 1 species in the final PS oligonucleotide product relies almost exclusively on the fidelity and reproducibility of the synthesis.

Process validation also involves demonstrating adequate and reproducible removal of contaminants. The most likely contaminants of oligonucleotide products are oligonucleotide-related species present in a multi-product production facility. In oligonucleotide synthesis, these contaminants would be expected in the solid phase bioreactor, the solid phase itself, and the cleavage/deprotection equipment. Likewise, in purification of oligonucleotides, chromatographic fittings, frits and resins are expected sites of contaminants. Therefore, cleaning processes are validated to demonstrate that they are effective at preventing cross-contamination between products sharing common process equipment. Equipment cleaning procedures are validated following an approved protocol. As an added measure, when cleaned equipment is devoted to making a different product, a changeover assay is routinely done to verify removal of the previous product before the next product is introduced. Fortunately, most therapeutically useful oligonucleotides are very soluble in water. This facilitates cleaning of equipment and reduces the use of cleaning agents such as detergents, which can leave a residue.

Some purification equipment items, such as chromatography column frits, are dedicated for single product use, due to the difficulty of validating adequate cleaning. Materials such as chromatography resins, filtration membranes, and peristaltic pump tubing are also dedicated to single-product use.

Some characteristics of the processes used in oligonucleotide manufacturing facilitate validation. Manufacture of oligonucleotides at large scale involves the use of relatively few methods. There are separate synthetic methods for PS, PO, hybrid, and chimeric compounds (Figure 8), methods that are specific for addition of particular phosphoramidites.

These methods can be rearranged to generate a large number of sequences, which facilitates process development and process validation. Tailoring the synthesis process for a particular product largely involves relatively minor alterations in contact times, column volumes, and molar equivalents delivered of the synthesis reagents. While each product must be independently validated, knowledge gained in validating one product can be applied to validation of others in the same class.

On the purification end of the process, we have successfully utilized the same three purification steps described above (HIC, anion exchange chromatography, and ultrafiltration, Figure 1) for a variety of types of compounds (PS, PO, hybrid, chimeric, oligonucleotides with linkers and/or modified bases).

While each product requires a unique purification process that must be independently validated for use, the differences between processes relate primarily to pH, salt concentration, and the number of column volumes of wash and elution buffers used in each step. This approach permits rapid scale-up from less than 1 mmol up to 100 mmol without compromising purity and yield.

Therefore, use of a small number of common raw materials coupled with rugged process designs facilitates process validation, cleaning validation and documentation (i.e., SOPs, batch records, and protocols and reports) necessary under cGMPs.

While purification equipment also permits in-process monitoring of pressure, conductivity, flow, and absorbance, the monitoring of the purification stream by HPLC (Figure 4) and CGE (Figure 5) analysis is routinely used to make pooling decisions, as a tool for controlling the process, and in building trend charts.

II.E. Second Generation Oligonucleotides

While current large scale manufacturing is targeted towards phosphodiesters and phosphorothioates, "second generation" chemistries are

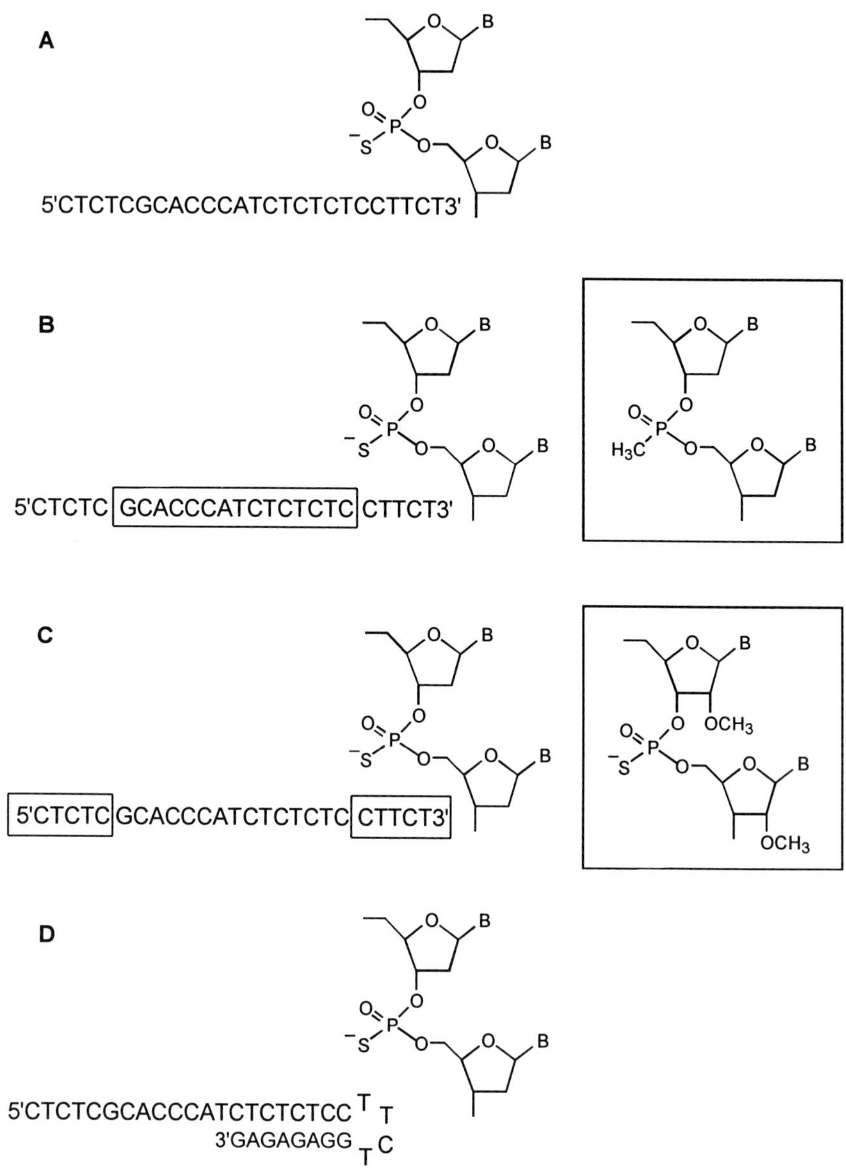

Figure 8. Structures of different modifications of GEM®91 oligonucleotide. **A:** phosphorothioate (PS); **B:** Chimeric phosphorothioate-methylphosphonate-phosphorothioate (PS/MP/PS); **C:** Hybrid 2'-O-methyl phosphorothioate-2'-deoxy-phosphorothioate-2'-O-methyl phosphorothioate (OM/PS/OM); **D:** Self-stabilized phosphorothioate 3'-hairpin.

rapidly leaving the lab bench and becoming candidates for use as therapeutic agents (Figure 8). In addition, oligonucleotides containing a variety of modified bases and linker molecules are of increasing interest for a variety of other applications.

II.E.1. Synthesis

Nonionic methylphosphonate oligonucleotides (Hogrefe, *et al.*, 1993) may be synthesized as a pure methylphosphonate, or as a product which combines regions containing methylphosphonates and standard phosphorothioates or phosphodiesters. Such products are referred to as "chimeric" molecules. These molecules present a challenge to cleavage, deprotection, and to some extent purification, since the methylphosphonate portions of a chimeric product are relatively less stable to basic pH than the non-methylphosphonate portions. For this reason, concentrated ammonium hydroxide, successfully used with first generation oligonucleotides (Figure 1), is not ideal for cleavage/deprotection. Using a gentler deprotection agent such as ethyl-enediamine in ethanol gives better results.

"Hybrid" molecules are composed of DNA and modified RNA-like portions. The 2'-O-methyl amidites used for the modified RNA portions of the molecule require longer coupling times to achieve acceptable coupling efficiencies. Thus, large-scale synthesizers having the capability to recycle coupling reagents are best suited to the production of such hybrid oligonucleotides.

II.E.2. Purification

The aqueous purification process described above is readily adapted for use in purification of oligonucleotides with modified bases and linkers, and for purification of chimeric and hybrid molecules. Purification is trouble-free as long as the process is designed to avoid conditions such as low pH, which affect the stability of particular modifications.

III. FACILITY DESCRIPTION

The design of Hybridon's cGMP API production facility required intensive architectural, engineering, and coordination efforts. Areas requiring special attention included layout and utilities, heating ventilation, and air conditioning (HVAC) for clean rooms, solvent distribution, water systems [reverse osmosis and deionized water (RO/DI) and water for injection (WFI)], product, and personnel flow. Figure 9 shows the layout of Hybridon's cGMP facility.

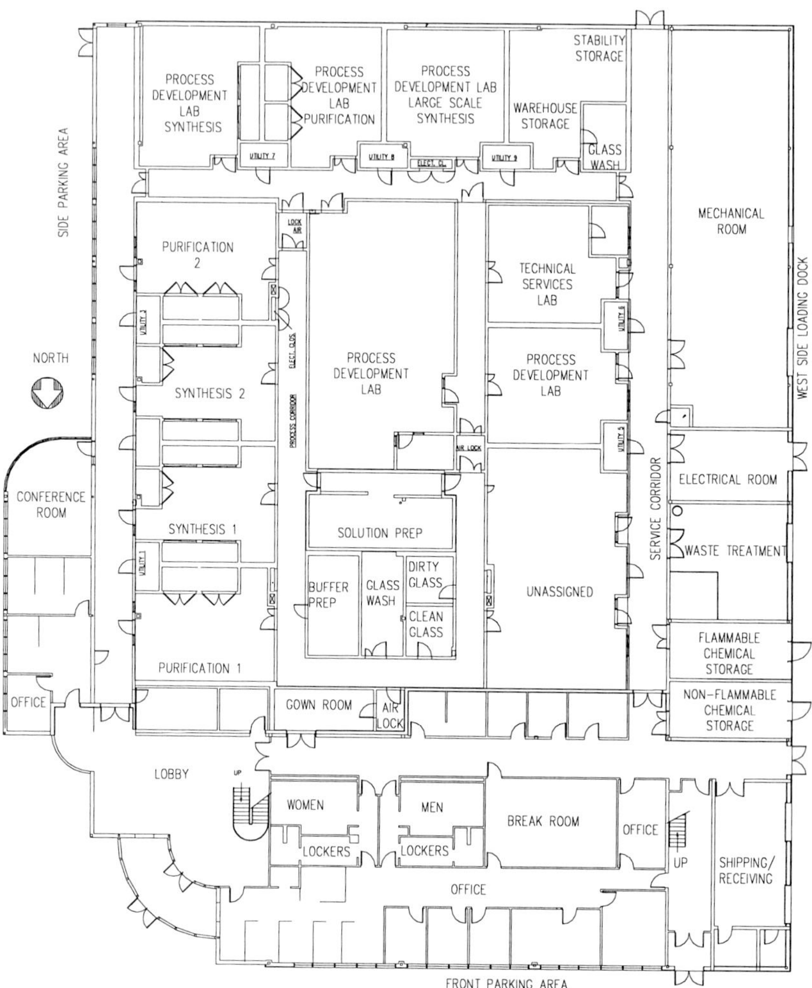

Figure 9. First Floor Layout of Hybridon's Integrated cGMP Bulk Oligonucleotide Manufacturing Facility. Besides Production, the building housing Hybridon's bulk manufacturing facility in Milford, MA contains all the requisite support groups for Production including Quality Control, Quality Assurance, Technical Support, Process Development, Engineering and Maintenance, Purchasing, Materials Control, and Administration. The production suites are located along the observation corridor. The solution preparation area is located near the center of the building. The area labeled "unassigned" is under construction to house separately large scale lyophilizer operations, and organic synthesis and purification of specialty products. See Table 1 for additional description of the plant.

Table 1. Statistics of Hybridon's cGMP Production Plant

- 36,000 square feet facility
- 15,000 square feet of class 100,000 controlled area
- Seamless, chemically resistant highly cross-linked vinyl ester floors
- HVAC 38,000 cubic feet per minute, stainless steel exhaust system
- 27 miles of pipe, 2 miles welded stainless steel piping for high purity solvent distribution
- 25,000 gallon tank farm with 6,000 cubic feet per minute exhaust, transfer station for solvent distribution
- 122 gallon per hour (maximum) still for Water For Injection, with 1,500 gallon storage tank, 35 gallon per minute distribution system
- 360 gallon per hour reverse osmosis/deionized water system with 1,000 gallon tank, 60 gallon per minute distribution system
- Class 1, Division II explosion proof solvent preparation area
- 600 gallon per hour Waste Water Treatment System
- Buffer Preparation Suite
- Air locks and gowning room
- Individual Air Handling Units for each Suite to prevent cross-contamination
- 175 kW emergency Generator with Uninterrupted Power Supply system
- 276 cubic feet per minute dry compressor system
- 600 ton chiller capacity
- 170 horsepower boiler system for Water For Injection and Heating Ventilation, and Air Conditioning

In designing and building such a facility, it was imperative that the engineering and construction firms work towards a GMP facility from the beginning of the project. This required extensive documentation of specifications, blue prints, "as built" drawings, engineering change orders, construction qualification, and ultimately facilities validation.

Facilities validation included a Master Validation Plan, blueprints, and drawings, specifications, change orders, commissioning reports, IQ, OQ, and PQ protocols, hazardous operations study, process description, material and personnel flows, preventative maintenance protocols, and environmental monitoring. In addition to FDA regulations, the facility was required to be in compliance with EPA, NFPA, and OSHA regulations. All the appropriate federal, state, and local permits had to be in place before completion of the construction. The process of obtaining these permits typically took 12 to 18 months to complete and facility validation required 18-24 months to complete.

A list of engineering and utility statistics for Hybridon's cGMP facility, designed for an annual throughput capacity of one thousand kilograms of synthetic oligonucleotides, is shown in Table 1. Hybridon's facility was designed such that multiple products could be produced simultaneously in separate suites. Similarly, storage areas were designed for separate quarantine and release of raw materials and products.

IV. OUTSOURCING MANUFACTURE OF FINISHED PHARMACEUTICAL DRUG PRODUCT

Several well established commercial contractors in the U.S. specialize in pharmaceutical manufacturing operations. For clinical trials, these contractors present an attractive option for the manufacture of a pharmaceutical product versus establishing an internal fill and finish capability. Contract manufacturing of products can reduce manufacturing risk and improve the use of available resources. These are desirable from both a cGMP and a sound business perspective.

Advantages of outsourcing include the deferral of capital expenses to build a production facility and avoiding the investment required to develop the operating expertise and attendant manpower. In addition, the ready availability of expert contractors in this field provides an opportunity to tap their technical strengths at a reasonable cost. Critical elements to the success of the contract manufacturing experience include selection of the contractor, technology transfer, R&D reports, and process validation.

IV.A. Selection of a Finished Drug Product Contractor

Hybridon's first step in the selection of a contract manufacturer for the dosage form is a thorough quality assurance audit of the contractor's facility. Deficiencies noted by the auditor are identified as either major or minor and submitted to the contractor in writing. cGMP violations require the contractor to commit to addressing these deficiencies before the facility is selected.

The contractor provides the quality assurance auditor with copies of the following: observations issued to the firm by the FDA over at least the last three years, the company's responses to these observations, and the FDA establishment inspection reports for each. Facilities that have failed one or more preapproval inspections are disqualified from further consideration.

The second tier of selection criteria includes contract price, production flexibility, production capacity, dosage form development capabilities, and client references. To evaluate manufacturing pricing, Hybridon requests from each contractor quotes for multiple scales spanning early clinical to commercial production scale because of the evolving needs of clinical programs.

The flexibility of a contractor's production schedule is also an important factor. A good measure of production flexibility is the lead time between requisition for production and the actual production date. Lead times of one month versus three or more months facilitate faster responses to the changing requirements of clinical programs.

IV.B. Technology Transfer Document

Hybridon prepares a Technology Transfer Document (Table 2) to communicate the manufacturing process conditions which are necessary for the successful manufacturing of the finished pharmaceutical drug product by the contractor.

Table 2. Table of Contents for Pharmaceutical Manufacturing Technology Transfer Document

- Development findings
- Product definition detailing the formulation, container, closure, and seal
- Manufacturing instructions
- Formulation precautions
- Certificate of analysis for the active ingredient
- Certificate of analysis for the reference standard
- Relevant analytical methods
- Specifications for raw materials
- Specifications for excipients
- Excipient release testing
- In-process testing
- Raw materials release testing
- Drug product release testing
- Stability data on drug substance
- Stability data on drug product
- MSDS (material safety data sheet)

Table 3. Table of Contents for the Pharmaceutical Research and Development Report

• Preformulation studies on the drug substance and drug product
• Development of prototype and the final dosage form
• Stability data on the drug substance and drug product
• Information on drug product development and production scale lots
• Analytical methods for assay of the drug substance and drug product
• Analytical methods for monitoring stability of the drug substance and drug product

IV.C. Research and Development Report

A Research and Development Report (Table 3) is also prepared to chronicle the development of the dosage form from early stage development through validation and commercial manufacture.

IV.D. Validation of the Manufacturing Process

As with API production, validation of pharmaceutical manufacturing assures reproducibility and control of the production process, which are key cGMP requirements. The validation process begins with a protocol that outlines the critical parameters of the manufacturing process, specification ranges necessary to the process, and how these parameters will be monitored during the validation process. Drug product contract manufacturers generally use a standard validation protocol that can be modified to suit most products.

The validation of the manufacturing process requires the manufacturing of three lots of drug product that meet the established manufacturing specifications. The lot size used to validate the process should approximate the intended commercial scale. Upon completion, a validation report is written to document the findings of the validation process.

V. QUALITY ASSURANCE

At Hybridon, Quality Assurance serves several functions: 1) maintenance of all the documentation necessary for manufacturing and control of the product, 2) control of the receipt and use of raw materials, packaging, and labeling so that only approved materials are used, 3) development, implementation, and maintenance of quality systems to ensure that the plant remains in a state of control, 4) performance of internal audits to measure compliance to internal quality systems, vendor audits, and the oversight of corrective action programs to maintain cGMP compliance.

Documentation is crucial to the success of the quality system in manufacturing. Documentation forms the basis for consistent manufacture, proof of quality, and the ability to evaluate process efficiency and quality trends.

V.A. Documents

V.A.1. Policies

Policies are higher level documents that describe in general terms how the various departments interact and how the quality systems function. These documents provide the necessary overview of how the GMP functions operate.

V.A.2. Standard Operating Procedures (SOPs)

SOPs provide "how to" instructions on equipment operation, completion of forms, formatting and writing documents, system operations, and change control.

V.A.3. Specifications

Specifications are written parameters that are absolutely necessary for the control of raw materials, containers/closures, and in-process and finished materials.

V.A.4. Test Methods

Test methods are written descriptions of the analytical procedures used to obtain consistent and accurate results for acceptance or release of raw materials, intermediates, and completed products.

V.A.5. Batch Records

Batch records provide instructions on performing operations and capturing the important production and control information from each batch produced. They form the basis of review and approval of each batch of product.

V.A.6. Forms

Many different types of data, other than manufacturing information, must be recorded and tracked. Forms provide standard reporting formats for environmental monitoring data, analytical results, process deviations, and change control. This ensures that all relevant data are included and presented in an easy to read manner.

V.B. Control Systems

V.B.1. Document Change Control

Change control over all of this documentation is critical to ensure that only approved changes are made, and that only the current versions are available for users. The Quality Assurance department handles document control and its accompanying change control. Documents are stored in locked fireproof cabinets, accessible only through QA documentation personnel. Superseded versions are identified as such and maintained in these files. This provides a history of how the processes were run and what specifications were in effect throughout the development of the oligonucleotide.

V.B.2. Raw Material Control

Control of raw materials is critical to compliance with cGMP. In the initial stages of product development, raw material is less controlled since it may be changed frequently in research applications. However, very early in the process, as material parameters are identified and development continues, ever tightening limits are established as raw material specifications. This requires a close relationship with the vendors of these materials, sharing solutions to quality issues, new information, and test methodology for proper characterization. The present tense is intentionally used here, because new materials and manufacturing methods are constantly being developed at a fairly rapid rate. Sharing of test methods and the establishment of vendor agreements to accept the client's test results as a basis for accepting or rejecting lots is crucial.

A raw material quarantine/approval system provides control over the use of acceptable materials. Records are kept of each lot of material received, its test results, whether accepted or rejected, and its ultimate use in batches of product.

Since a key tenet of cGMP states that the quality of a material cannot be increased by testing, it is important to audit vendors of critical materials to verify that they are operating under a state of control. This audit program eventually results in a vendor certification program, whereby certain vendors are certified to supply each raw material, and test results supplied are accepted with only periodic verification.

V.B.3. Facility Control

At Hybridon, Quality Assurance is responsible for ensuring that the facility installation qualifications and operation qualifications are performed adequately on the air handling systems, water systems, and major process equipment, and that only authorized changes are made. Unauthorized changes can invalidate these systems.

Separate air handlers are used for the manufacturing suites to prevent any possibility of cross contamination. The manufacturing areas are classified under FED STD 209E and are routinely monitored. This is a U.S. government standard for airborne particulate cleanliness classes in clean rooms and clean zones. City supplied water is used as a feed water for the RO/DI system. This water is processed to produce water that exceeds the USP requirements for "Purified Water". It is then used for non-critical processing and as feed water for a Water for Injection, USP (WFI) system. The WFI water is delivered to points of use via a constantly circulating hot loop. Both the RO/DI and the WFI water are monitored frequently during the week, so that all points of use and critical areas before and after filters are tested.

V.C. Product Release Control

Establishment of final release specifications requires a detailed look at process capability, safety issues, and test methodology. Oligonucleotides are novel compounds, and as such have little history with regulatory agencies. Guidance for product release, especially for early production lots, is very limited. As the process move towards commercialization, it becomes imperative to establish impurity profiles.

Specifications for the completed drug substance are initially set with wide tolerances for impurities and minimal characterization. As the process becomes more established, certainly by phase III trials, the specifications have become considerably tighter with much reduced impurity levels and characterization of the major impurities. Since any one method may lack the necessary specificity (Kambhampati, *et al.*, 1993) identification has been accomplished by applying several methods: ultraviolet (UV) absorbance spectroscopy, ion exchange HPLC (Figure 4), CGE (Figure 5), MALDI-TOF MS (Figure 6), and ^{31}P-NMR spectroscopy (Figure 7). Impurities may be encountered from reaction by-products, reactants, nucleotide or base deletions from the chain ($n - 1$, $n - 2$, etc.) and salt. Depending upon the contaminant, ^{1}H-NMR and reverse phase HPLC are also used. CGE has been the most effective method for measuring impurities (Figure 5).

V.D. Quality Committees

In the maintenance quality systems, it is necessary for QA to serve as a focal point for various committees. The Materials Review Board (MRB) consists of a core of individuals from QA, materials management, production, and technical services. These members have the expertise to judge the acceptability and criticality of material from the perspective of product quality, regulatory compliance, technical needs, and availability. Additional members can be brought in for issues requiring other expertise. The MRB meets to resolve non-conformances of any material, from raw materials and supplies through in-process and finished products. Specifications for materials, containers, closures, packaging, and labeling are handled through a subcommittee of the MRB.

VI. QUALITY CONTROL

The essentials of quality control (QC) for pharmaceutical manufacturing are described in 21 CFR parts 210 and 211, the cGMP regulations. Specific guidance is included in FDA documents (Food and Drug Administration, 1987,1990,1993,1994; Kambhampati, *et al.*, 1993) and professional association publications (Lazar, 1993,1995,1996; European Chemical Industry Council, 1994; International Society for Pharmaceutical Engineering, 1996). At Hybridon, the QC unit conducts analyses to evaluate the identity, strength, quality, and purity of raw materials, drug substance, and drug product. Based on these findings, QC recommends acceptance or rejection of the tested lots to QA, which decides on their final disposition.

VI.A. Raw Materials

The quality and purity of the active pharmaceutical ingredient and the drug product are impacted by the quality of the starting materials used in the synthesis of the oligonucleotide. An effect can be observed both when these materials are incorporated into the oligonucleotide and when they support the synthesis or purification processes. Therefore, appropriate specifications and analytical methods are important for every material that is ordered for large scale manufacturing. In contrast to classical small molecule pharmaceutical production, few of the raw materials used in oligonucleotide manufacture are specified in USP/NF monographs. Standard Operating Procedures (SOPs) describe what to do when the material arrives from the vendor to have a high degree of confidence that it will function in production as expected. Typical procedures address: 1) how to verify that the correct material has arrived; 2) where to quarantine the material and under what prescribed storage conditions; 3) the number of containers that must be sampled to obtain a statistically valid representation of the lot; and 4) what controls must be satisfied before the material is released for use by production. Other procedures detail warehouse controls, handling and disposition of rejected materials, and safety and health considerations.

VI.B. Oligonucleotide Testing Including Stability

During the production process, many in-process tests are conducted to monitor key steps. Typically the yield for the synthesis is estimated before purification by determining the oligonucleotide concentration from UV absorbance, and initial quality is monitored by chromatographic and electrophoretic methods.

Specifications are established for each of these steps; only samples that have analytical results within the specified limits are approved for further processing. When production is complete, the purified oligonucleotide is subjected to rigorous testing to ensure its identity, quality, purity, and strength. Table 4 contains a list of typical analyses that may be specified for a particular API.

Several of these tests have been developed specifically for antisense oligonucleotides intended to be used as therapeutic agents. New technology (Belenky, et al, 1995) was required to verify the absolute sequence of the oligonucleotide because the high specificity of antisense activity is due to its base sequence. Sophisticated analytical techniques such as CGE (Figure 5) and MALDI-TOF MS (Figure 6), as in II.C.2 above, are used to confirm the specific identity of these large molecules.

Expiry dating required by cGMP regulations for finished drug product is determined based on a stability testing program. The International Commission on Harmonization (ICH) has issued guidance on the essentials of an acceptable stability program, and the FDA published these in the Federal Register (1994) as requirements for all new drug substance and drug product stability programs.

It is Hybridon's responsibility to demonstrate that the product maintains its strength throughout the expected lifetime using well designed stability studies. Data from stability-indicating analytical methods are required to support the expiration period. These studies are conducted as part of the quality control evaluation.

Table 4. Typical Analytical Tests for Oligonucleotide Pharmaceutical Substance or Product

IDENTITY TESTS

- Appearance by color and morphology
- Molecular Weight by MALDI-TOF mass spectrometry
- Phosphorothioate backbone by ^{31}P-NMR
- Base Sequence by Capillary Gel Electrophoresis-LIF or MALDI-TOF MS
- Retention time by HPLC, CGE, or PAGE
- Elemental Analysis

PURITY TESTS

- Chromatographic Purity by ion-exchange HPLC
- Electrophoretic Purity by CGE
- Moisture Content by Karl Fischer coulometric titration
- Heavy Metals by specific atomic absorption
- Volatile Organic Compounds by head-space gas chromatography

QUALITY TESTS (Current USP <test method identification>)
- Endotoxin Level by USP<85>
- Pyrogen by USP<151>
- Sterility by USP<71>
- Particulate Matter by USP<1>
- Content Uniformity by USP<905>

STRENGTH TESTS

- Assay by UV absorbance with theoretical extinction coefficient
- Assay by CGE with authenticated reference standard
- Assay by ion-exchange HPLC with authenticated reference standard

VI. ENVIRONMENTAL QUALITY

Besides controlling the manufacturing processes and resulting products for quality, QC also tests for environmental quality. The environment within which the operations occur and the cleanliness of the production equipment impact the quality and purity of the drug products. Routine environmental monitoring programs for particulate counts are conducted to protect against incidental contamination. Cleaning processes are validated through sensitive and specific analytical tests.

Prevention of microbial contamination is particularly important during the water-based purification stage of production. Control of the water systems to the quality specified by the USP for use in production of drug substances requires an extensive monitoring program with realistic Alert and Action limits. In-process microbial load testing may be needed to monitor levels at those steps in the process that are vulnerable to microbial growth. There is a regulatory guidance (FDA, 1994) concerning control of endotoxins in drug substances intended for parenteral use. The endotoxin level measured in the substance is reported on the Certificate of Analysis.

VII. PROCESS DEVELOPMENT

During the last four years, process development of oligonucleotides focused on answering three crucial questions to the future feasibility of commercializing oligonucleotides as therapeutics. 1) Can oligonucleotides be produced at sufficiently large scale to support parenteral administration? 2) Can oligonucleotides be produced in a cost efficient fashion at large scale? 3) Can oligonucleotides be made at the purity required of therapeutic agents?

VII.A. Development of a Commercial Operation Scale

In addressing the first question, we demonstrated that flow-through technology with a fixed bed column reactor is suitable for scaling up the solid phase synthesis of oligonucleotides to commercial scale. At Hybridon, we designed, built, and applied for a patent for one such large scale instrument, the Hybridon 601. In Table 5, its performance is compared to the OligoProcess, a second instrument that Hybridon codeveloped with Pharmacia Biotech and which was first brought to full scale in the production of Hybridon's first clinical compound GEM®91 in our production facility in Milford, MA.

The crude GEM®91 throughput of either synthesizer averages 560 grams per batch, a productivity rate of 47 grams per hour of synthesizer operation. In purification we developed a two-step low pressure column chromatography process that is readily scaleable to produce about 2 kilograms of purified GEM®91 per batch, a productivity rate exceeding 250 grams per hour of operation. Based on these factors, we calculate an annual throughput for the Hybridon production plant of 1000 kilograms. The average recovery of GEM®91 during purification is 70%.

Table 5. Comparison of Solid Synthesis Support of GEM®91 (25mer PS) at Different Scales and Synthesizers

Machine	Scale, mmol	%Yield, DMT	%PO, NMR	%N, CGE	%N−1, CGE
Millipore 8800	2	NA	NA	58	NA
OligoPilot II	3	79	0.3	67	7
OligoProcess	10	76	0.3	72	4
OligoProcess	40	75	0.4	71	4
OligoProcess	100	78	0.3	77	4
Hybridon 601	10	78	0.3	74	4
Hybridon 601	50	76	0.3	68	7

Syntheses were carried out with CPG support except Pharmacia 30HL support was used with the OligoProcess. DMT %yield results were obtained from anion exchange HPLC. %PO was determined by ^{31}P-NMR. %N and %N−1 data resulted from CGE. CPG leads to higher n − 1 content. Compare OligoPilot II at 3 mmol and Hybridon 601 at 50 mmol with the OligoProcess runs.

VII.B. Development of Cost Efficient Production Processes

In the accompanying semilog plot (Figure 10), we show production costs for GEM®91 at the largest scale available at the time. Since 1991, production costs have decreased at a 50% compounded annual rate. Over the retrospective portion of the time period depicted in the graph, cost improvements are due to multiple factors, such as the improved efficiency in the use of amidites and solvents, due to the change from solution phase to solid phase synthesis, labor cost reductions due to over 1000-fold scale up, and substantial reductions in the price of starting materials. In 1995, scaling up the process reduced the labor contribution to the synthesis cost by a factor of ten. In addition, the substitution of the reverse phase purification process by the readily scaleable aqueous process has, so far, reduced labor cost in purification by a factor of four.

The costs for 1997 and 1998 were projected from the previous 6 years of cost history for GEM®91. These costs assume production of 50 kg/year at customary high yields, 3.4 g purified oligonucleotide per mmol of solid support, about 40%. Costs for other molecules will vary on the basis of chemistry, yield, and volume. However, the GEM®91 learning curve is readily applicable to other molecules. Further gains are likely in the future, when oligonucleotides will be produced at an even larger scale. Future cost improvements are expected on the basis of implementing novel, proprietary Hybridon technology including less expensive solid supports for synthesis that also result in higher yields and final product purity, and less expensive sulfurization reagents. We believe that production cost will continue to decrease at this rate for at least the next two years.

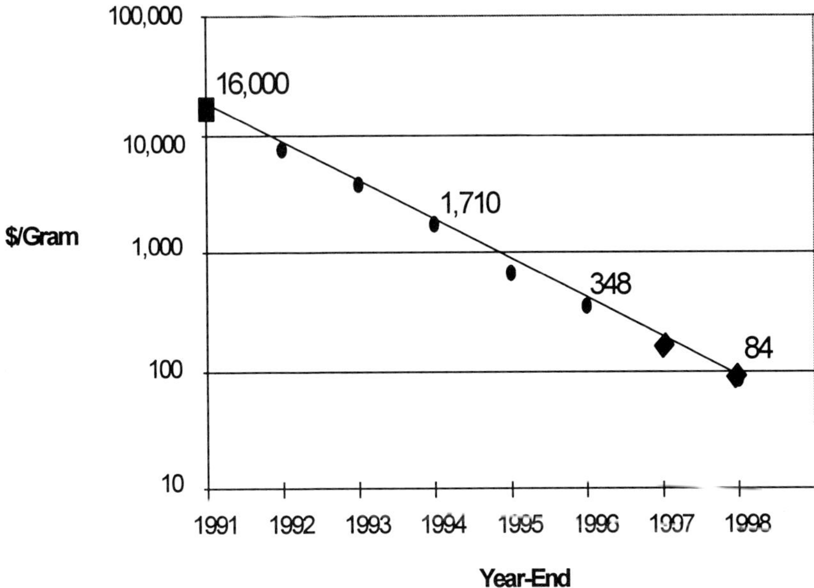

Figure 10. Decrease in Production Cost of GEM®91 over 6 Years. Solid line is a best fit linear regression of the fully allocated costs.

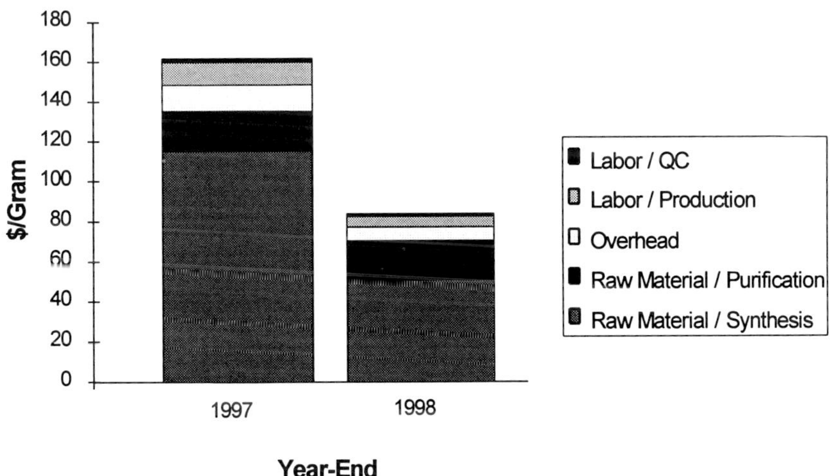

Figure 11. Standard Cost Distribution Analysis of GEM® 91. This cost analysis is based on 12 synthesis lots per month at the 100 mmol scale and 6 purification lots per month at the 800 gram scale.

As shown in the standard cost distribution graph (Figure 11), following the scale-up of these processes, 84% of the production cost of GEM®91 comes from the raw material cost. These costs are largely independent of operation scale, and can only be addressed by increasing the efficiency of conversion of raw materials into product, decreasing the cost of raw materials through invention, increasing manufacturing volume, and vertical integration of production.

Presently, the three major raw material contributors to bulk oligonucleotide production cost are the sulfurization reagent, synthesis solid support, and amidites. At Hybridon, we have directed our cost reduction efforts by focusing on the invention of novel sulfurization reagents, polymer supports, in situ generation of amidites, and nucleotide dimer block synthons.

For instance, HYB-12, our proprietary sulfurization reagent (patent pending) costs only 10% of the cost of Beaucage reagent we formerly used. Similarly, our proprietary polymer support (patent pending) costs a fraction of the commercially available supports and performs very favorably in comparison (Table 6).

Hybridon's method for *in situ* generation of phosphoramidites (Zhang and Tang, 1996) (patent pending) leads to a 50% reduction in the cost of amidites, the building blocks of the oligonucleotides. This cost improvement can be compounded by the use of dimer synthons, providing a near optimal combination of the advantages of solution phase production (in the synthesis of the dimer synthons) with solid phase synthesis (in the assembly of the dimer synthons into the final oligonucleotide product).

The use of dimer block synthons halves the oligonucleotide synthesis cycle time, and reduces the use of solvents and other reagents, while doubling the oligonucleotide plant synthetic throughput capacity, and increasing crude quality and yield. The key advantages of solution phase production are the possibility of very large scale operation and high efficiency in monomer use, and relatively low solvent consumption. However, its drawback is that it cannot maintain a high level of purity for oligonucleotides more than 8 bases long. In contrast, solid phase synthesis excels at making high purity, longer oligonucleotides.

VII.C. Development of Therapeutic Quality Products

The use of dimer synthons in synthesis creates the most exciting opportunities to produce therapeutic quality oligonucleotides. Our experience in the large scale production of GEM®91 (a 25-mer phosphorothioate) shows purity, assessed by either ion exchange

Table 6. Synthesis of GEM®91 (25mer PS) Using Different Supports

Support	Loading, µmol/g	%Yield, DMT	%PO, NMR	%N, CGE	%N−1, CGE
Hyb-polymer	60.2	79.4	0.3	77.5	3.3
Hyb-polymer	95	NA	NA	77.0	1.7
Hyb-polymer	134.4	67.4	0.4	73.5	2.1
CPG	82	78.8	0.3	66.8	5.2
ABI-HLP	186	77.0	0.6	77.0	4.6
Pharmacia 30HL	92	74.3	0.3	72.0	3.0

Syntheses were conducted with 1.5 eq. of amidites and tetrazole. However, ABI-HLP support needs 3 eq. of amidites and ethylthiotetrazole. Loading refers to the amount of protected nucleoside present in the support. %DMT results were obtained from anion exchange HPLC. %PO was determined by ^{31}P-NMR. %N and %N−1 data resulted from CGE.

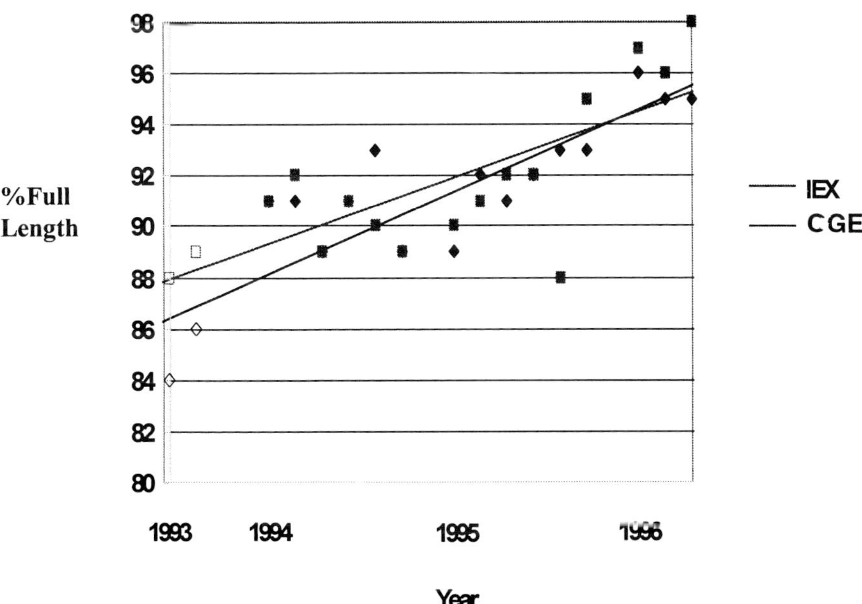

Figure 12. Continuous Improvement in the purity of all lots of GEM®91 produced since 1993. Percent purity based on ion exchange chromatography (IEX) analysis is represented by closed squares (■), where the dashed line represents best fit linear regression. Purity by capillary gel electrophoresis (CGE) is represented by closed diamonds (♦), where solid line represents best fit linear regression.

chromatography (IEX) (Figure 4), or capillary gel electrophoresis (CGE) (Figure 5), improving from 87% in 1993 to 98% in 1996 (Figure 12). The balance of the oligonucleotide species (2%) present in the product is the n – 1 family of GEM®91 related oligonucleotides that comprises the collection of 24-mers missing one nucleotide.

As discussed earlier, reduction of n – 1 species during purification of phosphorothioate oligonucleotides is largely ineffective. Since starting our work in 1993 in the dimer-based synthesis of phosphorothioate oligonucleotides (Tang, *et al.*, 1993), we have found that the use of dimer synthons eliminates the detectable occurrence of n – 1 species (assessed by CGE) in the crude product and, as a result, in the final product.

Thus, implementation of this chemistry at the large scale enables the production of oligonucleotides essentially free from n – 1 species, and the attainment of product purities exceeding 99%. In addition, we have been able to demonstrate that the combination of the extensive use of dimers in the synthesis of phosphorothioate oligonucleotides with anion exchange chromatography reduces the presence of phosphodiester bonds in the oligonucleotide product to undetectable levels, as evaluated by ^{31}P NMR (Figure 7).

VIII. SUMMARY

At the beginning of the decade, the feasibility of commercial manufacturing of oligonucleotides to support therapeutic application was in doubt. However, over the intervening years, the key manufacturing questions regarding scale of operation, product quality, and cost of goods have been answered. At Hybridon, these answers have taken physical form in the first commercial scale cGMP production facility in the world.

The Milford, MA facility has an annual capacity of 1000 kilograms of oligonucleotides. This capacity may be sufficient to support several products at commercial scale. To that end, the plant was designed to produce two products concurrently without cross-contamination risk. In addition, many more products can be made in the course of a year by campaigning production of each in the existing independent manufacturing suites. In contrast to the manufacturing of other drugs, oligonucleotides, made by assembling chemically similar building blocks, expand the manufacturing scope of a single, flexible cGMP plant to the production of many biologically distinct, but chemically similar therapeutic and diagnostic products. Currently, Hybridon Specialty Products manufactures more than ten different oligonucleotides used in commercial diagnostic applications, and clinical trials in one year.

Current Good Manufacturing Practices apply to the production of oligonucleotides under the oversight of the Center for Drug Evaluation and Research. Although oligonucleotides are structurally more complex than lower molecular weight chemicals, the same regulatory concepts of building quality into the product by means of controlling the manufacturing process and quality testing still hold true. At Hybridon Specialty Products, we have implemented cGMPs by taking a quality system approach. Logically, this approach grows from documentation, and change control of the production batch record to establishing in-process controls, raw material specifications, testing, and disposition, facility and equipment design, validation, and preventive maintenance, and personnel technical and quality systems training. The quality systems encompass the Quality Control function to ensure that testing is also performed under controlled, scientifically sound, and validated conditions. Within the Hybridon organization, Production and Quality Control functions are both subject to the regulatory guidance and compliance auditing of the Quality Assurance unit. Quality Assurance and senior business management are ultimately responsible for cGMP compliance.

Implementation of cGMP should lead to robust processes that result in consistent quality products. By combining the application of cGMP with continuous improvement of quality parameters, we have witnessed a steady increase in final product quality over the last five years. Moreover, we also observe very consistent final product quality among different clinical lots, supporting the notion that well-controlled processes are practiced. Presently, oligonucleotide quality standards are coming ever closer to the standards expected of smaller molecular weight compounds, and we foresee the probability of additional quality improvements through the introduction of new technologies in manufacturing, such as the use of dimer amidites, and better reaction solid supports.

Due to the inherent complexity and size of oligonucleotides, along with operation scale and product quality, lower cost of goods (COG) is a pivotal a factor in making therapeutic oligonucleotides feasible. Over the last seven years, we have reduced COG at a 50% compounded annual rate. Thus, commercial scale production of about 100 kilograms per year can be made presently at cost levels consistent with supporting an active therapeutic product for otherwise incurable, life threatening diseases. Cost reduction now also permits commercial production of much more potent derivatives for special applications, such as vaccine potentiation, specific disease states, or particular combinations of advanced chemistries. Considering both process development lead compounds undergoing scale-up, and development of advanced oligonucleotide derivatives with much greater potency, we believe that the scope of application of oligonucleotides as therapeutics will be extended within two years to more cost-sensitive markets. Our role in manufacturing is to stay ahead of therapeutic requirements in terms of scale, quality, and cost, to help advance the novel and seemingly limitless potential of oligonucleotide therapies.

ACKNOWLEDGEMENTS

We thank Bonnie J. Cauley for carrying out ^{31}P NMR spectroscopic analyses.

IX. REFERENCES

Agrawal, S. (ed.) (1996) *Antisense Therapeutics*. Humana Press, Totowa, NJ.

Beaucage, S.L. and Iyer, P.R. (1992) Advances in the synthesis of oligonucleotides by the phosphoramidite approach. *Tetrahedron* **48**:2223-2311

Beaucage, S. L. (1993) Oligodeoxyribonucleotides synthesis - phosphoramidite approach. In Agrawal, S., ed., *Protocols for Oligonucleotides and Analogs: Synthesis and Properties*. Humana Press, Totowa, NJ, 19-61.

Belenky, A., Smisek, D.L., and Cohen, A.S. (1995) Sequencing of antisense DNA analogues by capillary gel electrophoresis with laser-induced fluorescence detection. *J. Chromat.* A, **700**:137-149.

European Chemical Industry Council (1994) Good Manufacturing Practices Guidelines for Bulk Pharmaceutical Chemicals Manufacturers.

Caruthers, M.H. (1989) Synthesis of oligonucleotides and oligonucleotide analogues. In Cohen, J.S., ed., *Oligodeoxynucleotides - Antisense Inhibitors of Gene Expression*. CRC Press, Boca Raton, FL, 7-24.

Food and Drug Administration (1994) Guide to Inspection of Bulk Pharmaceutical Chemical Manufacturing.

Food and Drug Administration (1993) Guide to Inspections of Pharmaceutical Quality Control Laboratories.

Food and Drug Administration (1987) Guideline for Submitting Supporting Documentation in Drug Applications for the Manufacture of Drug Substances.

Food and Drug Administration (1996) Discussion Draft: Guidance for Industry - Manufacture, Processing or Holding of Active Pharmaceutical Ingredients.

Food and Drug Administration (1990) Compliance Program 7356.002F - Bulk Pharmaceutical Chemicals (BPCs).

Hogrefe, R. I., Reynolds, M. A., Vaghefi, M. M., Young, K. M., Riley, T. A., Klem, R. E., and Arnold, Jr., L. J. (1993) An improved method for the synthesis and deprotection of methylphosphonate oligonucleotides. In Agrawal, S., ed., *Protocols for Oligonucleotides and Analogs: Synthesis and Properties*. Humana Press, Totowa, NJ, 115-143.

International Society for Pharmaceutical Engineering (1996) Baseline Pharmaceutical Engineering Guide, Volume 1: Bulk Pharmaceutical Chemicals.

Kambhampati, R.V.B., Chen, C.-W., and Blumenstein, J.J. (1993) Regulatory Issues

Related to the Chemistry, Manufacturing, and Controls of Oligonucleotide Therapeutics for use in Clinical Studies, presented at Commercializing Oligonucleotide-Based Therapeutics, Coronado, CA.

Kambhampati, R.V.B., Chiu, Y.Y., Chen, C.-W., Blumenstein, J.J. (1993) Regulatory concerns for the chemistry, manufacturing, and controls of oligonucleotide therapeutics for use in clinical studies. *Antisense Res. Dev.* **3**:405-410.

Lazar, M. S. (1993) PMA concepts for the process validation of bulk pharmaceutical chemicals. *Pharm. Technol.* **17** (12):32-40.

Lazar, M. S. (1995) PhRMA guidelines for the production, packing, repacking, or holding of drug substances: Part I. *Pharm. Technol.* **19** (12):22-32.

Lazar, M. S. (1996) PhRMA guidelines for the production, packing, repacking, or holding of drug substances: Part II. *Pharm. Technol.* **20** (1):50-63.

Puma, P. (1996) Chromatography in process development. In Swadesh, J., ed., *HPLC: Practical and Industrial Applications.* CRC Press, Boca Raton, FL, 80-110.

Seliger, H. (1993) Scale-up of oligonucleotide synthesis: solution phase. In Agrawal, S., ed., *Protocols for Oligonucleotides and Analogs: Synthesis and Properties.* Humana Press, Totowa, NJ, 391-436.

Sinha, N. (1993) Large-scale oligonucleotide synthesis, using the solid-phase approach. In Agrawal, S., ed., *Protocols for Oligonucleotides and Analogs: Synthesis and Properties.* Humana Press, Totowa, NJ, 437-464.

Federal Register (1994) Stability testing of new drug substances and products. *Federal Register* **59**:48754.

Tang, J.Y., Iadorola, P., and Agrawal, S. (1993) Synthesis of oligonucleotide analogs using dimer blocks as synthons. *FASEB J.* **7**:A1088

Zamecnik, P. (1991) Oligonucleotide based hybridization as a modulator of genetic message readout. In Wickstrom, E., ed., *Prospects for Nucleic Acid Therapy of Cancer and AIDS.* Wiley-Liss. New York, 1-6.

Zhang, Z. and Tang, J. Y. (1996) A novel phosphitylating reagent for *in situ* generation of deoxyribonucleoside phosphoramidites. *Tet. Lett.* **37**:331-334.

Zon, G., and Geiser, T.G. (1991) *Anticancer Drug Design* **6**:539-568.

4

The Regulatory Process and Gene Therapy[1]

Suzanne L. Epstein

Food and Drug Administration, Rockville, Maryland

I. INTRODUCTION

Gene therapy is a rapidly evolving field in which a wide variety of therapeutic approaches are being considered and tested. Approaches to treatment of various lung diseases include study and use of adenovirus vectors (Rosenfeld *et al.*, 1991, 1992; Engelhardt *et al.*, 1993; Crystal *et al.*, 1994), adeno-associated virus vectors (Flotte *et al.*, 1992, 1993), and DNA/liposome vectors (Canonico *et al.*, 1994; Malone *et al.*, 1989) for delivery of therapeutic genes. Retroviral vectors have also been studied (Drumm *et al.*, 1990). In regulating product development in such an area, it is important to be flexible and attentive to practical limitations. If the barrier to starting exploratory clinical trials is too high, then only a few therapies will be tried and the ones that might have succeeded may be missed. On the other hand, the public should be protected from unnecessary or unreasonable risk, and also from the raising of unrealistic hopes. In its oversight of biological products for gene therapy, the Center for Biologics Evaluation and Research (CBER) of FDA seeks a balance between these considerations. As time goes on and a longer track record accumulates, it will likely be possible to relax recommendations for some

---

[1] Reprinted from Brigham, K., ed. (1997) *Gene Therapy for Diseases of the Lung*, Marcel Dekker, New York

types of testing which pioneering investigators had to perform. Sponsors should be aware that change in regulatory policy occurs more quickly now than in the past, especially in an area like gene therapy with rapidly evolving technology. However, when products reach later stages of development, standards for approval call for demonstration of efficacy, as with any other product.

While FDA regulation applies to human clinical trials rather than to laboratory research, CBER staff can assist during the lab research phase of product development prior to submission of an investigational new drug (IND) application. Early interaction with CBER staff can be helpful, especially for those, such as academic scientists, not used to viewing their work as manufacturing. Pre-IND interaction can help sponsors anticipate questions, facilitate later stages of the process, benefit from accumulated experience of others, and avoid pitfalls. In addition to telephone consultation, pre-IND meetings can be arranged. Guidance documents are also available, as listed in Table 1.

The chapter title includes "regulatory process" in order to emphasize the idea of regulation as a dynamic process involving exchange and dialogue between the investigator/sponsor and CBER staff. It is inappropriate to think of regulation of biologics in terms of a check list of tests every product must undergo, especially in an area such as gene therapy in which systems differ greatly and available information changes so rapidly. The fundamentals have not changed: demonstration of the safety, identity, purity, potency, and ultimately efficacy of the product. However, the details depend on the experimental system, patient population, and the knowledge base for related products. The goal should be a process during which the sponsor guides a product through development and clinical testing, consulting with CBER staff whenever appropriate.

Table 1. Points to Consider Documents Relevant to Gene Therapy Issued by the Center for Biologics Evaluation and Research.

Points to Consider in the Production and Testing of New Drugs and Biologicals Produced by Recombinant DNA Technology (1995)

Points to Consider in the Characterization of Cell Lines Used to Produce Biologicals (1993)

Points to Consider in the Manufacture and Testing of Monoclonal Antibody Products for Human Use (1995)

Points to Consider in the Collection, Processing, and Testing of Ex-Vivo-Activated Mononuclear Leukocytes for Administration to Humans (1989)

Points to Consider in Human Somatic Cell Therapy and Gene Therapy (1991)

II. MANUFACTURING AND QUALITY CONTROL OF GENE THERAPY PRODUCTS

Production and testing of products for gene therapy overlap in many respects with those for other biological products, as discussed previously (Epstein, 1991). In particular, there are resemblances to the handling of other products produced by recombinant DNA technology and to handling of cell lines for use in making other biologicals. For this reason, guidance documents in these areas are helpful (Table 1). For example, manufacturing of gene therapy products may involve bacterial cultures producing plasmids, mammalian packaging cell lines, and producer cell lines making viral vectors. Cell banks of these types are to be characterized and documented as described in the "Points to Consider in the Characterization of Cell Lines Used to Produce Biologicals (1993)." As with other products, in process testing and product quality control testing should use quantitative assays of demonstrated sensitivity and specificity, whenever possible.

Conventional concerns about toxic contaminants of course apply to these novel products, too. Unnecessary use of toxic materials during manufacturing should be avoided. If toxic solvents, metals, dyes, or other chemicals are used during production, the risk they pose should be considered, residual levels measured, and specifications set. One example would be chloroform used in manufacturing liposomes. Chloroform is readily removed, assays for residual levels are sensitive, and its removal should be verified by testing.

Gene therapy products also present novel issues. One example is the potential for recombination. Recombination can occur between sequences in a vector and other elements in a packaging cell or producer cell (Vanin *et al.*, 1994). Recombination may also occur between a vector and either endogenous sequences in a recipient (Gareis *et al.*, 1991) or viruses infecting the recipient. Recombination has indeed been reported between two retroviruses infecting a single patient (Diaz *et al.*, 1995). The significance of this type of concern would depend on the product and the patient group.

For retroviral products, either transduced cells or retroviral vectors for direct administration to patients, one major concern is exclusion of replication competent retroviruses (RCR). This concern was at first theoretical, as initial animal studies did not show adverse effects (Cornetta *et al.*, 1990). While retroviral vectors have been used in humans for several years without apparent adverse effects (Blaese *et al.*, 1995; Bordignon *et al.*, 1995), the concern was heightened by the development of fatal lymphomas in immunosuppressed monkeys exposed to replication-competent murine retrovirus (Donahue *et al.*, 1992). For this reason, testing for RCR in cell banks, testing of bulk and final products, and monitoring of patients are recommended. For detailed recommendations, consult current versions of "Points to Consider" documents or contact CBER for guidance.

For adenovirus vectors, the analogous issue of replication-competent adenovirus (RCA) is of concern. Adenovirus infections of the respiratory tract are often mild, but could be significant adverse events in patients already compromised by lung diseases. Quantitative testing for presence of RCA is needed, and levels of RCA in patient doses are to be minimized.

For plasmid gene therapies, one issue has been raised by the common use of ampicillin resistance as a selection marker. As stated in the "Points to Consider in the Characterization of Cell Lines Used to Produce Biologicals (1993)," penicillin and other beta-lactam antibiotics should be avoided in production cultures, due to the possibility of residual antibiotic or adducts in the final product and antibiotic hypersensitivity in some people. For licensed products, current regulations (21 CFR 610.61(m)) require product labeling to declare any antibiotics added during manufacturing, and use of penicillin may concern physicians. If a construct has already been made and contains the *amp*[r] gene, it may be possible to use it in some situations if ampicillin is not used in making the final product. If the antibiotic is used in production, contact CBER concerning the nature of the protocol and the proposed patient population.

Investigators with products under development should be aware of this problem and consider alternative selection markers, especially since simply omitting ampicillin from production runs may lead to poor yields. Choosing a different selection marker avoids the difficulties of attempting later to prove absence of residual beta lactam and/or derivatized products, and of labeling issues if the product proceeds to licensure. Currently, other antibiotic selection systems can be used. A variety of non-antibiotic selection systems are being developed. Such approaches would have the added advantage of avoiding possible transfer of antibiotic resistance genes to other bacteria.

III. PRECLINICAL ANIMAL STUDIES

Animal studies are usually needed and informative in testing of new products prior to human use, even though they provide imperfect predictions about effects in humans (though see Section IV below). An animal species should be chosen that can be infected by the vector class in use, and if possible an animal that provides a model of the biological activity of the product. It will sometimes be possible to achieve an animal model of the disease. The appropriate species will vary, and uniform use of particular animal species such as primates is not mandated by CBER. The route of administration intended for clinical trials should be used if possible, as well as intravenous administration in certain cases. The dose range tested should include at least one dose within and one dose above the proposed clinical range, and should attempt to determine a toxic dose.

One goal of animal studies is of course testing for toxicity. Formulation components can be responsible for toxicity, so the final product formulation intended for human use should be the one studied. For example, a variety of cationic lipids have been studied as possible delivery systems for plasmid gene therapy (Farhood *et al.*, 1994; Gao and Huang, 1991). One observed toxicity of DNA/liposome products is inflammation of the lungs, and this effect has been shown to be due to the lipid component (Dr. S. Cheng, Genzyme, personal communication with permission).

In addition to toxicity, the goals of animal studies should be to study where the vector localizes, where it is expressed, and effects of overexpression or ectopic expression. Tumorigenicity may also be a concern in some cases. Possible adverse events might include effects of expression at higher than normal levels, or effects of expression in unintended cells or tissues. Such effects could include immune responses to a gene product in hosts not normally expressing it, or in hosts expressing a mutated form of that product. Such responses could result in autoimmunity or in alteration of the safety and/or efficacy of repeat treatment.

Of special concern is whether vector localizes to particular sites, and whether it is found in the gonads. While the gonads and germline are not impervious to viral attack and alteration (Raina and Adams, 1995), it may be that gene therapy vectors in current use will not alter the germ line. In one model system, plasmid DNA administered to pregnant mice was found in the progeny but had not been integrated into the germline (Tsukamoto *et al.*, 1995). Nevertheless, due to the long term consequences of germline alteration and the lack of societal consensus on germ line genetic intervention, preclinical animal testing is used to examine potential effects of gene therapy products. The presence of vectors in the gonads without chromosomal integration could also have reproductive effects of concern. These questions call for well-designed animal trials analyzed with sensitive, specific assay methods such as validated PCR. Positive signals in gonadal tissue would call for follow-up studies excluding blood contamination and analyzing the cell type(s) involved.

Note, however, that the position of CBER does not absolutely prohibit clinical use of products with possible mutagenic effects. Even negative results from animal studies analyzed with the most sensitive techniques can only put an upper limit on the possible frequency of altered sperm, for example. Other accepted therapeutics such as some chemotherapy drugs are known to be mutagenic. The goal is to explore the issue carefully, to follow up on unexpected results, and to accumulate data that will allow full patient informed consent, including knowledgeable advice to patients in the future, and will permit an accurate risk-benefit assessment to be made.

Recombination of vectors with other viruses present in the host at the time of treatment or with viruses infecting the host later are possible sources of concern in gene therapy, and have been discussed earlier (Section III). Whether such possibilities can appropriately be addressed *in vitro* or in animal models depends on the system.

IV. VECTOR MODIFICATION

In the past, CBER has considered any change in a vector to result in a new product, necessitating a new IND. There is now an accumulating body of evidence that will permit flexibility in this policy, and in the future further changes in policy may be possible. The goal is to facilitate progress towards effective therapies by abbreviating testing and reducing documentation, when this can be done while preserving patient safety.

Ordinarily, a new biological product requires a new IND. For gene therapy vectors this is not always appropriate. Vector modification is a process involving many incremental changes potentially leading to safer and more effective vectors. Each modified vector must receive appropriate testing and must meet appropriate specifications. However, repeating all preclinical studies will not be necessary in all cases. If this were required, it could lead to inappropriate conservatism by investigators and to continued use of vectors that might not be as good as newer ones, but were considered by investigators to be "approved." Note that CBER does not approve vectors for investigational use in general, but rather permits particular clinical trial proposals to proceed based on adequate data. What is adequate may depend upon the intended use.

How much of a change in a vector causes it to be classed as a new product subject to full retesting? This question must be judged case by case, and decisions will be based in part on the data accumulating concerning the effects or lack of effects of particular structural changes on localization, expression, etc. Still, comments can be made about some examples. If the only change in a series of closely related vectors is substitution of a different (especially if closely related) therapeutic gene as the insert, with no change in the viral or plasmid structures, the vectors may in some cases be considered members of a panel. Another case would be replacement of one antibiotic selection marker with another. Other minor changes in vector structure may also fall into this category.

In such cases testing could be abbreviated. Preclinical animal testing such as tissue localization, germ line alteration, tumorigenicity, and pharmacology/ toxicology studies may not need to be repeated. Instead, safety testing for each vector will focus on safety issues presented by the insert and its expressed product, plus quality control testing. If the inserted genes in a series of vectors are the same or closely related and would be expected to have the same potential for adverse effects, even such testing could be abbreviated.

Other modifications may consist of changes in lipids or other formulation components used with vectors. Different lipid components, different ratios of two lipids, or different ratios of lipid to DNA may be used to formulate a vector for administration to patients. The properties of the particular lipids used will affect the activity of liposomes in enhancing transfection (Felgner *et al.*, 1994). Use of a new lipid requires safety information on that new lipid, but not necessarily in conjunction with the precise vector proposed for use if the new lipid has been used with other vectors. If tissue localization, germ line alteration, and classical animal pharmacology/toxicology data are compared for use of varied lipid ratios, compositions, or ratios to DNA, this data can be used to evaluate whether such studies are necessary for additional combinations.

Characterization of modified vectors should include confirmation of the intended changes in structural features. If the same facility is producing multiple related vectors, it would be advisable to perform a quality control test capable of distinguishing the various vectors, such as mapping with appropriately chosen restriction enzymes.

V. CLINICAL TRIALS

Gene therapy clinical trials involve special unresolved issues, as well as usual issues of design and ethics affecting all trials. The concerns come about due to possible infectious spread or intrapatient recombination or replication of agents used in the trial, possible irreversible effects, potential for germ line alteration, and the uncertain time frame and nature of any adverse effects. Such issues have been analyzed by ethicists and by investigators in the field (Ledley, 1995). Other issues arise mainly due to the excitement surrounding this technology and the potential for unrealistic hopes and exploitation of patients or their families. Some of these issues can be addressed by appropriate conduct of recruiting and the informed consent process, some by long term patient monitoring, and some by additional attention to animal studies even while clinical trials are underway.

Regarding recruitment, an important issue is definition of diagnostic criteria to define patient subgroups in which genotype or diagnosis permits prediction of disease severity. Such criteria permit more realistic risk/benefit assessment. Another issue is the accuracy of laboratory tests used for diagnosis. Next, there is need for balanced presentation to patients or families of the options they have, whether or not these lie within the area of expertise of the investigator.

Access to clinical trials should be the same for patients of both genders unless there is a valid biological or medical reason to exclude one gender. Reproductive risks should also be handled comparably for both genders unless there are biological or medical reasons to do otherwise. Exclusion of pregnant women may be appropriate in many trials, while intentional *in utero* gene therapy brings up a host of issues that are currently under discussion nationally and internationally. However, requirements for use of birth control or for reproductive sterility will usually apply to both genders if they apply at all. Any hazard to germ cells may be greater in the male due to the rapid turnover of gametes. In addition to use of birth control, reproductive risks can be dealt with by informed consent procedures that honestly state the limits of our knowledge about risks to future offspring.

Patient monitoring in gene therapy trials will need to be long term, often life-long. Adverse events cannot be assumed unrelated to therapy simply because they are not immediate. Many consequences, for example immunological ones, might be considerably delayed. Consent forms should mention the need for long-term follow-up, including the possible need for samples. Planning is underway for a national information network to track such data. Patients should be advised not to donate blood, organs, tissues, and gametes until more is known about genetic alterations. Consent can also include mention of the importance of autopsy, and autopsy protocols should include germ cell analysis.

If there is reasonable risk that a vector used in a clinical trial may be subject to environmental release and spread or may mutate or recombine to release an infectious agent, then clinical trial design should include provision for appropriate isolation of the subject and for monitoring for such agents.

Questions arise as to the risk of interactions if one patient is exposed to multiple gene therapy agents. As with interactions among other drugs, this possibility should be examined case-by-case. There is no general prohibition on patients who have already received one gene therapy later receiving another gene therapy intervention, particularly if cells have been treated with vector *ex vivo*.

VI. "WELL-CHARACTERIZED PRODUCTS"

At a CBER-sponsored workshop, December 11-13, 1995, on well-characterized biotechnology products, a breakout session on plasmid DNA products was held. The recommendation of this session was that plasmid DNA products be considered eligible for designation as "well-characterized." A "well-characterized product" would have well-defined physicochemical characteristics, could be manufactured reproducibly, and would not require a separate establishment license or lot release submission to CBER if licensed. Note that a product can be considered well characterized by virtue of its structure and mode of manufacture and testing, and yet the product is not eligible for licensure unless data collected under an IND demonstrate safety or efficacy in an appropriate patient population.

As announced in the Federal Register, Vol. 61, No. 94, of May 14, 1996, "Elimination of Establishment License Application for Specified Biotechnology and Specified Synthetic Biological Products," the term "well-characterized" has been superceded by the term "specified." Plasmid DNA therapeutic products are included as "specified products," and so are not subject to a requirement for a separate establishment license or lot release submission to CBER.

VII. CONCLUSION

Application of the new technology of gene therapy to treatment of lung diseases will yield the greatest benefits if approached on a sound scientific and medical basis. CBER welcomes suggestions from the research and manufacturing communities as to how this process can best be facilitated in the regulatory arena.

Guidance documents including Points to Consider documents and other policy documents are available from CBER as follows: Division of Congressional and Public Affairs, CBER, HFM-12, 1401 Rockville Pike, Suite 200N, Rockville MD 20852-1448.

Investigators are invited to call CBER at (301) 594-1800 or 800 835-4709, or at the CBER FAX Information System (301) 594-1939 (call from a FAX machine with touchtone phone and then follow the instructions), or send E-mail to gsta@a1.cber.fda.gov. You will receive an automatic reply listing documents available by return E-mail. CBER staff are also available for questions about what products FDA regulates, for consultation, or for arrangement of pre-IND meetings at (301) 594-0830.

VIII. REFERENCES

Blaese, R. M., Culver, K. W., Miller, A. D., Carter, C. S., Fleisher, T., Clerici, M., Shearer, G., Chang, L., Chiang, Y., Tolstoshev, P., Greenblatt, J. J., Rosenberg, S. A., Klein, H., Berger, M., Mullen, C. A., Ramsey, W. J., Muul, L., Morgan, R. A., and Anderson, W. F. (1995) T lymphocyte-directed gene therapy for ADA-SCID: Initial trial results after 4 years. *Science* **270**:475-480.

Bordignon, C., Notarangelo, L. D., Nobili, N., Ferrari, G., Casorati, G., Panina, P., Mazzolari, E., Maggioni, D., Rossi, C., Servida, P., Ugazio, A. G., and Mavilio, F. (1995) Gene therapy in peripheral blood lymphocytes and bone marrow for ADA-immunodeficient patients. *Science* **270**:470-475.

Canonico, A. E., Plitman, J. D., Conary, J. T., Meyrick, B. O., and Brigham, K. L. (1994) No lung toxicity after repeated aerosol or intravenous delivery of plasmid-cationic liposome complexes. *J. Appl. Physiol.* **77**:415-419.

Cornetta, K., Moen, R. C., Culver, K., Morgan, R. A., McLachlin, J. R., Sturm, S., Selegue, J., London, W., Blaese, R. M., and Anderson, W. F. (1990) Amphotropic murine leukemia retrovirus is not an acute pathogen for primates. *Hum. Gene Ther.* **1**:15-30.

Crystal, R. G., McElvaney, N. G., Rosenfeld, M. A., Chu, C. S., Mastrangeli, A., Hay, J. G., Brody, S. L., Jaffe, H. A., Eissa, N. T., and Danel, C. (1994) Administration of an adenovirus containing the human CFTR cDNA to the respiratory tract of individuals with cystic fibrosis. *Nature Genet.* **8**:42-51.

Diaz, R. S., Sabino, E. C., Mayer, A., Mosley, J. W., Busch, M. P., and Transfusion Safety Study Group (1995) Dual human immunodeficiency virus type 1 infection and recombination in a dually exposed transfusion recipient. *J. Virol.* **69**:3273-3281.

Donahue, R. E., Kessler, S. W., Bodine, D., McDonagh, K., Dunbar, C., Goodman, S., Agricola, B., Byrne, E., Raffeld, M., Moen, R., Bacher, J., Zsebo, K. M., and Nienhuis, A. W. (1992) Helper virus induced T cell lymphoma in nonhuman primates after retroviral mediated gene transfer. *J. Exp. Med.* **176**:1125-1135.

Drumm, M. L., Pope, H. A., Cliff, W. H., Rommens, J. M., Marvin, S. A., Tsui, L.-C., Collins, F. S., Frizzell, R. A., and Wilson, J. M. (1990) Correction of the cystic fibrosis defect *in vitro* by retrovirus-mediated gene transfer. *Cell* **62**:1227-1233.

Engelhardt, J. F., Simon, R. H., Yang, Y., Zepeda, M., Weber-Pendleton, S., Doranz, B., Grossman, M., and Wilson, J. M. (1993) Adenovirus-mediated transfer of the CFTR gene to lung of nonhuman primates: Biological efficacy study. *Hum. Gene Ther.* **4**:759-769.

Epstein, S. L. (1991) Regulatory concerns in human gene therapy. *Hum. Gene Ther.* **2**:243-249.

Farhood, H., Gao, X., Son, K., Yang, Y., Lazo, J. S., Huang, L., Barsoum, J.,

Bottega, R., and Epand, R. M. (1994) Cationic liposomes for direct gene transfer in therapy of cancer and other diseases. *Ann. NY Acad. Sci.* **716**:23-34.

Felgner, J. H., Kumar, R., Sridhar, C. N., Wheeler, C. J., Tsai, Y. J., Border, R., Ramsey, P., Martin, M., and Felgner, P. L. (1994) Enhanced gene delivery and mechanism studies with a novel series of cationic lipid formulations. *J. Biol. Chem.* **269**:2550-2561.

Flotte T. R., Solow R., Owens R. A., Afione S., Zeitlin P. L. and Carter B. J. (1992) Gene expressions from adeno-associated virus vectors in airway epithelial cells. *Am. J. Respir. Cell Mol. Biol.* **7**:349-356.

Flotte, T. R., Afione, S. A., Conrad, C., McGrath, S. A., Solow, R., Oka, H., Zeitlin, P. L., Guggino, W. B., and Carter, B. J. (1993) Stable *in vivo* expression of the cystic fibrosis transmembrane conductance regulator with an adeno-associated virus vector. Proc. *Natl. Acad. Sci. USA* **90**:10613-10617.

Gao, X., and Huang L. (1991) A novel cationic liposome reagent for efficient transfection of mammalian cells. *Biochem. Biophys. Res. Commun.* **179**:280-285.

Gareis, M., Harrer, P., and Bertling, W. M. (1991) Homologous recombination of exogenous DNA fragments with genomic DNA in somatic cells of mice. *Cell. Molec. Biol.* **37**:191-203.

Ledley, F. D. (1995) After gene therapy: Issues in long-term clinical follow-up and care. *Adv. Genet.* **32**:1-16.

Malone, R. W., Felgner, P. L., and Verma, I. M. (1989) Cationic liposome-mediated RNA transfection. *Biochemistry* **86**:6077-6081.

Raina, A. K., and Adams, J. R. (1995) Gonad-specific virus of corn earworm. *Nature* **374**:770

Rosenfeld, M. A., Siegfried, W., Yoshimura, K., Yoneyama, K., Fukayama, M., Stier, L. E., Pääkkö, P. K., Gilardi, P., Stratford-Perricaudet, L. D., Perricaudet, M., Jallat, S., Pavirani, A., Lecocq, J.-P., and Crystal, R. G. (1991) Adenovirus-mediated transfer of a recombinant α1-antitrypsin gene to the lung epithelium *in vivo. Science* **252**:431-434.

Rosenfeld, M. A., Yoshimura, K., Trapnell, B. C., Yoneyama, K., Rosenthal, E. R., Dalemans, W., Fukayama, M., Bargon, J., Stier, L. E., Stratford-Perricaudet, L., Perricaudet, M., Guggino, W. B., Pavirani, A., Lecocq, J.-P., and Crystal, R. G. (1992) *In vivo* transfer of the human cystic fibrosis transmembrane conductance regulator gene to the airway epithelium. *Cell* **68**:143-155.

Tsukamoto, M., Ochiya, T., Yoshida, S., Sugimura, T., and Terada, M. (1995) Gene transfer and expression in progeny after intravenous DNA injection into pregnant mice. *Nature Genet.* **9**:243-248.

Vanin, E. F., Kaloss, M., Broscius, J. and Nienhuis, A. W. (1994) Characterization of replication-competent retroviruses from nonhuman primates with virus-induced T-cell lymphomas and observations regarding the mechanism of oncogenesis. *J. Virol.* **68**:4241-4250.

5

Production of Clinical Lots of Gene Therapy Vectors Using Good Manufacturing Practice: Experience in a University Setting

Alan R. Davis and Colleen Baker
University of Pennsylvania, Philadelphia, Pennsylvania

I. INTRODUCTION

Gene therapy has potential for the treatment of genetic disease, infectious disease, cancer and many others. The premises upon which this field are based are now being developed. To rationally move forward, however, it is critical that concepts which are set forth in the basic research arena using animal models are validated and tested in human clinical studies. A university medical center is an ideal place to test the concepts of gene therapy since basic researchers are in close proximity to clinical practice. However, the production of clinical lots of drug substance for human clinical studies is, for the most part, performed within the pharmaceutical or biotechnology industry. We describe here a specialized university laboratory for the production of clinical materials used in somatic cell or gene therapy.

II. THE PRODUCTION FACILITY AND ITS DESIGN

The Human Applications Laboratory (HAL) is a specialized clean room manufacturing facility located at the University of Pennsylvania in the Institute for Human Gene Therapy (IHGT). The IHGT is dedicated to the efficient translation of basic research in the areas of gene transfer technology and gene therapy into pilot human trials. The HAL is a 2200 ft^2 laboratory that was developed to provide gene therapy reagents to investigators interested in performing preclinical safety and toxicology studies in appropriate animal models as well as clinical grade reagents for human clinical trials. Its regulatory intent is to conform and operate in accordance with Food and Drug Administration (FDA) guidelines for Good Manufacturing Practice (GMP) and Good Laboratory Practices (GLP). HAL is equipped to work with biohazardous material at the BL2+ level.

It houses five separate tissue culture rooms with the manufacturing ability to provide a diverse array of gene therapy reagents at any given time. This design facilitates the quick and cost-efficient translation of new gene transfer technologies to clinical trials. It also allows for the isolation and cultivation of human cells that will be genetically modified *ex vivo*.

The HAL was the first pilot manufacturing facility that successfully brought two gene therapy technologies (i.e., retroviruses and adenoviruses) through the FDA approval process in the form of investigator-initiated Investigational New Drug (IND) applications. The initial project was the production and validation of recombinant retroviruses containing the LDL receptor for the familial hyper-cholesterolemia trial. The HAL has also brought clinical grade Cystic Fibrosis Transmembrane Regulator (CFTR) recombinant adenovirus through the regulatory process and into patients. HAL is being developed as a national resource facility through the National Gene Vector Laboratory (NGVL). This organization enables coordinated interactive research and interpretation of clinical data and makes gene therapy clinical trials more cost-effective. For example, the initial investment to establish and validate an adenovirus packaging cell line is spread out over multiple disease-based protocols (e.g. the same type of virus will be used in liver metabolic diseases and cancer) and within a disease, such as CF, among multiple institutions.

The development of HAL was based on several premises, including the fact that gene transfer technology represents the common thread that pulls together the development and application of human gene therapy throughout all disciplines. By definition, therefore, the programmatic implementation of a laboratory involved in production of gene transfer technologies is cross-cutting, spanning multiple disease states and medical disciplines. In this setting there is true economy of scale when supporting a diverse array of clinical protocols that are based around a common gene transfer technology.

Some of the physical features of HAL include an airlock, gowning room, for entrance and exit, a main laboratory for reagent preparation and testing, a pass-through autoclave as well as a computerized system that monitors equipment.

III. PRODUCTION OF CLINICAL LOTS OF GENE THERAPY VECTORS

Production of clinical lots of gene therapy vectors must conform to the current Good Manufacturing Process (cGMP) guidelines as elaborated in the Code of Federal Regulations (21 CFR 58) (Food and Drug Administration, 1995). These guidelines cover a number of specific requirements, not only for the establishment producing the gene therapy vectors (see above) but also the processes used within the establishment. These processes include the use of standard operating procedures which include all steps performed during any procedure. For instance, the purification of an adenoviral vector in a cGMP laboratory is performed using a standard operating procedure. The operation should also be managed carefully in terms of quality control procedures such as the quality of materials, manufacturing controls, and equipment validation and monitoring. In addition to the regulations within 21 CFR 58 regarding cGMP standards, such a production laboratory must also pay close attention to certain Points to Consider (Food and Drug Administration, 1991,1993). The 1993 document describes standards for the collection of cells for *in vivo* somatic cell therapy protocols, the quality control and culture medium used for such procedures, the handling of such cultures to minimize contamination with adventitious agents and the testing of cultures for these agents. For *in vivo* gene therapy, the cell bank used to grow the gene therapy vector should be extensively characterized and a Master Working Cell Bank prepared. Packaging cell lines should be similarly characterized and their construction described in detail. Materials used during these cell culture procedures need to be indicated and the specification for these materials, such as serum, antibiotics, or other chemicals, need to be provided. In addition, limits for their concentration in the final product, the methods used to remove them, and the effectiveness of their removal needs to be given.

For gene therapy a molecular and genetic characterization of the construct needs to be provided. This includes not only the method of construction, but also the sequencing of the transgene and flanking areas.

Viral vectors must be tested for replication competent viruses. For instance, adenovirus vectors need to be free of replication competent adenovirus (RCA) at a level of one plaque forming unit (PFU) of RCA per patient dose. This testing is done by passage of the recombinant adenovirus vector on a cell line that does not support the replication of vector virus such that only RCA can replicate. In the case of recombinant adenovirus vectors deleted in region E1, since such vectors can grow only in the 293 packaging cell line that supplies E1 function, the vector is passaged on a cell line such as A549 (human lung carcinoma) which does not supply E1 function, to test for RCA.

As an example of the manufacturing process we describe below a scheme for the production of clinical lots of recombinant adenovirus. The process begins with the construction of a recombinant plasmid containing the transgene. The plasmid is transfected with an adenovirus backbone to generate the recombinant virus.

The virus is then amplified by infection of a Master Cell Bank to generate a lysate that is purified to a master seed stock virus. Both the lysate and purified virus are subjected to quality assurance testing (see below). After testing, the master seed stock virus is used to infect a Master Cell Bank and generate the production lot lysate that is purified to yield the production lot virus. Quality assurance is performed on both the production lot lysate and the purified virus.

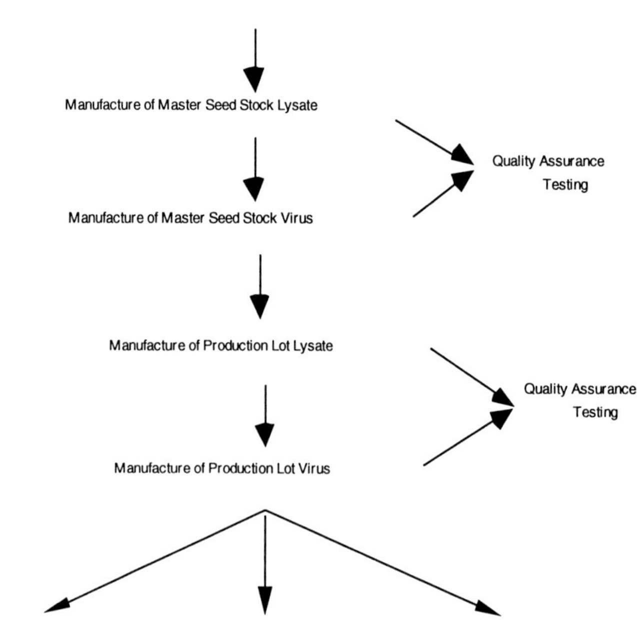

Production Lot Virus is the final product used for preclinical toxicology and clinical studies. It is this virus that is examined for stability over time, transgene function, and sequence analysis. The production lot is released by quality control testing (see below) which assures that the production process has been performed correctly and that the final product is within a range of specified values. The concentration of vector, in particles/ml, is determined as well as its infectivity, in plaque forming units/ml. Transgene activity is then assessed in a functional assay to measure the ability of the specific production lot of vector to produce a specified amount of functional transgene.

IV. NOMENCLATURE FOR RECOMBINANT ADENOVIRUS PRODUCED WITHIN HAL

A specific nomenclature has been devised for recombinant adenovirus vectors produced within the HAL. First a letter designation specifies the species followed by a number specifying the serotype. Second, a series of three numbers specifies the vector modification in the E2, E3, and E4 regions of the viral genome respectively. The specific number defines a specific mutation or deletion that has been engineered in the region. Next the promoter and transgene are specified. Therefore in this nomenclature the recombinant adenovirus H5.001CMVlacZ would be a human adenovirus type 5 that is unchanged from wild type in regions E2 and E3 but has a modification in region E4. The vector produces *lacZ* β-lactamase driven by a CMV promoter.

V. QUALITY ASSURANCE AND QUALITY CONTROL TESTING

Quality assurance testing is performed at various stages in the production process. It is performed on both the Master Seed Stock Lysate and Master Seed Stock Virus as well as the Production Lot Lysate and Production Lot Virus. In addition quality control and release testing is performed on the Production Lot Virus. A battery of quality assurance tests is also required for the Master Cell Bank. In general some of the quality assurance measures for the recombinant adenovirus Master Seed Stock may include tests for the presence of mycoplasma, a sterility test for the presence of bacterial and fungal contaminants, *in vitro* assays for the presence of viral contaminants, transmission electron microscopic evaluation, a test to detect adeno-associated virus (AAV), and an assay for the presence of replication competent adenovirus (RCA assay). Finally, a general safety test may be conducted. The general safety test involves inoculation of guinea pigs and mice intraperitoneally with the Master Seed Stock. These animals are observed for overt signs of ill health or unusual response. Their weights are measured just prior to and upon completion of the test period.

The quality assurance tests for the Production Lots may include some of the same tests and may also include the *Limulus* amebocyte lysate test. The purpose of this test is to detect and quantitatively determine the gram-negative bacterial endotoxin level of a test article or extract using the Limulus Amebocyte Lysate (LAL) gel-clot method for testing. Quality assurance tests can also be performed on the recombinant adenovirus lysates. Establishment of a Master Cell Bank requires extensive testing. Quality assurance tests performed on the Master Cell Bank might include mycoplasma and sterility testing as well as cell culture identification.

Once these tests are passed satisfactorily, 5% of the Master Cell Bank is tested for the various adventitious agents, for example bovine and porcine viruses, hepatitis B and C viruses, HIV1 and HIV2, HTLV1 and HTLV2, EBV, AAV, and cytomegalovirus.

As described above, quality control release testing using a recombinant adenovirus vector involves a determination of the structure of the viral DNA by restriction analysis and verification that the transgene is functional. Additionally, the vector must be titered, in the case of an adenovirus vector this is done by plaque assay and particle number determination. In the case of the adenovirus expressing CFTR (H5.020CBCFTR), functional analysis was performed using SPQ analysis.

VI. CASE STUDIES OF THE USE OF VIRAL VECTORS IN A GMP LABORATORY

As examples of production of viral vectors for gene therapy in a GMP laboratory we present two case studies. In the first, we describe the *ex vivo* administration of a viral vector expressing the LDL receptor for potential correction of the familial hypercholesterolemia (FH) genetic defect. In the second, we present the potential correction of the cystic fibrosis genetic disease with *in vivo* gene therapy using a viral vector expressing the cystic fibrosis transmembrane conductance regulator (CFTR).

VI.A. *Ex vivo* administration of a retrovirus vector expressing the LDL receptor

Familial hypercholesterolemia (FH) is an autosomal dominant disease caused by mutations in the LDL receptor gene. Patients with two mutant alleles develop massive accumulations of LDL early in childhood. In *ex vivo* gene therapy autologous hepatocytes were transduced with a retroviral vector expressing LDL receptor. These cells were then re-introduced intrahepatically. These studies were all performed in the HAL. In this paradigm, however, it was also necessary to first perform three critical pre-clinical studies. In the first, performed in an authentic animal model for FH, the Watanabe heritable hyerlipidemic (WHLL) rabbit, pre-clinical efficacy was demonstrated (Wilson *et al.*, 1990). In the next studies, two models of pre-clinical safety, the dog (Grossman *et al.*, 1993) and the baboon (Grossman *et al.*, 1992) were done before a human study was conducted. It has been our practice to perform such pre-clinical toxicology studies with vector produced within HAL but not subject to the full battery of tests used for actual human lots.

VI.B. *In vivo* administration of a recombinant adenovirus expressing the cystic fibrosis transmembrane conductance regulator

Cystic fibrosis is a primary cause of death in the Caucasian population from genetic disease. The most damage from this disease is caused in the lung although other organ systems are affected. Poor mucociliary clearance in the lung results in obstructive lung disease and chronic bacterial infections. The genetic basis of cystic fibrosis was recently uncovered when the gene which is mutated in this disease, CFTR, was found (Rommens *et al.*, 1989). After this, it was discovered that CFTR is a chloride channel (Anderson *et al.*, 1991).

It is however, extremely difficult to perform *ex vivo* gene therapy in the lung, therefore vectors were sought in which an *in vivo* approach could be utilized. Adenovirus-based vectors have an ability to transfer genetic material efficiently into lung epithelial cells and therefore has led them to be the vehicles of choice for the first trials of human gene therapy for cystic fibrosis. As was the case with FH, first pre-clinical efficacy studies demonstrated that this approach could be successful (Rosenfeld *et al.*, 1992). Next, pre-clinical toxicity studies were conducted in non-human primates (baboons). In these studies (Simon *et al.*, 1993) two adenovirus vectors: H5.020CBCFTR and H5.010 CMVlacZ were instilled through a bronchoscope into limited regions of lungs of baboons. Although minimal inflammation was seen at 10^7 and 10^8 pfu/ml, at 10^9 and 10^{10} pfu/ml a perivascular lymphocytic and histocytic infiltrate was seen. Therefore it was concluded that with this generation of the adenoviral vector, transfer into the lungs is associated with development of alveolar inflammation at high doses of virus.

Based on these pre-clinical toxicology studies, human trials using H5.020CBCFTR were initiated. In the first (Wilson *et al.*, 1994) patients were treated with CFTR expressing virus and followed for evidence of CFTR gene transfer and expression, immunological responses to CFTR or adenoviral proteins, and toxicity. The extensive preclinical studies preceding this trial included 1) proof of the ability of the vector to complement the CFTR defect *in vitro* using CFPAC cells derived from a pancreatic adenocarcinoma of a CF patient 2) demonstration that the CFTR producing vector could infect human airway xenografts in a *nu/nu* mouse and 3) demonstration that the CFTR producing vector could complement human CF xenografts.

In the most recent of trials (Knowles, *et al.*, 1995), twelve patients with cystic fibrosis were administered this vector in groups of three at doses of 2×10^7 to 2×10^{12} pfu. The adenoviral vector was administered to the nasal epithelium and gene transfer was quantitated by a number of molecular techniques. The safety of the treatment was monitored by nasal lavage and biopsy to assess inflammation as well as any replication of H5.020CBCFTR. Three lots of adenoviral vector with particle:plaque-forming unit ratios ranging from 20:1 to 50:1 were produced under current Good Manufacturing Practices in the HAL using the production and quality control procedure outlined above.

Since this time several other adenoviral vectors produced in the HAL have either entered or are about to enter clinical studies. These include studies in the use of an adenoviral vector expressing the herpes simplex virus thymidine kinase in combination with the normally non-toxic nucleoside analog ganciclovir for treatment of mesothelioma (Treat *et al.*, 1996). In cells expressing the viral thymidine kinase, ganciclovir is phosphorylated, then incorporated into nucleic acids, which leads to cell death. Malignant mesothelioma has been clearly linked to prior asbestos exposure. In addition this same vector is also being used in treatment of CNS malignancies (Eck *et al.*, 1996; this volume, Chap. 11).

Finally, clinical studies are beginning that utilize a new adenoviral vector in which region E4 is deleted for potential treatment of cystic fibrosis and ornithine transcarbamylase deficiency. Such vectors have the potential for increased safety and duration of transgene expression in that the immune response to the vector itself may be diminished. These "third generation" vectors are produced in the HAL according to the same cGMP production and quality assurance procedures that are elaborated above. All these studies would be far more difficult and costly if laboratories like the HAL did not exist. The HAL gives investigators the ability to place novel vectors in phase I/II clinical trials coupled with the flexibility to more rapidly modify these vectors without incurring extraordinary cost.

VII. REFERENCES

Anderson, M.P. Gregory, R.J., Thompson, S., Souza, D.W., Paul, S., Mulligan, R.C., Smith, A.E., and Welsh, M.J. (1991) Demonstration that CFTR is a chloride channel by alteration of its anion sensitivity. *Science* **253**:202-205.

Eck, S.L., Alavi, J.B., Alavi, A., Davis, A., Hackney, D., Judy, K., Mollman, J., Phillips, P.C., Wheeldon, E. B., and Wilson, J.M. (1996) Treatment of Advanced CNS Malignancies with the Recombinant Adenovirus H5.010RSVTK: A Phase I Trial. *Hum. Gene Ther.* **7**:1465-1482.

Food and Drug Administration (1991) Points to Consider in Human and Somatic Cell and Gene Therapy.

Food and Drug Administration (1993) Points to Consider in the Characterization of Cell Lines Used to Produce Biologicals.

Food and Drug Administration (1995) Good laboratory practices for nonclinical laboratory studies. 21 *Code Federal Regulations* **58**:265-278.

Grossman, M., Wilson, J.M., and Raper, S.E. (1993) A novel approach for introducing hepatocytes into the portal circulation. *J. Lab. Clin. Med.* **121**:472-478.

Grossman, M., Raper, S.E., and Wilson, J.M. (1992) Transplantation of genetically modified autologous hepatocytes into nonhuman primates: feasibility and short-term toxicity. *Hum. Gene Ther.* **3**:501-510.

Knowles, M.R., Hohneker, K.W., Zhou, Z., Olsen, J.C., Noah, T.L., Hu, P.-C., Leigh, M.W., Engelhardt, J. F., Edwards, L.J., Jones, K.R., Grossman, M., Wilson, J.M., Johnson, L.G., and Boucher, R.C. (1995) A controlled study of adenovrial-vector-mediated gene transfer in the nasal epithelium of patients with cystic fibrosis. *New Engl. J. Med.* **333**:871-873.

Rommens, J.M. Iannuzzi, M.C., Kerem, B., Drumm, M.L., Melmer, G., Dean, M., Rozmahel, R., Cole, J.L., Kennedy, D., Hidaka, N., *et al.* (1989) Identification of the cystic fibrosis gene: chromosome walking and jumping. *Science* **245**:1059-1065.

Rosenfeld, M.A., Yoshimura, K , Trapnell, B.C., Yoneyama, K., Rosenthal, E.R., Dalemans, W., Fukayama, M., Bargon, J., Stier, L.E., Stratford-Perricaudet L., *et al.* (1992) *In vivo* transfer of the human cystic fibrosis transmembrane conductance regulator gene to the airway epithelium. *Cell* **68**:143-155.

Simon, R.H., Engelhardt, J.F., Yang, Y., Zepeda, M., Weber-Pendleton, S., Grossman, M., and Wilson, J.M. (1993) Adenovirus-mediated transfer of the CFTR gene to the lung of nonhuman primates: toxicity study. *Hum. Gene Ther.* **4**.771-780.

Treat, J. Kaiser, L. R., Sterman, D. H., Litzky, L. Davis, A., Wilson, J. M., and Albelda, S.M. (1996) Treatment of advanced mesothelioma with the recombinant adenovirus H5.010RSVTK: A phase I trial (BB-IND 6274). *Hum.*

Gene Ther. 7:2047-2057.

Wilson, J.M., Chowdhury, N.R., Grossman, M., Wajsman, R., Epstein, A., Mulligan, R.C., and Chowdhury, J.R. (1990) Temporary amelioration of hyperlipidemia in low density lipoprotein receptor-deficient rabbits transplanted with genetically modified hepatocytes. *Proc. Natl. Acad. Sci. USA* **87**:8437-8441.

Wilson, J.M., Engelhardt, J.F., Grossman, M., Simon, R.H., and Yang, Y. (1994) Gene Therapy of Cysitic Fibrosis Lung Disease Using E1 Deleted Adenoviruses: A phase I trial. *Hum. Gene Ther.* **5**:501-519.

6

Gene Therapy Clinical Trials for Adenosine Deaminase Deficiency/Severe Combined Immunodeficiency

Erlinda M. Gordon and W. French Anderson
University of Southern California, Los Angeles, California

I. INTRODUCTION

Severe combined immunodeficiency due to adenosine deaminase deficiency (ADA-SCID) results from accumulation of the toxic purine metabolites, deoxyadenosine (dAdo) and its triphosphate form, dATP. Toxic levels of dATP cause a depletion of circulating T and B cells (Hershfield and Mitchell, 1994.) The clinical course of ADA-SCID is characterized by severe lymphopenia, impaired cellular and humoral immune responses, and recurrent infections, usually from opportunistic organisms, which are often fatal. Allogeneic bone marrow transplantation using an HLA-matched donor is the treatment of choice for ADA-SCID with a ~70% cure. However, only 20-30% of children with SCID will have an HLA-matched donor, and bone marrow transplantation using T-cell-depleted haplo-identical donors (i.e., parents) is much less successful (O'Reilly *et al.*, 1989.)

Without a bone marrow transplant, enzyme replacement therapy with PEG-ADA has been the mainstay of therapy (Hershfield *et al.*, 1987.) For patients with less severe deficiency and a more benign clinical course, PEG-ADA is often sufficient and is a less toxic alternative to bone marrow transplantation with cytoablation. Since dAdo diffuses between the extracellular and the intracellular compartments, PEG-ADA in the serum reduces the total level of dAdo. Hence, patients who receive PEG-ADA at a dose of 10-15 µg/kg intramuscularly twice a week often are able to enhance their immune responses, and are less susceptible to overwhelming opportunistic infections. Even patients with severe ADA deficiency can benefit from enzyme replacement therapy, although some exhibit only transient improvement. Patients who do not have the option of bone marrow transplantation, and who fail PEG-ADA therapy, are eligible candidates for currently active *ex vivo* gene therapy protocols.

II. THE FIRST HUMAN GENE THERAPY CLINICAL PROTOCOL

The advent of improved retroviral vectors with enhanced gene transfer efficiency and stable integration into the chromosome (Armentano *et al.*, 1989; Gilboa *et al.*, 1986; Miller *et al.*, 1987; Miller *et al.*, 1989), and the demonstration that this procedure of gene transfer could be effectively and safely used in humans (Kasid *et al.*, 1989; Rosenberg *et al.*, 1990; Anderson, 1992), allowed the development of gene therapy clinical trials in the United States and in Europe.

The first human gene therapy protocol, entitled "Treatment of Severe Combined Immunodeficiency Disease (SCID) due to Adenosine Deaminase (ADA) Deficiency with Autologous Lymphocytes Transduced with a Human ADA Gene," was submitted to the Recombinant DNA Advisory Committee (RAC) of the National Institutes of Health (NIH) on March 13, 1990 (Blaese *et al.*, 1990.) The protocol addressed issues raised in the document "Points to Consider in the Design and Submission of Protocols for the Transfer of Recombinant DNA into the Genome of Human Subjects."

II.A. Objectives

The stated objectives of the protocol were: 1) to evaluate the possible therapeutic efficacy of the administration of autologous lymphocytes transduced with a normal hADA gene in an effort to reconstitute the function of the cellular and humoral immune system in patients with ADA⁻SCID, and 2) to evaluate the *in vivo* survival of culture-expanded autologous T cells and the duration of expression of the inserted genes (Blaese *et al.*, 1990.) The procedure was to isolate the T lymphocytes of patients with ADA deficiency, grow these cells in culture in the presence of growth factors, insert the normal ADA gene into them via retroviral-mediated gene transfer, and return the gene-engineered T cells to the patient. A third objective was to evaluate and enhance the development of gene therapy for life-threatening disorders. ADA deficiency was chosen as the prototype disorder for the initial gene therapy clinical trial (Anderson, 1984) since T cells that integrate the ADA gene should have a distinct survival advantage over nontransduced T cells that have a shortened life span in the circulation. The natural course of ADA deficiency is such that the beneficial effects of gene therapy could be evaluated, although concomitant treatment with PEG-ADA introduced a confounding biologic variable.

The retroviral vector system for the ADA trial (Blaese *et al.*, 1990) was identical to that used in the previous Phase I/II gene marker protocol wherein a retroviral vector (LNL6) bearing only the neo[R] gene was introduced into tumor infiltrating lymphocytes (TIL) without adverse effects (Rosenberg, *et al.*, 1990.)

For the ADA trials, a corrective gene, the ADA gene, was inserted into the same vector backbone. The vector was called LASN, where the L stands for long terminal repeat (LTR), A for ADA, S for the SV40 promoter, and N for the neo[R] gene. The long terminal repeat sequences (LTR) of the murine leukemia virus (MuLV) served as the promoter-enhancer that controlled the expression of the ADA gene. This vector had safety features to greatly reduce the possible occurrence of replication-competent virus by homologous recombination (Miller and Buttimore, 1987; Miller, 1990.)

II.B. Safety Issues

The primary concerns involving use of retroviral vectors for gene delivery were 1) the potential for the production of a pathological recombinant retrovirus, and 2) insertional mutagenesis leading to a malignancy. Preclinical studies of monkeys showed no evidence of pathology or cancer (Cornetta, *et al.*, 1990,1991.) Further, there were no side effects or pathology attributable to gene transfer noted in patients who participated in the gene marker (N2 TIL) study (Rosenberg *et al.*, 1990; Anderson *et al.*, 1992.)

More than 2000 mice had received retroviral-mediated gene transfer, and no mouse developed a tumor bearing a vector genome insertion. The NIH concluded that "the use of amphotropically packaged retroviral vectors does not pose a public health risk to patients or to health care personnel, even in the event of accidental exposure to experimental material" (Wyngaarden, 1990.) In both *in vitro* and *in vivo* experiments, the corrected T cells were able to detoxify high levels of dAdo, and the gene-corrected T cells from patients had a greater survival advantage than the noncorrected cells when infused into nude mice (Bordignon *et al.*, 1990.) However, since there are no animal models for ADA deficiency, the growth advantage of gene-corrected murine or monkey T cells over nontransduced T cells could not be elucidated in animals.

II.C. Entry Criteria

Eligible patients were children with documented ADA-SCID who were not eligible to receive allogeneic bone marrow transplantation, as determined by the patient's physician in concert with the child's parents, and who failed to regain or maintain complete immune reconstitution with PEG-ADA. To assess the degree of immune responsiveness to PEG-ADA before gene therapy, the patients needed to have received PEG-ADA, 15 µg/kg, for at least nine months before entry into the study. Patients with significant pulmonary or hepatic dysfunction were not initially considered eligible until patients with normal pulmonary or hepatic function were evaluated. Gene therapy was postponed for children with acute infections. HIV seropositivity or malignancy were exclusion criteria.

Immunologic inclusion criteria included at least three of the following features: an absolute lymphocyte count of $<1500/\text{mm}^3$, a T-cell lymphocyte count of $<1000/\text{mm}^3$, <3 of 7 positive CMI (cell mediated immunity) skin test panel, $<30\%$ of control response to *in vitro* antigen-stimulated lymphocyte proliferation, $<50\%$ of control response to *in vitro* proliferation response to allogeneic cells, $<50\%$ of control *in vitro* lymphocyte proliferation response to mitogens, $<50\%$ of control *in vitro* T-helper function response, $<50\%$ of control *in vitro* T-cell cytotoxicity response, $<40\%$ of control immunoglobulin production *in vitro*, >2 SD below normal mean serum IgG, IgA, or IgM, an isohemagglutinin titer of $<1:16$ by age 3 years, and deficient specific antibody responses to challenge with nonviable vaccine antigens.

II.D. Clinical Course

Patients were admitted to the Clinical Center of the National Institutes of Health, Bethesda, MD, for entry into study. After appropriate clinical evaluation and documentation of protocol eligibility, the child underwent lymphopheresis at intervals (Figure 1) with a maximum of 7 ml/kg of blood withdrawn per lymphopheresis.

Table 1. Results of ADA-SCID Deficiency Gene Therapy Clinical Trials

Patient; reference	Cell number plus [peak] PEG-ADA response before gene therapy	Cell number after gene therapy	Vector DNA persistence	Vector gene expression (RT-PCR)	Immunologic response after gene therapy
1; Blaese *et al.*, 1995; Mullen *et al.*, 1996	500 [1250]	1000 - 1500	1 copy/cell @ 18 mo.	@ 18 mo. - viral specific sequences	++++ during treatment - 18 mo.
2; Blaese *et al.*, 1995; Mullen *et al.*, 1996	200 [1000]	1000	not done	not done	++++ during treatment - 12 mo. [18 mo. not yet reported]
3; Kohn *et al.*, 1995	N/A	N/A	6/10 clonogenic cells were LASN$^+$ @ 1 yr.	0.03-0.3% control LASN$^+$ cells	normal
4; Kohn *et al.*, 1995	N/A	N/A	all tested clones were LASN$^+$ @ 1 yr.	not done	normal
5; Kohn *et al.*, 1995	N/A	N/A	all tested clones were LASN$^+$ @ 1 yr.	not done	normal
6; Bordignon *et al.*, 1995	2400 [4000]	2800	@ 6 mo. PBC contained PBL-origin vector; @35 mo. PBC, BM, and early PB contained BM-origin vector	in PBC for 1 yr., then slight in BM for 1 yr., then	++++ during treatment - 3 yr.
7; Bordignon *et al.*, 1995	2000 [4000]	3000	" "	" "	++++ during treatment - 20-24 mo., then to peak PEG ADA level
8; Hoogerbrugge *et al.*, 1996	100 (1000)	350	PBC @ 3 mo., no later; not found in BM	not done	not done
9; Hoogerbrugge *et al.*, 1996 [PEG-ADA after gene therapy]	800 (3000)	800	PBC @ 3 mo., no later BM @ 2 yr., not before or after	not done	not done
10; Hoogerbrugge *et al.*, 1996	1000 [very erratic]	200	PBC @ 2 mo., no later	not done	not done

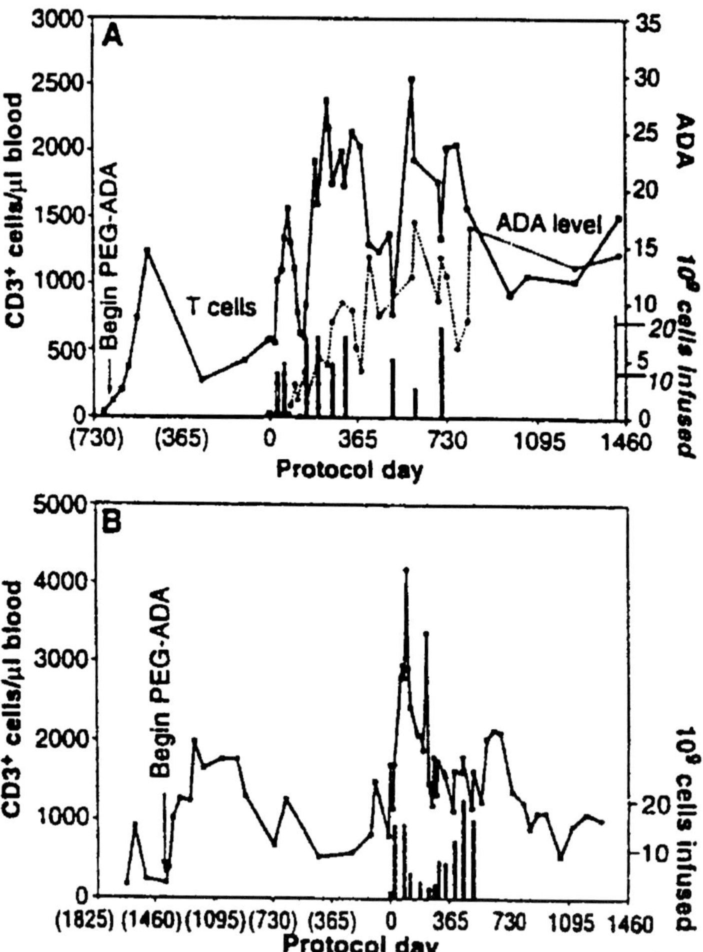

Figure 1. Peripheral blood T cell counts for Patients 1 and 2 since the time the diagnosis of ADA deficiency was made, as well as protocol days and the total number of cells infused for each patient. ADA level is measured in nanomoles of adenosine deaminase per minute per 10^8 cells. Vertical bars indicate the dates of cell infusion, and their height represents the total number of nonselected cells infused at each treatment. The T cell numbers represent total CD3-bearing T cells determined by standard flow cytometric analysis. **A.** Patient 1 began gene therapy on 14 September 1990 (protocol day 0) and received a total of 11 infusions. Cellular ADA enzyme level is indicated by the dashed line. ADA activity was determined as described (Kohn *et al.*, 1989; Ferrari *et al.*, 1991.) Values shown are the mean of duplicate samples and represent EHNA-sensitive ADA enzyme activity. **B.** Patient 2 began gene therapy on 31 January 1991 (protocol day 0) and received a total of 12 infusions *(reprinted, with permission, from Blaese, et al., 1995.)*

The peripheral blood mononuclear cells were isolated by Ficoll-Hypaque density gradient centrifugation, washed, counted and cultured in AIM-V nutrient medium containing IL-2, a T-cell growth factor, and OKT3 monoclonal antibody to activate the T cells. Proliferating T lymphocytes were incubated with the LASN vector supernatant, in the presence of protamine and IL-2, two times daily for 2-3 days. The cells were replenished with fresh AIM-V medium, harvested, washed and resuspended in normal saline, filtered through a platelet filter and infused through a peripheral vein after a test dose was given. An initial infusion of ~6 × 10^7 cells/kg body weight was given in a volume of 10 ml/kg body weight administered over a period of 2 hrs. Later (see Figure 5 for details), the number of ADA gene-corrected T lymphocytes was progressively increased and given at intervals.

The followup included a complete blood count with differential, platelet count, chemistry panel, T and B cell phenotype analysis, lymphocyte proliferative responses, cellular ADA concentration, and analysis for vector DNA by PCR or Southern analysis 18 to 72 hrs after completion of each treatment. At the end of the study, additional tests included a skin test panel for delayed type hypersensitivity (DTH), immunoglobulin levels, isohemagglutinin titers, functional evaluation of B-cell responsiveness, helper and suppressor T cell function, evaluations for natural killer cell, lymphocyte-activated killer cell, cell-mediated lympholysis, and antiviral cytotoxic activity in peripheral blood lymphocytes (PBL), and antibody titer in response to immunization with tetanus toxoid, diphtheria toxoid, keyhole limpet hemocyanin, pneumovax, *Hemophilus influenzae* polysaccharide, and *Brucella abortus*.

Designated response criteria included development of more than one new positive delayed-type hypersensitivity skin test to immunogens, antigen-specific proliferative responses *in vitro* to antigens where no responsiveness previously existed, production of specific antibodies in response to antigenic challenge, a normal lymphocyte count in a patient who was previously lymphopenic, or resolution of a chronic infection without specific antibiotic therapy.

The first patient began gene therapy on September 14, 1990. The patient's response to PEG-ADA therapy had diminished after an initial increase in peripheral blood lymphocyte numbers. Within six months of beginning gene therapy, lymphocyte counts, ADA enzyme activity, and immunologic responses improved (Figures 1-3, Table 1.) Tests of immunologic function continue to be conducted up to the present time, and the significant improvement in the patient's immune responses have been maintained. Furthermore, normal levels of peripheral lymphocytes have been maintained over this time period with reduced dosage of PEG-

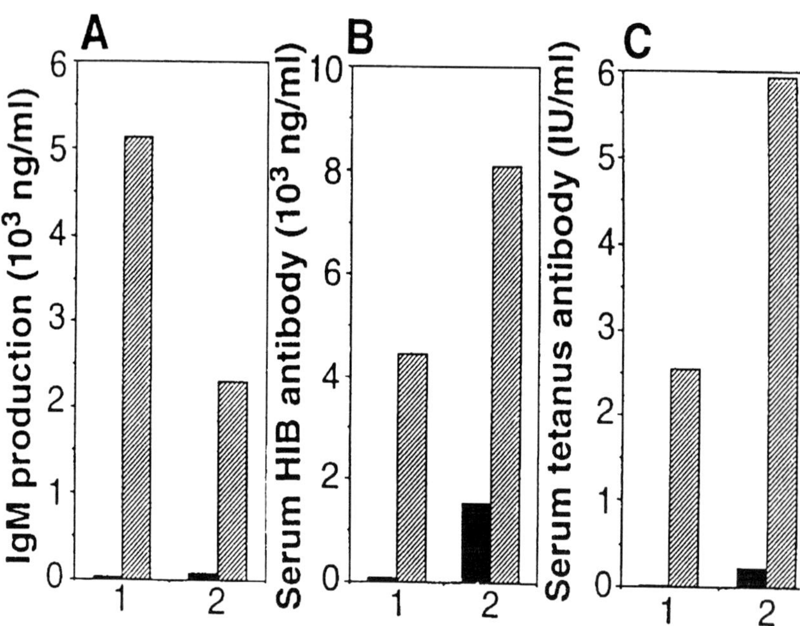

Figure 2. Humoral immune function of patients 1 and 2 before (solid bars) and after (hatched bars) gene therapy. **A.** IgM production by the patient's peripheral blood mononuclear cells in cultures stimulated with the T cell-dependent polyclonal activator PWM performed as described (Waldmann *et al.*, 1974.) "Before" samples were from D(-9.) Follow-up cultures were at D500 (patient 1) and D560 (patient 2.) In each case, the patient's cells, stimulated with the T cell-independent B cell stimulant EBV (Kirchner *et al.*, 1979), produced normal amounts of IgM (not shown), indicating intact B cell function before and after gene therapy, as expected. At least two normal subjects were included concurrently in each assay. **B.** Serum antibody response to *Hemophilus influenzae* B. Patient 1 had failed to respond to two immunizations while on PEG-ADA alone (D-9 shown.) Her response at protocol D591 is shown, after immunization. Patient 2 had some HIB-specific antibodies present before therapy (D-122), whose amounts increased without additional immunization during the protocol (D560.) **C.** Serum tetanus antibody. Patient 1 had negligible response to five separate tetanus immunizations before gene therapy (D-48 shown), but responded briskly at D731, 24 days after re-immunization. Serum titers for patient 2 are shown for D-9, 140 days after immunization while on PEG-ADA alone, and after receiving gene therapy (D592), 32 days after a booster tetanus immunization *(reprinted, with permission, from Blaese, et al., 1995.)*

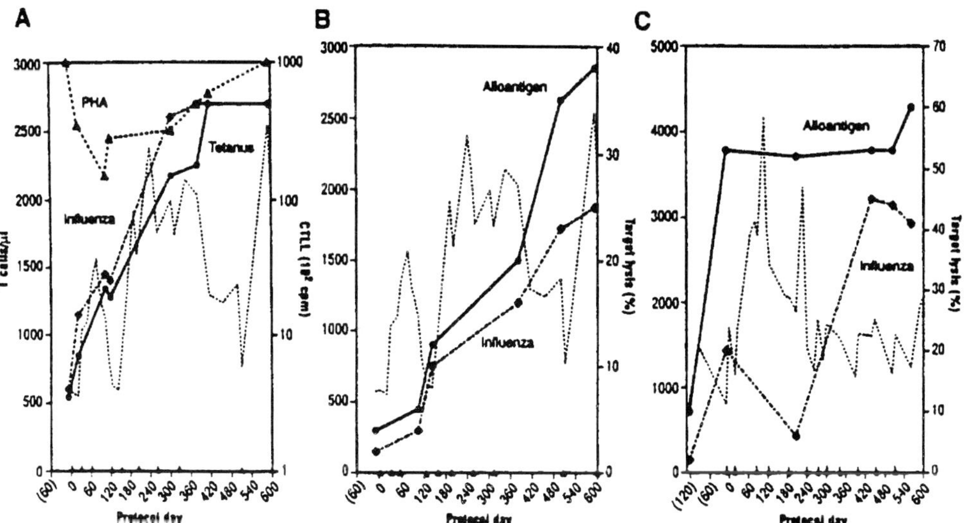

Figure 3. Evaluation of the *in vitro* cellular immune responses of blood T cells from patients 1 and 2 at various times before and during the gene therapy trial. At least two normal subjects were included concurrently in each assay. **A.** Production of IL-2 by cultured cells from patient 1 after stimulation with the mitogen PHA and with the specific antigens tetanus toxoid and influenza A virus as described (Clerici *et al.*, 1989.) IL-2 was quantitated by bioassay measuring the proliferation of the IL-2-dependent T cell line CTL at a 1:2 dilution of the lymphocyte culture supernatant. The fine dashed line indicates the patient's T cell count for reference. Solid triangles along the base line indicate the dates of cell infusion. **B.** *In vitro* killing of an allogeneic target B cell line by blood T cells from patient 1 as described (Biddison *et al.*, 1980; Clerici *et al.*, 1991.) Lysis (as percent specific isotope release during a 6-hour incubation of effector and target cells at a ratio of 60:1) was measured after *in vitro* pre-stimulation for 7 days. Solid triangles along the base line indicate the dates of cell infusion. **C.** *In vitro* killing of a ^{51}Cr-labeled, influenza A-infected autologous B cell and a ^{51}Cr-labeled allogeneic target B cell line by blood T cells from patient 2 as described above *(reprinted, with permission, from Blaese, et al., 1995.)*

ADA per kg body weight. Peripheral blood cells (PBC) were assessed for ADA activity, and ADA transcripts, as well as for vector DNA (Figures 4-5.) RT-PCR analysis demonstrated vector-derived transcripts in PBC about 3 months and 18 months after the last infusion of transduced lymphocytes. Southern blot and PCR analyses indicated that the ADA cDNA copy number/cell rose during the course of treatment, then stabilized at about 1 copy per cell. This level has persisted.

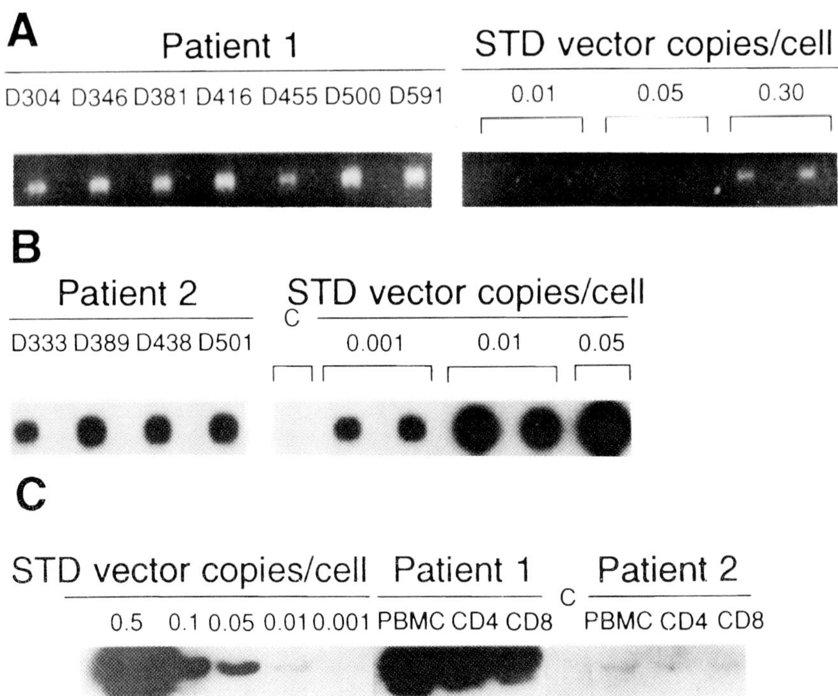

Figure 4. PCR evaluation of the frequency of LASN vector-positive cells in the blood of patients 1 and 2 at various protocol days. **A.** Cells from patient 1 for protocol days D304 to D591. PCR analysis was performed as described (Morgan *et al.*, 1990) in an ethidium-stained gel. **B.** Cells from patient 2 for protocol days D333 to D501. PCR products were probed with ^{32}P-labeled *neo* gene as described (Morgan *et al.*, 1990.) **C.** Purified CD4[+] and CD8[+] cell subpopulations from patient 1 (D1480) and patient 2 (D1198) prepared by separation of peripheral blood mononuclear cells (PBMCs) by fluorescence-activated cell sorting (FACS.) The purity of the separated T cell subpopulations from which DNA was extracted exceeded 98%, as confirmed by FACS analysis. Direct PCR with ^{32}P deoxycytosine triphosphate was performed as described (Bunnell *et al.*, 1995.) Standards (STD) were prepared from DNA obtained from cell mixtures of a known proportion of LASN-transduced cells containing a single vector insert mixed with vector-negative cells *(reprinted, with permission, from Blaese, et al., 1995.)*

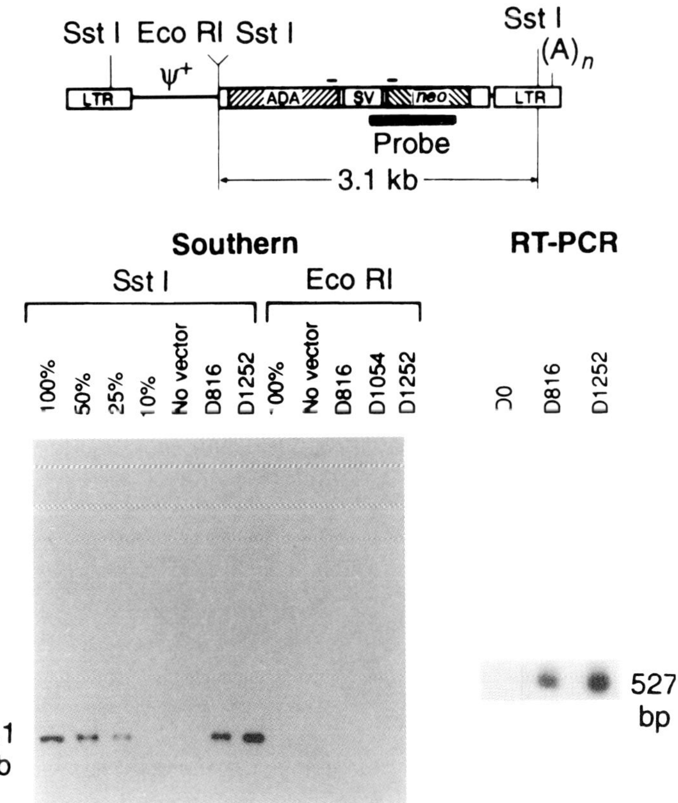

Figure 5. Quantitative Southern hybridization analysis of DNA prepared from the blood mononuclear cells of patient 1 on protocol days D816 and D1252. DNA digested with SstI should yield a single restriction fragment of 3.1 kb containing both the vector neo and ADA genes. EcoRI cuts only once within the vector sequence, and therefore a detectable band would indicate that a predominant clone with a single unique vector integration site was present in that blood sample. None was detected. Polyadenylated mRNA was extracted from the patient cells on days D0, D816, and D1252 and analyzed for vector message by RT-PCR. The primer locations used are indicated as short solid lines above the vector diagram. SV, SV40 early promoter; (A)$_n$, polyadenylation site; Psi$^+$, extended retrovirus packaging signal. Hatched regions indicate protein coding regions *(reprinted, with permission, from Blaese, et al., 1995.)*

Endogenous sequences and transcripts were distinguished from vector sequences using vector-specific markers (Mullen *et al.*, 1996.) Since such longevity was not expected, attempts were made to ascertain the clonal origin of these sequences. Among the 6 successful clones from 7-8 months after the final gene therapy infusion, 5 clones had demonstrable LASN sequences. All 6 clones had measurable ADA activity, suggesting that ADA^+ cells have a significant selective advantage *in vivo* (Table 2.) The genetic source of the endogenous ADA activity in the sixth clone could not be determined. Plausibly, there is another enzyme, normally exhibiting very low adenosine deaminase activity, whose activity may have been induced. Further, there is the possibility of spontaneous revertants of the ADA gene. Both occurrences would be extremely rare, but this observation emphasized the need to use additional tests besides enzymatic activity to test for efficacy of the gene therapy. PCR, RT-PCR, and restriction analyses of DNA and transcript RNAs provide more reliable identification of the product. In the other 5 clones, RT-PCR verified the vector origin of the transcript (Figures 6,7.)

The second patient presented with less severe symptoms, and responded only transiently to PEG-ADA. Patient 2's response to gene therapy was not as strong and convincing as the response of patient 1. Nonetheless, both patients tolerated the gene therapy infusions without acute adverse events, followed by a decrease in the incidence of infectious episodes and an overall improvement in clinical status. The laboratory responses of patient 2 had been quite variable before gene therapy, and the transduction efficiency of her peripheral blood lymphocytes was only ~10% that of patient 1. Despite the difficulty in interpreting the observations of patient 2, her immune response has been maintained even on a reduced PEG-ADA dose after the infusions were discontinued. RT-PCR analysis detected vector-derived transcripts ~2 years after the last treatment and the level of vector-containing cells persisted during a 4-year follow-up (see Figure 5 legend for details.)

III. SUBSEQUENT HUMAN GENE THERAPY TRIALS

Four other groups have since attempted human gene therapy for ADA-SCID patients. The gene therapy protocols of the subsequent American group and the two European groups differed somewhat from this original protocol. Table 1 shows the results of the gene therapy experience in 10 patients. In addition, two patients in Japan have received gene therapy infusions, using the original ADA protocol (Kawamura *et al.*, 1995.) No data on the patients' response to treatment are available at the present time.

Table 2. ADA Enzyme Activity in Uncultured PBMCs and in Longterm Lymphoctye Cell Lines

Cells	Source	Vector DNA*	Vector RNA[†]	ADA, %[‡]
Primary PBMCs	Normal donor	-	-	100
	Patient pretreatment	-	-	1
	Patient day 115	+	-	3
	Patient day 500	+	+	11
	Patient day 1252	+	+	18
Cultured cell lines	Patient clone 1.1	+	NT[§]	61
	Patient clone 1.3	+	+	60
	Patient clone 1.7	+	+	141
	Patient clone 1.14	+	+	74
	Patient clone 1.20	+	NT	43
	Patient clone 1.21	-	-	24
	Patient pretreatment	-	-	10
	TJF-2	-	-	1
	TJF-2-LASN	+	+	407
	K562	-	-	98
	K562-LASN	+	+	3387

*Primary uncultured PBMCs were obtained from Ficoll-Hypaque fractionated venipuncture blood. Patient clones were obtained from the patient on day 985, 9 months after the final infusion of treated cells. Cultured patient pretreatment cells were PBMCs obtained prior to the first treatment, that were stimulated with PHA *in vitro* and grown *in vitro* for greater than 7 weeks in tissue culture before analysis.

[†]Vector DNA +: cells were found positive for LASN vector DNA by Southern hybridization analysis of genomic DNA and by PCR using primers specific for LASN vector

[‡] Vector RNA +: cells were found positive for LASN vector RNA by RT-PCR of DNase-treated mRNA using primers specific for LASN vector, and that did not amplify cellular ADA mRNA.

[§]Percentage of normal ADA activity: EHNA-sensitive adenosine deaminating activity values of cells as a percentage of normal peripheral blood PBMC activity. Percentage of normal ADA activity of cloned cells represents total adenosine deaminating activity, because the limited number of cloned cells precluded repetition of the assay including EHNA as a specific inhibitor of ADA. *(Reprinted, with permission, from Mullen, et al., 1996.)*

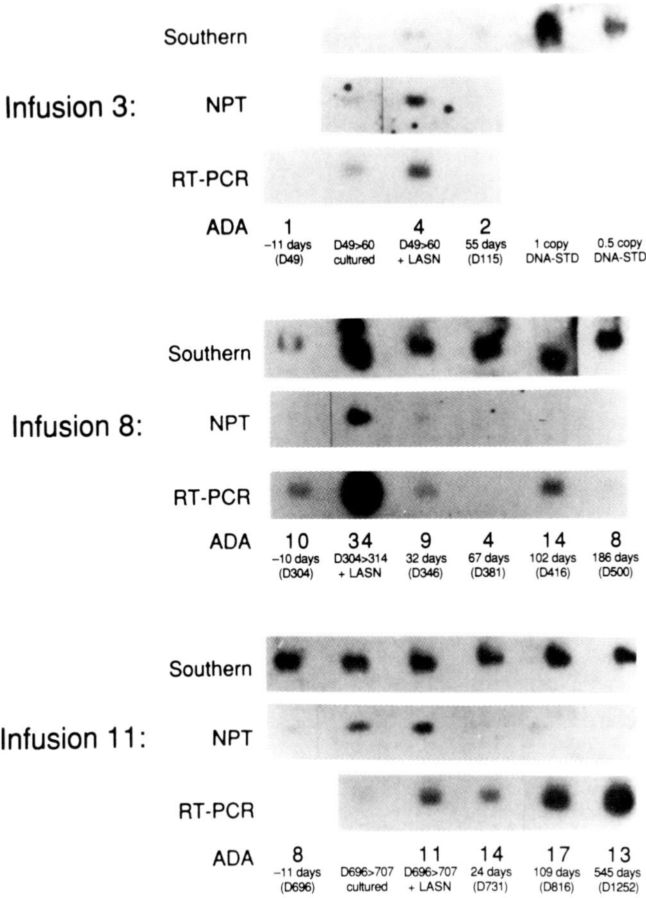

Figure 6. LASN vector expression in patient PBMCs. Patient PBMC samples from infusion cycles 3, 8, and 11 were studied. The study date is indicated under each column. The number of days from the initiation of the study is indicated by "D" (e.g. D115.) The number of days relative to the infusion date for the infusion cycle is also indicated (e.g. "55 days" relative to infusion 3.) PBMCs from blood were directly assayed and are identified with a "D" number in parentheses, e.g. (D115.) Cells grown in culture without additional vector are designated "cultured." Those designated "LASN" were exposed to additional vector and were aliquots of the cells infused into the patient in the infusion cycle. For each point in time, a Southern blot was performed to identify the LASN vector. DNA from K562-LASN cells provided the positive control. NPT is neomycin phosphotransferase activity in lysates of 5×10^6 cells. RT-PCR represents LASN vector message in DNase-treated RNA. Positive patient signals from 0.3 µg of cDNA were similar in intensity to 0.3 ng of cDNA from K562-LASN. EHNA-sensitive ADA activity is described for each PBMC sample *(reprinted, with permission, from Mullen, et al., 1996.)*

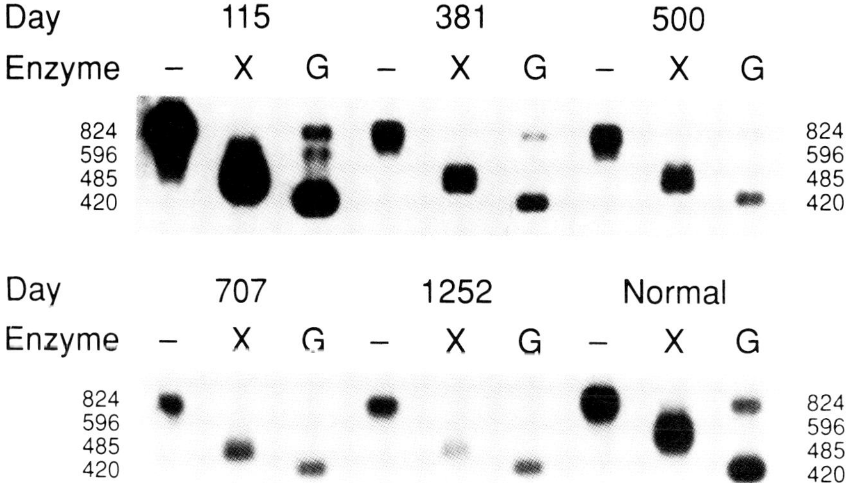

Figure 7. RT-PCR analysis of endogenous (nonvector) ADA-mRNA. A 0.3 µg amount of cDNA was amplified with PCR using vectors that were specific for endogenous ADA and that fail to amplify LASN vector-derived mRNA. Each PCR product was split into three aliquots and digested with: no enzyme (-), Bst XI (X) or Bgl II (G.) Normal and mutant ADA genes should both produce an 824-bp band that reduces to 420 bp with Bgl II digestion. (The larger bands in the "G" lanes represent incomplete digestion with Bgl II.) Bst XI digestion of the PCR product from a normal gene yields a 596-bp band, whereas the mutant ADA gene which contains an additional Bst XI site yields a 485-bp band. Samples are from the patient on the study day indicated or from a normal human donor. The bands corresponding to days 115, 381, and 500 represent an overnight exposure of the blot, whereas the bands from days 707, 1,252, and the normal donor are from a 3 hr exposure of the same blot *(reprinted, with permission, from Mullen, et al., 1996.)*

The Italian group (Bordignon *et al.*, 1995) used a human ADA minigene (promoter and full length cDNA) which was inserted in the LTR U3 region of the Moloney murine leukemia virus-derived vector (DCA.) Two variant vectors with unique marker restriction sites inserted into a nonfunctional region of the retroviral LTR were used to transduce PBL (DCAl) or BM (DCAm) independently, to distinguish the cellular origin of subsequently isolated vector-containing cells. One patient received 9 injections over 24 months, while the other patient received 5 injections over 9 months. In most respects, the patient selection criteria and treatment regimen were essentially the same as in the Blaese *et al.* (1990) protocol. The patients were given a series of transfusions of transduced autologous cells after ADA-PEG treatment had been transiently successful. These transfusions consisted of a mixture of PBL and BM cells, which had been separately transduced with one of the two vectors. ADA activity was measured in these patients for 7-35 months after treatment. At 6 months, 5-30% of BM cells and 0.8-8.5% of PBL expressed the corresponding vector transcripts. By 16 months, transductants accounted for approximately 5% of peripheral blood T (PBT)-cells and 25% of BM cells in patient 1, 2% of PBT-cells and 17% BM cells in patient 2. No revertant ADA^+ vector cells were observed.

Since the PBC and BM cells had been transduced with otherwise identical vector distinguished only by unique markers, the cellular source of the persistent vector could be ascertained. In patient 1, PBL at 7 and 19 months contained vector DCAl, which had been transduced into PBC, but one year after the end of therapy BM and circulating granulocytes contained DCAm, originally transduced into BM. At 35 months, both PBL and BM cells contained DCAm exclusively in patient 2 and almost exclusively in patient 1. Thus, persistence of gene expression was noted in BM cells transduced with the DCAm vector.

To confirm this result, individual clones derived from cells collected at these later times were analyzed with more precision. From patient 1, 26/35 clones had vector sequences derived from BM; from patient 2, 37/42 clones had BM-derived vector. Thus, short term transgene expression observed for ~12 months after gene therapy appears to be supported by transduced PBL. In contrast, the transgene expression in transduced BM cells increased more slowly, but continued to persist or even increase over time, up to 35 months after the infusions were discontinued. Furthermore, ADA activity was detected in a wider variety of cell types, including erythrocytes. Hence, long term transgene expression appears to be supported by transduced BM cells.

Hoogerbrugge *et al.* (1996) reported BM transduction and gene transfer in three patients. Their vector was not described in detail. The course of treatment for two of these patients (#1 and #3) is similar to that in the Blaese *et al.* (1990) protocol, but patient 2 received PEG-ADA starting 4 months after gene therapy. Furthermore, and importantly, only a single infusion of transduced BM cells was administered to each patient. BM aspirates were enriched for $CD34^+$ using an immunoaffinity system. Less than 2% of the original cell population was recovered. Three months after the treatment, no change in PBL count was seen in patient #1 or #3; patient #2 experienced a transient rise in cell number after PEG-ADA treatment began which did not persist. PCR analysis detected vector sequences in PBC for all three patients up to 14 weeks after gene therapy, and in BM of patient 2 only, at 24-26 weeks but not subsequently. However, this vector did not have a selective marker, and the ability of these workers to distinguish endogenous ADA sequences from vector-derived sequences is not clear. No RT-PCR for detection of gene products is reported.

Subsequently, three neonates, prenatally diagnosed as having homozygous ADA⁻ gene deficiency, were infused with autologous cord blood-derived cells, including $CD34^+$ progenitor/stem cells, transduced with LASN (Kohn *et al.*, 1995.) This protocol differs from the other protocols in that: 1) the patients received transduced $CD34^+$ cells enriched from umbilical cord blood, rather than peripheral blood lymphocytes, with the hope that these stem cells would be capable of self-renewal, and would provide a continuous *in vivo* source of the enzyme, and 2) the newborn infants were not symptomatic when gene therapy was given. The patients received PEG-ADA enzyme replacement therapy during the first four days after birth, and on the fourth day, the transduced cord blood cells were reinfused.

The same vector, LASN, was transduced into umbilical cord blood cell preparations enriched for $CD34^+$ progenitor/stem cells by treating the cord blood with biotin-CD34 antibody conjugate, and then using an avidin immunoaffinity column to separate biotinylated cells. The process of separating and enriching the $CD34^+$ cells, and three rounds of LASN transduction to increase yield, was conducted in four days. On the fourth day, the transduced autologous cells (which were not subjected to G418 selection) were infused into the patients. The patients were also given and maintained on PEG-ADA therapy. ADA expression (i.e., enzyme activity and RT-PCR amplified transcript) were assessed for the following 18 months. In these three patients, 12.5 - 21.5% of the clonogenic cells were neo resistant, indicating that they had been stably transduced. Semi-quantitative PCR, using vector-specific primers, detected LASN ADA sequences in peripheral blood mononuclear cells and granulocytes over the entire 18 months of the study. BM aspirates were examined at 1 year.

Vector DNA was detected in granulocytic and nongranulocytic cells in all three patients. Among the clonogenic cells with detectable vector sequences, about 5% were neomycin-resistant. When specific clones were tested for the presence of vector, 6/10 were LASN[+] in patient 1, and all tested clones were LASN[+] in the other two patients. This again may reflect the selective survival advantage of cells expressing the ADA gene. Interestingly, while 4-6% of BM cells were LASN[+], only 0.03-0.001% of PBC were LASN[+]. Therefore, the isolated BM CD34[+] cells were examined directly by PCR. About 1% of these cells were LASN[+], which agrees with the order of magnitude of neo-resistance populations. Since integration sites are variable, each original transduced cell should have a distinct restriction pattern of integrated vector sequences. Nine or ten CFUs were examined from each patient, and 3-5 distinct patterns were found in each patient. This result suggests that the current population of CD34[+] - LASN[+] cells originated from a small number of clonal sources in each individual.

RT-PCR of PBC showed activity 0.03-0.3% of control standard LASN-expressing cells. This result meant that ADA activity was too low to be tested directly in PBC, but cells expanded *in vitro* produced sufficient enzyme to test for biologic activity. These values corresponded well to the theoretical yield, i.e., if 1% of total cell number were transplanted originally, at 1-10% efficiency of transduction, this would result in about 0.1-0.01% of BM cells transduced. Clinically, all three infants remained asymptomatic on supplemental PEG-ADA therapy. Results of the ADA gene therapy protocol conducted in Japan are not yet available.

IV. CONCLUSIONS

The first four human gene therapy trials for ADA-SCID have been successfully conducted. While a single infusion of transduced cord blood CD34[+] cells may be sufficient for neonatal gene therapy, multiple infusions of transduced cells appear to be required for patients with symptomatic SCID. Immunologic parameters improved during the course of gene therapy. These included peripheral blood lymphocyte cell count, ADA enzyme activity, ADA transcripts, response to mitogens, and a variety of immunological responses. All of these activities appeared to stabilize at about the levels attained at the cessation of gene therapy, except in older children who received only one infusion of transduced cells. In those patients who received multiple infusions of ADA-transduced cells, immune responses improved progressively over the course of gene therapy. Taken together, the results of the first ADA gene therapy clinical trials indicate that *ex vivo* ADA gene therapy is safe and may be effective as supplemental therapy for patients with severe ADA-SCID, or curative in patients with less severe symptomatology.

V. REFERENCES

Anderson, W.F. (1984) Prospects for human gene therapy. *Science* **226**:401-409.

Anderson, W.F. (1992) Human gene therapy. *Science* **256**:808-813.

Anderson, W.F. (1994) Genetic engineering and our humanness. *Hum. Gene Ther.* **5**:755-760.

Biddison, W.E., Ward, F.E., Shearer, G.M., and Shaw, S. (1980) The self determinants recognized by human virus-immune T cells can be distinguished from the serologically defined HLA antigens. *J. Immunol.* **124**:548-552.

Blaese, R.M., Anderson, W.F., Culver, K.W., and Rosenberg, S.A. (1990) Treatment of severe combined immune deficiency disease (SCID) due to adenosine deaminase (ADA) deficiency with autologous lymphocytes transduced with a human ADA gene. *Hum. Gene Ther.* **1**:327-362.

Blaese, R., Culver, K., Miller, A., Carter, C., Fleisher, T., Clerici, M., Shearer, G., Chang, L., Chiang, Y., Tolstoshev, P., Greenblatt, J., Rosenberg, S., Klien, H., Berger, M., Muller, C., Ramsey, J., Muul, L., Morgan, R., and Anderson, W.F. (1995) T lymphocyte gene therapy for ADA deficiency (SCID): Results of the initial trial with 4 years of observation. *Science* **270**:475-480.

Bunnel, B.A., Muul, L.M., Donahue, R.R., Blaese, R.M., and Morgan, R.A. (1995) High-efficiency retroviral-mediated gene transfer into human and nonhuman primate peripheral blood lymphocytes, *Proc. Natl. Acad. Sci. USA* **92**:7739-7743.

Bordignon, C., Notarangelo, L.D., Nobili, N., Ferrari, G., Casorati, G., Panina, P., Mazzolari, E., Maggioni, D., Rossi, C., Servida, P., Ugazio, A.G., and Mavilio, F. (1995) Gene therapy in peripheral blood lymphocytes and bone marrow for ADA-deficient patients. *Science* **270**:470-475.

Clerici, M., Stocks, M.I., Zajac, R.A., Boswell, R.N., Lucey, D.R., Via, C.S., and Shearer, G.M. (1989) Detection of three distinct patterns of T helper cell dysfunction in asymptomatic, human immunodeficiency virus-seropositive patients. Independence of CD4 cell numbers and clinical staging. *J. Clin. Invest.* **84**: 1892-1989.

Clerici, M., Lucey, D.R., Zajac, R.A., Boswell, R.N., Gebel, H.M., Takahashi, H., Berzofsky, J.A., and Shearer, G.M. (1991) Detection of cytotoxic T lymphocytes specific for synthetic peptides of gp160 in HIV-seropositive individuals. *J. Immunol.* **146**:2214-2219.

Cornetta, K., Moen, R.C., Culver, K., Morgan, R.A., McLachlin, J.R., Sturm, S., Selegue, J., London, W., Blaese, R.M., and Anderson, W.F. (1990) Amphotropic murine Leukemia Retrovirus is not an acute pathogen for primates. *Hum. Gene Ther.* **1**:15-30.

Cornetta, K., Morgan, R.A., Gillio, A., Sturm, S., Baltrucki, L., O'Reilly, R.O., and Anderson, W.F. (1991) No retroviremia or pathology in long-term follow-up of monkeys exposed to a murine amphotropic retrovirus. *Hum. Gene Ther.* **2**:215-219.

Culver, K.W., Morgan, R.A., Osborne, W.R.A., Lee, R.T., Lenschow, D., Able, C., Cornetta, K., Anderson, W.F., and Blaese, R.M. (1990) *In vivo* expression and survival of gene-modified T lymphocytes in Rhesus monkeys. *Hum. Gene Ther.* **1**:399-410.

Culver, K.W., Anderson, W.F., and Blaese, R.M. (1991) Lymphocyte gene therapy. *Hum Gene Ther.* **2**:107-109.

Ferrari, G., Rossini, S., Giavazzi, R., Maggioni, D., Nobili, N., Soldati, M., Ungers, G., Mavilio, F., Gilboa, E., and Bordignon, C. (1991) An *in vivo* model of somatic cell gene therapy for human severe combined immunodeficiency. *Science* **251**:1363-1366.

Fong, S.W., Qaqundah, B.Y., and Taylor, W.F. (1974) Developmental patterns of ABO isoagglutinins in normal children correlated with the affect of age, sex and maternal isoagglutinins. *Transfusion* **14**:551-559.

Gilboa, E., Eglitis, M.A., Kantoff, P.W., and Anderson, W.F. (1986) Transfer and expression of cloned genes using retroviral vectors. *BioTechniques* **4**:504-512.

Gillio, A., Bordignon, C. J., Kerman, N., Kantoff, P., Eglitis, M., McLachlin, J., Karson, E., Yu, S.F., Zwiebel, J., Nienhuis, A., Karlsson, S., Blaese, M., Kohn, D., Armentano, D., Gilboa, E., Anderson, W.F., and O'Reilly, R.J. (1987) Retroviral vector-mediated gene transfer and expression in nonhuman primates following autologous bone marrow transplantation. *Ann. NY Acad. Sci.* **511**:406-417.

Hershfield, M.S. (1993) The role of polyethylene glycol adenosine deaminase in the evolution of therapy for adenosine deaminase deficiency. In Gupta S., and Griscelli, C. (eds.), *New Concepts in Immunodeficiency Diseases.* Wiley, New York, 417-426.

Hershfield, M.S., and Mitchell, B.S. (1994) Immunodeficiency diseases caused by adenosine deaminase deficiency and purine nucleoside phosphorylase deficiency In Scriver, C.R., Beaudet, A.L., Sly, W.S. and Valle, D., (eds.), *The Metabolic and Molecular Bases of Inherited Disease.* McGraw-Hill, New York, 1725-1768.

Hoogerbrugge, P.M., Van Beusechem, V.W., Fischer, A., Debree, M., le Deist, F., Perignon, J.L., Morgan, G., Gaspar, B., Fairbanks, L.D., Skeoch, C.H., Moseley, A., Harvey, M., Levinsky, R.J., and Valerio, D. (1996) Bone marrow gene transfer in three patients with adenosine deaminase deficiency. *Gene Ther.* **3**:179-183.

Kantoff, P.W., Kohn, D.B., Mitsuya, H., Armentano, D., Sieberg, M., Zwiebel, J.A., Eglitis, M.A., McLachlin, J.R., Wiginton, D.A., Hutton, J.J., Horowitz, S.D., Gilboa, E., Blaese, R.M., and Anderson, W.F. (1986) Correction of adenosine deaminase deficiency in human T and B cells using retroviral-mediated gene transfer. *Proc. Natl. Acad. Sci. USA* **83**:6563-6567.

Kasid, A., Morecki, S., Aebersold, P., Cornetta, K., Culver, K., Freeman, S., Director, E., Lotze, M.T., Blaese, R.M., Anderson, W.F., and Rosenberg, S.A. (1990) Human gene transfer: characterization of human tumor-infiltrating lymphocytes as vehicles for retroviral-mediated gene transfer in man. *Proc. Natl. Acad. Sci. USA* **87**:473-477.

Kawamura, N., Tame, A., Furuta, H., Kobayashi, I., Ariga, T., Okano, M., Sakiyama, Y., Blaese, R.M., Matsumoto, S. (1995) Gene therapy for a patient with adenosine deaminase deficiency: a retrovirus vector-mediated adenosine deaminase gene transfer protocol for human gene therapy. *Gene Ther.* **2**:684.

Kirchner, H., Tosato, G., Blaese, R.M., Broder, S., Magrath, I.T. (1979) Polyclonal immunoglobulin secretion by human B lymphocytes exposed to Epstein-Barr virus *in vitro. J Immunol.* **122**:1310-1313.

Kohn, D.B., Mitsuya, H., Ballow, M., Selegue, J.E., Barankiewicz, J., Cohen, A., Gelfand, E., Anderson, W.F., and Blaese, R.M. (1989) Establishment and characterization of adenosine deaminase-deficient human T cell lines. *J Immunol*, **142**:3971-3977.

Kohn, D.B., Weinberg, K.I., Nolta J.A., Heiss, L.N., Lenarsky, C. Crooks ,G.M., Hanley, M.E., Annett, G., Brooks, J.S., El-Khoureiy, A., Lawrence, K., Wells, S., Moen, R.C., Bastian, J., Williams-Herman, D.E., Elder, M., Wara, D., Bowen, T., Hershfield, M.S., Mullen, C.A., Blaese, R.M., and Parkman, R. (1995) Engraftment of gene modified umbilical cord blood cells in neonates with adenosine deaminase deficiency. *Nature Med.* **1**:1017-1023.

Morgan, R.A., Cornetta, K., and Anderson, W.F. (1990) Applications of the polymerase chain reaction in retroviral-mediated gene transfer and the analysis of gene-marked human TIL cell. *Hum. Gene Ther.* **1**:135-149.

Miller, A.D. (1990) Retrovirus packaging cells. *Hum. Gene Ther.* **1**:5-14.

Miller, A.D., and Buttimore, C. (1987) Redesign of retrovirus packaging cell lines to avoid recombination leading to helper virus production. *Mol. Cell Biol.* **6**:2895-2902.

Miller, A.D., and Rosman, G.J. (1989) Improved retroviral vectors for gene transfer and expression. *BioTechniques* **7**:980-984.

Mullen, C.A., Snitzer, K., Culver, K.W., Morgan, R.A., Anderson, W.F., and Blaese, R.M. (1996) Molecular analysis of T lymphocyte-directed gene therapy for adenosine deaminase deficiency: long-term expression *in vivo* of genes introduced with a retroviral vector. *Hum. Gene Ther.* **7**:1123-1129.

O'Reilly, F.J., Keever, C.A., Small, T.N., and Brochstein, J. (1989) The use of HLA-non-identical T-cell-depleted marrow transplants for the correction of severe combined immunodeficiency disease. *Immunodeficiency Rev.* **1**:273-309.

Rosenberg, S.A., Blaese, R.M., and Anderson, W.F. (1990) The N2-TIL human gene therapy clinical protocol. *Hum. Gene Ther.* **1**:73-92.

Waldmann, T.A., Durm, M., Broder, S., Blackman, M., Blaese, R.M., and Strober, W. (1974) Role of suppressor T cells in pathogenesis of common variable hypogammaglobulinaemia. *Lancet* **2**(7881):609-613.

Walker, R.H. (1993) *AABB Technical Manual*, 11th Ed., American Association of Blood Banks, Bethesda, MD, 622-623.

Wyngaarden, J.B. (1990) Human gene transfer protocol. *Federal Register* **54** (March 13, 1989):10508, reprinted in *Hum. Gene Ther.* **1**:91-92.

7

Development of an Oligodeoxynucleotide Pharmaceutical for the Treatment of Human Leukemia

Alan M. Gewirtz and Deborah Lee Sokol
University of Pennsylvania School of Medicine, Philadelphia, Pennsylvania

I. INTRODUCTION

In recent years, a large number of extracellular growth factors which regulate hematopoietic cell development have been molecularly cloned, expressed as active proteins, and utilized in the clinic. The intracellular events that are triggered when these growth factors interact with their receptors have also begun to be defined. Nevertheless, most of the molecular machinery that regulates blood cell development remains enigmatic and difficult to access. This situation applies particularly to normal blood cells because of the difficulty of applying modern molecular analytic techniques to the small numbers of cells that are ordinarily available for such investigations.

To approach the problem of understanding gene function in hematopoietic cells two main strategies have been applied. One involves infecting cells with a vector engineered to express the gene of interest at high levels (Clarke *et al.*, 1988; Liebermann and Hoffman-Liebermann, 1989; Prochownik *et al.*, 1990). If the cell's phenotype/behavior changes in the infected cell, but not a sham transfected cell, one can tentatively attribute the change in phenotype to the newly expressed gene. Function may therefore be inferred. This approach, while potentially informative, has a number of drawbacks. First, it is not certain that the changes observed are directly related to the gene's function, as any effect observed may be an indirect one. Second, the experiment is not really "physiologic" since overexpression of the gene may exaggerate its normal function or impart new functions by leading to an excess of the encoded protein. Finally, this approach is often limited to leukemic cell lines since normal cells are often much more problematic in terms of both infection and expression of the vector.

An alternative experimental approach, and one which may yield more physiologically relevant data, is to either physically "knock-out" the target gene or to interfere with its function by perturbing the use of the gene's encoded mRNA. Homologous recombination remains the "gold standard" for experiments of this type since this methodology physically destroys the gene of interest (Galli-Taliadoros *et al.*, 1995; Heyer and Kohli, 1994; Morrow and Kucherlapati, 1993; Osman *et al.*, 1994; Willnow and Herz, 1994). However, because this strategy ultimately depends on selecting cells in which a rare crossover event has occurred, it would appear to have limited therapeutic practicality at this time. Ribozymes, antisense RNAs and antisense DNAs, all of which interfere with use of the targeted mRNA, appear to have more immediate relevance from a therapeutic point of view. For these reasons, these approaches have become an increasingly popular strategy for exploring gene function in cells of virtually any type.

For the past several years, we have been engaged in trying to develop an effective strategy of disrupting specific gene function with antisense oligodeoxynucleotides (ODN). We have also been actively engaged in attempting to utilize this strategy in the clinic. This latter pursuit has focused on finding appropriate gene targets that can be successfully targeted using an antisense approach, and then developing "scale-up" methods so that techniques developed in the laboratory could be applied in the clinic. It was our opinion that human leukemias would be particularly amenable to this therapeutic strategy. They can be successfully manipulated *ex vivo*, the tumor is "liquid" *in vivo*, therefore more likely to successfully take up ODN, and a great deal is known about their cell and molecular biology. The latter in particular facilitates choice of a gene target. Accordingly, if ODN were going to be developed as therapeutics, the hematopoietic system seemed an ideal model system.

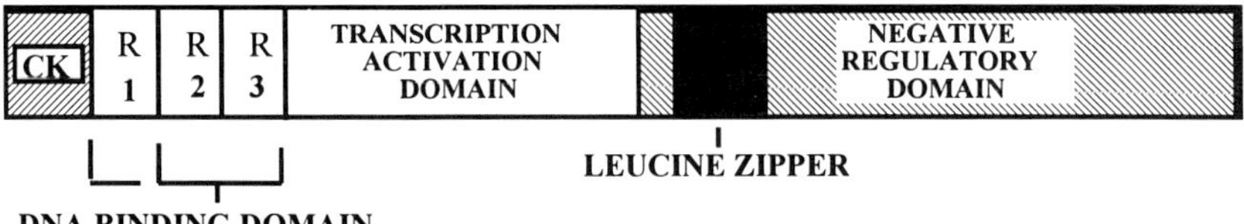

Figure 1. Functional map of the c-Myb protein. See text for details.

II. THE C-*MYB* PROTO-ONCOGENE

Of the genes that we have targeted for disruption using the antisense ODN strategy (Gewirtz and Calabretta, 1988; Luger *et al.*, 1996; Ratajczak *et al.*, 1992abc; Small *et al.*, 1994; Takeshita *et al.*, 1993) one that has been of particular interest to our laboratory, and one where therapeutically motivated disruptions are now in clinical trial, is the c-*myb* gene (Lyon *et al.*, 1994). C-*myb* is the normal cellular homologue of v-*myb*, the transforming oncogene of the avian myeloblastosis virus (AMV) and avian leukemia virus E26. It is a member of a family composed of at least two other highly homologous genes designated A-*myb* and B-*myb* (Nomura *et al.*, 1993). Located on chromosome 6q in humans, c-*myb* expresses a predominant transcript encoding a ~75 kDa nuclear binding protein (Myb) which recognizes the core consensus sequence 5'-PyAAC(G/Py)G-3' (Biedenkapp *et al.*, 1988). Myb consists of three primary functional regions (Sakura *et al.*, 1989) (Figure 1).

At the NH$_2$ terminus is the DNA binding domain. This region consists of three imperfect tandem repeats (R1, R2, R3) each consisting of 51-52 amino acids. Three perfectly conserved tryptophan residues are found in each repeat. Together they form a cluster in the hydrophobic core of the protein that maintains the DNA binding helix-turn-helix structure. The mid-portion of the protein contains an acidic transcriptional activating domain. The DNA binding portion of the protein is required for these transcriptional effects to be observed. The protein also contains a negative regulatory domain that has been localized to the carboxy terminus. Interestingly, the carboxy terminus is deleted in v-Myb and this has been thought to contribute to v-Myb's transforming ability. Recently reported experiments have been among those to confirm this hypothesis and have further demonstrated that amino terminal deletions give rise to a protein with even more potent transforming ability (Dini *et al.*, 1995). Deletions of both the amino and carboxy termini create a protein with the greatest transforming ability, and one which induces the formation of hematopoietic cells that are more primitive than those produced by amino terminal deletions alone (Dini *et al.*, 1995). These data suggest that simultaneous loss of Myb's ability to bind DNA and interact with as yet unidentified proteins are potent transforming stimuli. Nevertheless, this simple hypothesis is complicated by the observation that overexpression of the C-terminal portion of c-Myb can also be oncogenic (Press *et al.*, 1994) whereas overexpression of the whole protein is not (Dini *et al.*, 1995). At the least, one may conclude that sequestration of certain potential Myb binding proteins may also be an oncogenic event.

Recently, a putative leucine zipper structure was described within the amino terminal portion of Myb's carboxy terminal domain (Kanei-Ishii *et al.*, 1992). Leucine zippers, such as those found in the transcription factors Jun, Fos, and Myc are thought to facilitate the protein-protein interactions which permit heterodimerization of DNA binding proteins. Such dimerization is thought to play a key role in regulating the transcriptional activity of these factors. A Myb dimerizing binding partner has yet to be identified but Myb-Myb homodimerization, which likely occurs through its leucine zipper, does lead to loss of DNA binding and transactivation ability (Nomura *et al.*, 1993). Accordingly, one could reasonably postulate that Myb driven transactivation and/or transformation might be regulated by the binding of additional protein partners in the leucine zipper domain (Kanei-Ishii *et al.*, 1992). Alternatively, loss of the ability of Myb to dimerize with a putative regulatory partner might also contribute, directly of indirectly, to cellular transformation and leukemogenesis. Point mutations in the Myb negative regulatory domain might be one mechanism for bringing about such a loss (Kanei-Ishii *et al.*, 1992). Finally, interaction (not physical dimerization) with other nuclear binding proteins such as the CCAAT enhancer binding protein (C/BEP) (Burk *et al.*, 1993), and the related myeloid nuclear factor NF-M (Ness *et al.*, 1993) may also regulate Myb's transactivation or repressor functions.

The above discussion suggests that the c-*myb* gene might play a role in leukemogenesis. Additional, albeit indirect, evidence also support this contention. For example, *c-myb* amplification in AML and overexpression in 6q⁻ syndrome has been reported (Barletta *et al.*, 1987). The mechanism whereby overexpressed Myb might be leukemogenic is uncertain but points out the important difference of working with primary cells as opposed to cell lines. As noted above, it has been reported that overexpressing Myb protein is not by itself leukemogenic (Dini *et al.*, 1995) but this work was carried out in cell lines which may give results that are valid only for the lines tested. As was also noted above, one could reasonably postulate that Myb driven transformation might be regulated by the binding of additional protein partners in the leucine zipper domain (Kanei-Ishii *et al.*, 1992). Recent evidence demonstrating that Myb interacts with other nuclear binding proteins, and that Myb's carboxy-terminus may interact with a cellular inhibitor of transcription supports this hypothesis (Burk *et al.*, 1993; Vorbrueggen *et al.*, 1994). Other potential mechanisms might relate to Myb's ability to regulate hematopoietic cell proliferation (Gewirtz *et al.*, 1989), perhaps by its effects on important cell cycle genes including c-*myc* (Cogswell *et al.*, 1993), and *cdc2* (Ku *et al.*, 1993). Finally, Myb also plays a role in regulating hematopoietic cell differentiation (Weber *et al.*, 1990). It functions as a transcription factor for several cellular genes, including the neutrophil granule protein *mim*-1 (Ness *et al.*, 1989), CD4 (Nakayama *et al.*, 1993), IGF-1 (Travali *et al.*, 1991), CD34 (Melotti *et al.*, 1994), and possibly other growth factors (Szczylik *et al.*, 1993), including *c-kit* (Ratajczak *et al.*, 1992cd). The latter is of particular interest since it has been shown that when hematopoietic cells are deprived of *c-kit* ligand (Steel Factor) they undergo apoptosis (Yu *et al.*, 1993). Accordingly, Myb is clearly an important hematopoietic cell gene that may, directly or indirectly, contribute to the pathogenesis or maintenance of human leukemias. For this reason it is a rational target for therapeutically motivated disruption strategies.

III. TARGETING THE C-*MYB* GENE

Our investigations were initially designed to elucidate the role of Myb protein in regulating hematopoietic cell development. Because the results obtained from these studies had obvious clinical relevance, more translationally oriented studies were also undertaken. These have now culminated in clinical trials that are presently ongoing at the Hospital of the University of Pennsylvania. Below we summarize the steps carried out in the clinical development of the c-*myb* targeted antisense ODN. In addition, we will also allude briefly to our initial clinical experience with the *myb* targeted ODN.

IV. *IN VITRO* EXPERIENCE IN THE HEMATOPOIETIC CELL SYSTEM

IV.A. Role of c-*myb* encoded protein in normal human hematopoiesis

Attempts to exploit the c-*myb* gene as a therapeutic target for antisense ODN began as an outgrowth of studies which were seeking to define the role of Myb protein in regulating normal human hematopoiesis (Gewirtz *et al.*, 1989; Gewirtz and Calabretta, 1988). During the course of these studies it was determined that exposing normal bone marrow mononuclear cells (MNC) to a c-*myb* ODN antisense to codons 2-7 resulted in a decrease in cloning efficiency and progenitor cell proliferation. The effect was lineage indifferent since c-*myb* antisense DNA inhibited granulocyte-macrophage colony forming units (CFU-GM), CFU-E (erythroid), and CFU-Meg (megakaryocyte). Inhibition of colony formation was dose related, and inhibition of the targeted c-*myb* mRNA was also demonstrated. In contrast, a c-*myb* ODN with the corresponding sense sequence had no consistent effect on hematopoietic colony formation when compared to growth in control cultures.

Sequence specific, dose related biologic effects accompanied by a specific decrease or total elimination of the targeted mRNA were strong pieces of evidence to suggest that the effects we were observing were due to an "antisense" mechanism. It should be added that the effects we observed were largely confirmed using homologous recombination (Mucenski *et al.*, 1991). In other investigations, it was also determined that hematopoietic progenitor cells appeared to require Myb protein during specific stages of development, in particular when they were actively cycling (Caracciolo *et al.*, 1990), as might be expected given the above functional description of Myb protein.

IV.B. Myb protein is also required for leukemic hematopoiesis

Since the c-*myb* antisense ODN inhibited normal cell growth, we were also interested in determining their effect on leukemic cell growth. While one could reasonably postulate that aberrant c-*myb* expression or Myb function might play a role in carcinogenesis, demonstrating this was another matter. To address this question, we employed a variety of leukemic cell lines, including those of myeloid and lymphoid origin. In addition, we also employed primary patient material. We first determined the effect of *myb* sense and antisense ODN on the growth of HL-60, K562, KG-1, and KG-1a myeloid cell lines (Anfossi *et al.*, 1989). The antisense ODN inhibited the proliferation of each leukemia cell line, although the effect was most pronounced on HL-60 cells. Specificity of this inhibition was demonstrated by the fact that the sense ODN had no effect on cell proliferation, nor did "antisense" sequences with 2 or 4 nucleotide mismatches.

To determine whether the treatment with *myb* antisense ODN modified cell cycle distribution of HL-60 cells, we measured the DNA content in exponentially growing cells exposed to either sense or antisense *myb* ODN. Control cells, and cells treated with c-*myb* sense ODN had twice the DNA content of HL-60 cells exposed to the antisense ODN. The majority of these cells appeared either to reside in the G1 compartment or were blocked at the G1/S boundary. To examine the effect of the c-*myb* ODN on lymphoid cell growth we employed a lymphoid leukemia cell line, CCRF-CEM. As we noted in the case of normal lymphocytes (Gewirtz *et al.*, 1989), the CCRF-CEM cells were extremely sensitive to the anti-proliferative effects of the c-*myb* antisense ODN.

When exposed to the sense ODN, we found negligible effects on CEM cell growth in short term suspension cultures. In contrast, exposure to c-*myb* antisense DNA resulted in a daily decline in cell numbers. Compared to untreated controls, antisense DNA inhibited growth by ~2 logs. Growth reduction was not just a cytostatic effect, since cell viability was reduced only ~70% after exposure to the antisense ODN, and CEM cell growth did not recover when cells were left in culture for an additional nine days.

Results obtained from primary patient material were equally encouraging (Table 1) (Calabretta *et al.*, 1991). We began by attempting to determine if CFU-L from AML patients could be inhibited by exposure to c-*myb* antisense ODN. Of the 28 patients we initially studied, colony and cluster data were available in 16 and 23 cases respectively. After exposure to relatively low doses of c-*myb* antisense ODN (60 µg/ml) colony formation was inhibited in a statistically significant manner in 12/16 (~75%). Inhibition of cluster formation fell in a similar range. Of equal importance the numbers of residual colonies in the antisense treated dishes was ~10%.

An obvious problem with interpreting these results however, was determining the nature of the residual cells, i.e. were they the progeny of residual normal CFU or CFU-L. To try to answer this question in a rigorous manner we turned out attention to CML where the presence of the t(9:22) or *bcr/abl* neogene provided an unequivocal marker of the malignant cells (Ratajczak *et al.*, 1992a).

Exposure of CML cells to c-*myb* antisense ODN resulted in inhibition of CFU-GM derived colony formation in >50% cases evaluated and thus far we have studied in excess of 40 patients. Representative data are shown in Figure 2 and are presented as a function of oligomer effect on cells with "greater" cloning efficiency (control colonies >250/plate) (Figure 2A) versus "lesser" cloning efficiency (control colonies <250/plate) (Figure 2B).

In this particular study, colony formation was observed in eight of eleven cases evaluated and was statistically significant ($p \leq 0.03$) in seven. The amount of inhibition seen was dose dependent and ranged between 58% and 93%. In two cases the effect of the c-*myb* oligomers on CFU-GEMM colony formation was also determined to assess the effect of the oligomers on progenitors more primitive than CFU-GM. In each case, significant inhibition of CFU-GEMM derived colony formation was noted.

Table 1. Effect of c-*myb* Oligomers on Primary AML Cell Colony and Cluster Formation

	Colonies (% Control)		Colonies (% Control)	
Case	Sense	Antisense	Sense	Antisense
1	86	18 (0.058)	60	37 (0.080)
4	NG	NG	90	28 (0.036)
5	NG	NG	70	22 (0.101)
6	NG	NG	79	22 (0.026)
7	170	100 (0.423)	76	128 (0.502)
8	92	11 (0.008)	96	46 (0.020)
10	NG	NG	190	216 (0.034)
11	45	14 (0.021)	58	21 (0.084)
14	68	01 (0.152)	90	53 (0.071)
15	66	81 (0.736)	100	100 (0.896)
16	NG	NG	66	24 (0.001)
17	NG	NG	16	8 (0.023)
18	NG	NG	110	77 (0.164)
19	113	116 (0.717)	91	91 (0.763)
20	92	09 (0.051)	100	50 (0.009)
21	94	00 (0.006)	90	06 (0.004)
22	80	13 (0.001)	103	11 (0.015)
23	63	06 (0.001)	74	27 (0.004)
24	87	17 (0.002)	91	26 (0.018)
25	100	00 (0.019)	107	38 (0.364)
26	76	00 (0.009)	89	00 (0.001)
27	79	21 (0.014)	59	18 (0.043)
28	88	20 (0.009)	94	152 (0.096)

Blast cells were isolated from the peripheral blood of AML patients and exposed to sense or antisense oligomers. Colonies and clusters were enumerated and values were compared with growth in control cultures, which contained no oligomer. For each case, the number of colonies or clusters arising in the untreated control dishes was assumed to represent maximal (100%) growth for that patient. The numbers of colonies or clusters arising in the oligomer-treated dishes are expressed as a percentage of this number. NG, no growth. The statistical significance (determined by Student's t test for unpaired samples) of the change observed in the antisense-treated dishes relative to the untreated control is give as a p value in parentheses *(reprinted, with permission, from Calabretta et al., 1991)*.

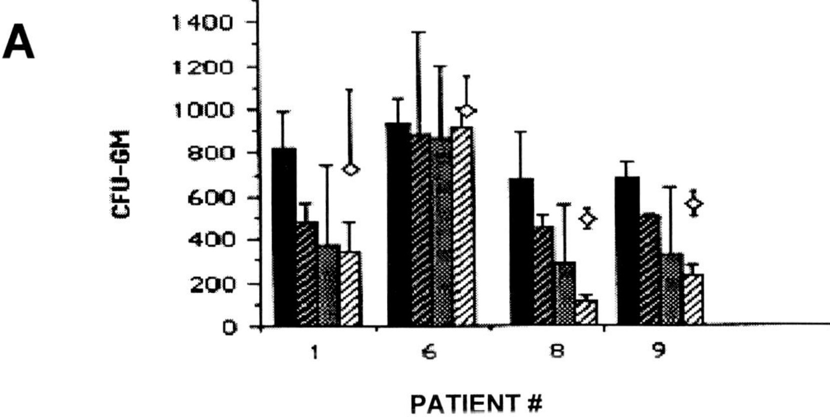

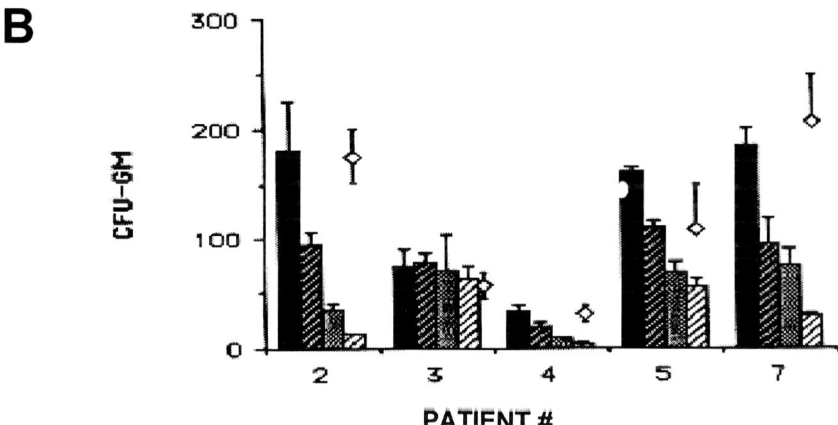

Figure 2. Effect of c-*myb* oligomers on chronic myeloid leukemia cell colony formation by cells with "high" (**A**) and "low" (**B**) cloning efficiency. Colony forming cells were enriched from patient peripheral blood or bone marrow and exposed to oligomers as detailed in the text. At 24 hours cells were plated and resulting colonies were enumerated in plates containing untreated control cells [■]; antisense (▨-20 μg/ml then 10 μg/ml; ▦-40 μg/ml then 20 μg/ml; ▨-100 μg/ml then 50 μg/ml); and sense (◇-100 μg/ml then 50 μg/ml) treated cells. Values plotted are mean ±SD of actual colony counts compared to growth in control cultures which contained no oligomers *(reprinted, with permission, from Ratajczak et al., 1992a).*

It is also important to note that, colony inhibition was sequence specific. For example, c-*myb* sense sequence ODN failed to significantly inhibit colony formation when employed at the highest antisense doses utilized (Figure 2).

IV.C. Normal and leukemic progenitor cells rely differentially on c-Myb function

In order to be useful as a therapeutic target, leukemic cells would have to be more dependent on Myb protein than their normal counterparts. To examine this critical issue we incubated phagocyte and T cell depleted normal human marrow mononuclear cells (MNC), human T lymphocyte leukemia cell line blasts (CCRF-CEM), or 1:1 mixtures of these cells with sense or antisense oligodeoxynucleotides to codons 2-7 of human c-*myb* mRNA (Calabretta *et al.*, 1991). ODN were added to liquid suspension cultures at Time 0 and at Time +18 hrs. Control cultures were untreated. In controls, or in cultures to which "high" doses of sense ODN were added, CCRF-CEM proliferated rapidly, whereas MNC numbers and viability decreased <10%.

In contrast, when CCRF-CEM were incubated for four days in c-*myb* antisense DNA, cultures contained $4.7 \pm 0.8 \times 10^4$ cells/ml (mean $\pm$ SD; n = 4) compared to $285 \pm 17 \times 10^4$/ml in controls. At the effective antisense dose, MNC were largely unaffected. After four days in culture, remaining cells were transferred to methylcellulose supplemented with recombinant hematopoietic growth factors. Myeloid colonies/clusters were enumerated at day ten of culture inception. Depending on cell number plated, control MNC formed from 31 ± 4 to 274 ± 18 colonies. In dishes containing equivalent numbers of untreated or sense ODN exposed CCRF-CEM, colonies were too numerous to count. When MNC were mixed 1:1 with CCRF-CEM in antisense oligomer concentrations ≤ 5 µg/ml, only leukemic colonies could be identified by morphologic, histochemical and immuno-chemical analysis.

However, when antisense oligomer exposure was intensified, normal myeloid colonies could now be found in the culture while leukemic colonies could no longer be identified with certainty using the same analytic methods. Finally, at antisense DNA doses used in the above studies, AML blasts from eighteen of twenty-three patients exhibited ~75% decrease in colony and cluster formation compared to untreated or sense oligomer treated controls. When 1:1 mixing experiments were carried out with primary AML blasts and normal MNC, we were again able to preferentially eliminate AML blast colony formation while normal myeloid colonies continued to form.

IV.D. Use of c-*myb* ODN as bone marrow purging agents

The above experiments suggested that leukemic cell growth could be preferentially inhibited after exposure to c-*myb* antisense ODN. In contemplating a clinical use for our findings application in the area of bone marrow transplantation seemed compelling. In this application, exposure conditions are entirely under the control of the investigator. In addition, the patient's exposure to the antisense DNA is minimal. This circumstance would also make approval by regulatory agencies less difficult. We therefore determined if the antisense ODN could be utilized as *ex vivo* bone marrow purging agents.

To examine this issue, normal MNC were mixed (1:1) with primary acute myelogenous leukemia (AML) or chronic myelogenous leukemia (CML) blast cells and then exposed to the ODN using a slightly modified protocol designed to test the feasibility of a more intensive antisense exposure. With this in mind, an additional ODN dose (20 µg/ml) was given just prior to plating the cells in methylcellulose. In control growth factor-stimulated cultures leukemic cells formed 25.5±3.5 (mean ± SD) colonies and 157±8.5 clusters (per 2×10^5 cells plated). Exposure to c-*myb* sense ODN did not significantly alter these numbers (19.5±0.7 colonies and 140.5±7.8 clusters; p>0.1). In contrast, equivalent concentrations of antisense ODN totally inhibited colony and cluster formation by the leukemic blasts. Colony formation was also inhibited in the plates containing normal MNCs, but only by ~50% in comparison to untreated control plates (control colony formation, 296±40 per 2×10^5 cells plated; treated colony formation, 149 ±15.5 per 2×10^5 cells).

To assess the potential effectiveness of an antisense purge, we carried out co-culture studies with cells obtained from CML patients in blast crisis and in chronic phase of their disease (Ratajczak *et al.*, 1992a). CML was a particularly useful model because cells from the malignant clone carry a tumor specific chromosomal translocation which can be easily identified in tissue culture by looking for *bcr/abl*, the mRNA product of the gene produced by the translocation (Witte, 1993). RNA was therefore extracted from cells cloned in methyl-cellulose cultures after exposure to the highest c-*myb* antisense ODN dose. The RNA was then reverse transcribed and resulting cDNA amplified. For each patient studied, mRNA was also extracted from a comparable number of cells derived from untreated control colonies using the same technique. Eight cases were evaluated and in each case *bcr/abl* expression as detected by RT PCR correlated with colony growth in cell culture. In cases which were inhibited by exposure to c-*myb* antisense ODNs (7/11), *bcr/abl* expression was also greatly decreased or non-detectable (Figure 3A).

These results suggested that *bcr/abl* expressing CFU might be substantially or entirely eliminated from a population of blood or marrow mononuclear cells by exposure to the antisense oligodeoxynucleotides. To explore this possibility further, replating experiment were carried out on samples from two patients (Figure 3B). We hypothesized that if CFU belonging to the malignant clone were present at the end of the original 12 day culture period, but not detectable because of failure to express *bcr/abl*, they might re-express the message upon re-growth in fresh cultures. Accordingly, cells from these patients were exposed to ODN and then plated into methylcellulose cultures formulated to favor growth of either CFU-GM or CFU-GEMM.

As was found with the original specimens, untreated control cells and cells exposed to sense ODN had RT-PCR detectable *bcr/abl* transcripts. Those exposed to the c-*myb* antisense ODN had none. One of the paired dishes from these cultures was then solubilized with fresh medium, and all cells contained therein were washed, disaggregated, and re-plated into fresh methylcellulose cultures *without* re-exposing the cells to ODN. After 14 days, CFU-GM and CFU-GEMM colony cells were again probed for *bcr/abl* expression. Control and sense treated cells had RT-PCR detectable mRNA but none was found in the antisense treated colonies. These results suggest that elimination of *bcr/abl* expressing cells and CFU was highly efficient and perhaps permanent.

IV.E. Why does downregulating Myb kill leukemic cells preferentially? A Hypothesis

Our initial studies on the function of the c-*kit* receptor in hematopoietic cells suggested that c-*kit* might be a Myb regulated gene (Ratajczak *et al.*, 1992d). Since c-*kit* encodes a critical hematopoietic cell tyrosine kinase receptor (Ratajczak *et al.*, 1992c), we hypothesized that dysregulation of c-*kit* expression may be an important mechanism of action of Myb AS ODN.

In support of this hypothesis it has been shown that when hematopoietic cells are deprived of c-*kit* R ligand (Steel Factor) they undergo apoptosis (Yu *et al.*, 1993). It has also recently been shown that when CD56[bright] NK cells, which express c-*kit*, are deprived of their ligand (Steel Factor), they too undergo apoptosis, perhaps because *bcl*-2 is downregulated (Carson, *et al*, 1994). Malignant myeloid hematopoietic cells, in particular CML cells, also express c-*kit* and respond to Steel Factor. Accordingly, we postulated that perturbation of Myb expression in malignant hematopoietic cells may force them to enter an apoptotic pathway by downregulating c-*kit*. Preliminary studies of K562 cells exposed to c-*myb* antisense ODN demonstrates that such cells do in fact

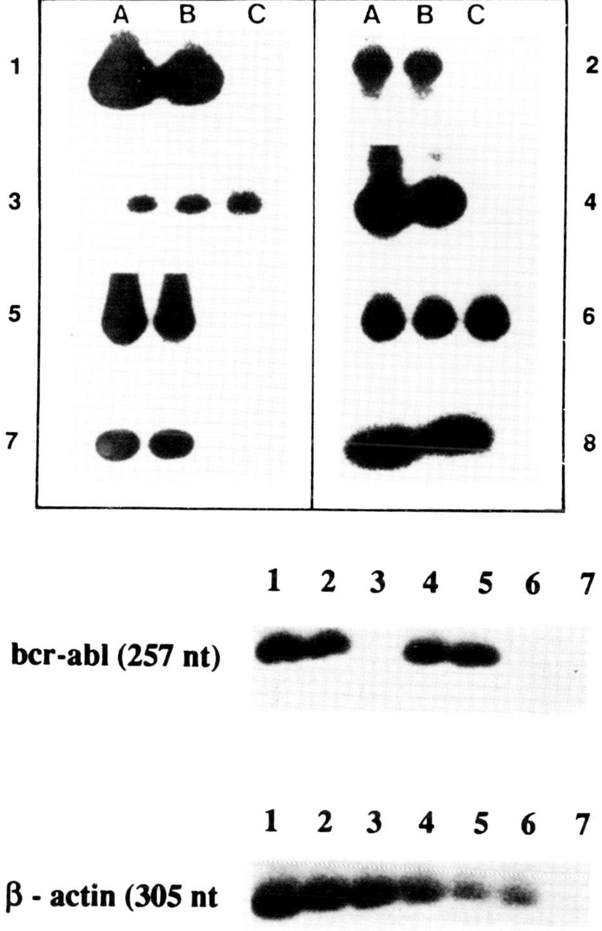

bcr-abl (257 nt)

β - actin (305 nt

Figure 3. Top: Detection of *bcr/abl* transcripts in CFU-GM derived colonies from marrow of 8 patients (#1 8) whose marrow was either not treated (**A**), treated with c-*myb* sense ODN (**B**), or treated with c-*myb* antisense ODN (**C**). All colonies present in the variously treated methylcellulose cultures were harvested and subjected to analysis. Colony selection bias was therefore avoided. **Bottom:** Detection of *bcr/abl* transcripts in CFU-GM (lanes 1-3) and CFU-GEMM (lanes 4-6) derived colonies obtained from re-seeded primary colonies of patient #8. Note that while β-actin transcripts are clearly detected in all colony samples, *bcr/abl* in only detectable in colonies derived from untreated control colonies (lanes 1 & 4) and colonies previously exposed to Myb sense ODN (lanes 2 & 5). Cells derived from colonies originally exposed to c-*myb* AS ODN do not have detectable *bcr/abl* expressing cells. Lane 7 is a control lane for the PCR reactions and is appropriately empty *(reprinted, with permission, from Ratajczak et al., 1992a).*

Figure 4. c-*Myb* antisense ODN causes K562 cells to undergo apoptosis. K562 cells were exposed to c-*myb* sense or antisense ODN (~20 μM) for 36 hours. After this time DNA was extracted from the cells, electrophoresed and stain with ethidium bromide. Lane 1: Untreated control; nuclear DNA is intact. Lane 2: Cells exposed to c-*myb* anti-sense ODN; note characteristic "laddering" of DNA. Lane 3: Cells exposed to c-*myb* sense ODN.

V. EFFICACY OF C-*MYB* OLIGODEOXYNUCLEOTIDES *IN VIVO*: DEVELOPMENT OF ANIMAL MODELS

The studies described above were carried out primarily with unmodified DNA. Such molecules are subject to endonuclease and exonuclease attack at the phosphodiester bonds and are therefore of little utility *in vivo*. We therefore needed to address two questions at this point. First, we needed to know if a more stable, chemically modified ODN would give similar results. Second, we needed to know if these materials would have effectiveness in an *in vivo* system against human leukemia cells. Since we could not give this material to patients we established a human leukemia/SCID mouse model system which would allow us to address both questions simultaneously (Ratajczak *et al.*, 1992b).

To carry out these experiments, SCID mice were injected IV with K562 chronic myeloid leukemia cells after cyclophosphamide conditioning. K562 cells express c-*myb*, the antisense oligodeoxynucleotide target, and the tumor specific *bcr/abl* oncogene that was utilized for tracking the human leukemia cells in the mouse host. After tumor cell injection, animals developed blasts in the peripheral blood within four to six weeks. After peripheral blood blast cells appeared, mean (±SD) survival of untreated mice (n = 20) was 6 ± 3 days. Dying animals had prominent central nervous system infiltration, marked infiltration of the ovary, and scattered abdominal granulocytic sarcomas. Infusion of either sense or scrambled sequence c-*myb* phosphorothioate

ODN (24 bp; codons 2-9) for three, seven, or fourteen days had no statistically significant effect on sites of disease involvement, or animal survival in comparison to control animals.

In contrast, animals treated for 7 or 14 days with c-*myb* AS ODN survived 3.5 to 8 times longer (p < 0.001) than the various control animals (n = 60) (Figure 5). In addition, animals receiving c-*myb* AS DNA had either rare microscopic foci or no obviously detectable CNS disease and a >50% reduction of ovarian involvement. A 3 day infusion of *myb* AS (100 μg/day) was without effect. Infusing mice (n = 12) with AS ODN (200 μg/day × 14 days) complementary to the c-*kit* proto-oncogene, which K562 cells do not express, also had no effect on disease burden or survival (n = 12). These results suggested that phosphorothioate modified c-*myb* antisense DNA might be efficacious for the treatment of human leukemia *in vivo*.

VI. MURINE C-*MYB* P-AS ODN TOXICITY STUDIES

An important consideration in the design of any clinical trial is anticipation of toxic effects of the candidate ODN. Two broad categories of potential toxicities should be considered.

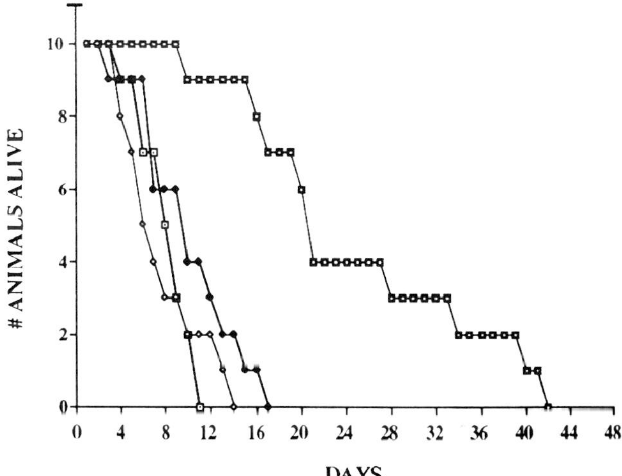

Figure 5. Survival curves of SCID-human chimeric animals transplanted with K562 chronic myelogenous leukemia cells. Animals received a 14 day infusion of oligomers at a dose of 100 μg/day. Legend- [⎯☐⎯ control, ⎯♦⎯ sense, ⎯■⎯ antisense, ⎯◇⎯ scrambled] *(reprinted, with permission, from Ratajczak et al., 1992b).*

The first is the direct result of perturbing the targeted gene's function in tissues where it is expressed. In tumor tissue, this potential "toxic" effect is in fact the desired therapeutic effect. However, inhibition of hematopoiesis by a *myb* directed ODN in patients being treated for melanoma for example (Hijiya *et al*, 1994) would clearly be a "toxic" and therefore undesirable side effect. The second broad category of potential side effects relates to toxicities which are not gene specific but rather inherent to phosphorothioate compounds (P-ODN). Every *in vivo* study described above employed a *human* c-*myb* sequence in a murine host. Accordingly, toxicity inherent to P-AS ODN might be observed but effects of perturbing c-*myb* function in the murine host could not. This follows from the knowledge that human and murine sequences differ at least 5 bases in the targeted regions.

Therefore, expression of the murine gene should not be affected. To address this issue we targeted the murine *myb* gene in normal Balb/c mice and infused animals with doses up to ten times those employed in SCID mice implanted with K562 cells (Figure 5).

Effects of murine c-*myb* P-AS ODN on body weight and blood cell counts at doses up to 300 µg/day × 14 days are shown in Table 2 above. Doses up to 1 mg/day × 14 days have been administered to normal Balb/c mice without major toxic effects. At this dose, some matting of fur was noted in the animals, and some skin ulceration at the site of pump implantation was also observed. Animals also developed slight anemia with hematocrits ~40%.

Table 2. Effect of *Murine* c-*myb* Infusion on Normal Balb/c mice

Oligomer	Mass, g			WBC, 10^3/mm^3			Hematocrit,%			PLT, 10^3/mm^3		
	Day 0	Day 14	Day 21	Day 0	Day 14	Day 21	Day 0	Day 14	Day 21	Day 0	Day 14	Day 21
Sense, 100 µg/day	24.8	24.6	24.0	3.2	7.7	3.2	53	52	52	554	635	215
Sense, 300 µg/day	27.1	25.7	25.3	3.4	6.7	2.9	52	51	51	642	619	218
Antisense, 100 µg/day	25.7	25.1	26.5	2.6	4.5	2.6	53	47	48	740	412	188
Antisense, 300 µg/day	25.2	24.6	26.1	2.1	5.1	2.7	52	44	48	595	397	183

Animals (n = 10) received up to 300 µg/day of P-AS ODN × 14 days. Animals were sacrificed at day 21. Complete autopsies were carried out which included histologic examination of major organs. No apparent toxicities were noted other than thrombocytopenia which appeared to be non-sequence specific. Animals had no clinically manifest bleeding abnormalities.

The animals otherwise behaved normally and appeared to be in excellent condition.

VII. USE OF ANTISENSE OLIGONUCLEOTIDES IN A CLINICAL SETING

Based on the type of data presented above, we have begun to evaluate the *myb*-targeted antisense ODN in the clinic (Gewirtz et al., 1996). Towards this end, we initiated clinical trials to evaluate the effectiveness of phosphorothioate modified ODN antisense to the c-*myb* gene as marrow purging agents for chronic phase (CP) or accelerated phase (AP) chronic myelogenous leukemia (CML) patients, and a phase I intravenous infusion study for blast crisis (BC) patients, and patients with other refractory leukemias. ODN purging was carried out for 24 hrs on CD34$^+$ marrow cells.

Patients received busulfan and cytoxan, followed by re-infusion of previously cryopreserved P-ODN purged MNC. In the pilot marrow purging study 7 CP and 1 AP CML patients were treated, and 7/8 engrafted. In 4/6 evaluable CP patients, PBL metaphase chromosome spreads appeared 85-100% normal 3 months after engraftment, suggesting that a significant purge had taken place in the marrow graft. Five CP patients have demonstrated marked, sustained, hematologic improvement with essential normalization of their blood counts. Follow-up periods ranged from 6 months to ~ 2 years. In an attempt to further increase purging efficiency we incubated patient MNC for 72 hours in the P-ODN.

Though PCR and long term culture initiating cells (LTCIC) studies suggested a very efficient purge had occurred, engraftment in 5 patients was poor. In the phase I systemic infusion study, 18 refractory leukemia patients (2 patients were treated at 2 different dose levels; 13 had AP or BC CML). c-*Myb* antisense ODN was delivered by continuous infusion at dose levels ranging between [0.3 mg/kg/day × 7 days] to [2.0 mg/kg/day × 7 days]. No recurrent dose related toxicity has been noted.

On the other hand, two idiosyncratic toxicities, not clearly drug related, were observed (1 transient renal insufficiency; 1 pericarditis). One BC patient survived ~14 months with transient restoration of CP disease. These studies show that ODN may be administered safely to leukemic patients. Whether or not patients treated on either study derived definite clinical benefit is not certain. Nevertheless, the results of these studies suggest to us that ODN may eventually demonstrate therapeutic utility in the treatment of human leukemias.

VIII. CONCLUSIONS

The ability to block gene function with antisense oligodeoxy-nucleotides has become an important tool in many research laboratories. Since activation and aberrant expression of proto-oncogenes appears to be an important mechanism in malignant transformation, targeted disruption of these genes and other molecular targets with oligodeoxynucleotides could have significant therapeutic utility as well. In this regard, the potential therapeutic usefulness of oligodeoxynucleotides has been demonstrated in many systems and against a number of different targets including viruses, oncogenes, proto-oncogenes, and an increasing array of cellular genes. These studies, in aggregate, suggest that synthetic ODN have the potential to become an important new therapeutic agent for the treatment of human cancer. Nevertheless, it is clear that considerable optimization will be required before antisense oligonucleotides will emerge as an effective agent for treating human disease. Progress will need to occur on several fronts. These include issues related to the chemistry of the molecules employed. For example, how different chemical modifications affect uptake, stability, and hybridization efficiency of a synthetic DNA molecule must be determined for each derivative. A clearer understanding of the mechanism of antisense mediated inhibition, including where such inhibition takes place, will also be required. Finally, cellular "defense" mechanisms, such as increasing transcription of the targeted message, may also be factors to consider in planning effective treatment strategies with these agents. It is also clear that choice of target is an important issue. Nevertheless, while many issues remain to be resolved, we remain optimistic that this approach will one day prove useful for the treatment of patients with a variety of hematologic malignancies.

IX. REFERENCES

Anfossi, G., Gewirtz, A. M., and Calabretta, B. (1989) An oligomer complementary to c-*myb*-encoded mRNA inhibits proliferation of human myeloid leukemia cell lines. *Proc. Natl. Acad. Sci. USA* **86**:3379-3383.

Barletta, C., Pelicci, P. G., Kenyon, L. C., Smith, S. D., and Dalla-Favera, R. (1987) Relationship between the c-*myb* locus and the 6q-chromosomal aberration in leukemias and lymphomas. *Science* **235**:1064-1067.

Biedenkapp, H., Borgmeyer, U., Sippel, A. E., and Klempnauer, K. H. (1988) Viral *myb* oncogene encodes a sequence-specific DNA-binding activity. *Nature* **335**:835-837.

Burk, O., Mink, S., Ringwald, M., and Klempnauer, K. H. (1993) Synergistic activation of the chicken *mim*-1 gene by v-*myb* and C/EBP transcription factors. *EMBO J.* **12**:2027-2038.

Calabretta, B., Sims, R. B., Valtieri, M., Caracciolo, D., Szczylik, C., Venturelli, D., Ratajczak, M., Beran, M., and Gewirtz, A. M. (1991) Normal and leukemic hematopoietic cells manifest differential sensitivity to inhibitory effects of c-*myb* antisense oligodeoxynucleotides: an *in vitro* study relevant to bone marrow purging. *Proc. Natl. Acad. Sci. USA* **88**:2351-2355.

Caracciolo, D., Venturelli, D., Valtieri, M., Peschle, C., Gewirtz, A. M., and Calabretta, B. (1990) Stage-related proliferative activity determines c-*myb* functional requirements during normal human hematopoiesis. *J. Clin. Invest.* **85**:55-61.

Carson, W.E, Haldar, S., Baiocchi, R.A., Croce, C.M., and Caligiuri, M.A. (1994) The c-*kit* ligand suppresses apoptosis of human natural killer cells through the upregulation of *bcl*-2. *Proc. Natl. Acad. Sci. USA* **91**:7553-7557.

Clarke, M. F., Kukowska-Latallo, J. F., Westin, E., Smith, M., and Prochownik, E. V. (1988) Constitutive expression of a c-*myb* cDNA blocks Friend murine erythroleukemia cell differentiation. *Mol. Cell Biol.* **8**:884-892.

Cogswell, J. P., Cogswell, P. C., Kuehl, W. M., Cuddihy, A. M., Bender, T. M., Engelke, U., Marcu, K. B., and Ting, J. P. (1993) Mechanism of c-*myc* regulation by c-Myb in different cell lineages. *Mol. Cell Biol.* **13**:2858-2869.

Dini, P. W., Eltman, J. T., and Lipsick, J. S. (1995) Mutations in the DNA-binding and transcriptional activation domains of v-Myb cooperate in transformation. *J. Virol.* **69**:2515-2524.

Galli-Taliadoros, L. A., Sedgwick, J. D., Wood, S. A., and Korner, H. (1995) Gene knock-out technology: a methodological overview for the interested novice. *J. Immunol. Methods* **181**:1-15.

Gewirtz, A. M., and Calabretta, B. (1988) A c-*myb* antisense oligodeoxynucleotide inhibits normal human hematopoiesis *in vitro*. *Science* **242**:1303-1306.

Gewirtz, A. M., Anfossi, G., Venturelli, D., Valpreda, S., Sims, R., and Calabretta, B. (1989) G1/S transition in normal human T-lymphocytes requires the nuclear protein encoded by c-*myb*. *Science* **245**:180-183.

Gewirtz, A. M., Luger, S., Sokol, D., Gowdin, B., Stadtmauer, E., Reccio, A., and Ratajczak, M. Z. (1996) Oligodeoxynucleotide therapeutics for human myelogenous leukemia: interim results. *Blood* **88**(Supplement 1*)*:270a.

Heyer, W. D., and Kohli, J. (1994) Homologous recombination. *Experientia* **50**:189-191.

Hijiya, N., Zhang J, Ratajczak M.Z., Kant, J.A., DeRiel, K., Herlyn, M., Zon, G., and Gewirtz, A.M. (1994) Biologic and therapeutic significance of Myb expression in human melanoma. *Proc. Natl. Acad. Sci. USA* **91**:4499-4503.

Kanei-Ishii, C., MacMillan, E. M., Nomura, T., Sarai, A., Ramsay, R. G., Aimoto, S., Ishii, S., and Gonda, T. J. (1992) Transactivation and transformation by Myb are negatively regulated by a leucine-zipper structure. *Proc. Natl. Acad. Sci. USA* **89**:3088-3092.

Ku, D. H., Wen, S. C., Engelhard, A., Nicolaides, N. C., Lipson, K. E., Marino, T. A., and Calabretta, B. (1993) c-*myb* transactivates *cdc2* expression via Myb binding sites in the 5'-flanking region of the human *cdc2* gene [published erratum appears in *J. Biol. Chem.* (1993) **268**:13010]. *J. Biol. Chem.* **268**:2255-2259.

Liebermann, D. A., and Hoffman-Liebermann, B. (1989) Proto-oncogene expression and dissection of the myeloid growth to differentiation developmental cascade. *Oncogene* **4**:583-592.

Luger, S. M., Ratajczak, J., Ratajczak, M. Z., Kuczynski, W. I., DiPaola, R. S., Ngo, W., Clevenger, C. V., and Gewirtz, A. M. (1996) A functional analysis of proto-oncogene *Vav's* role in adult human hematopoiesis. *Blood* **87**:1326-1334.

Lyon, J., Robinson, C., and Watson, R. (1994) The role of Myb proteins in normal and neoplastic cell proliferation. *Crit. Rev. Oncog.* **5**:373-388.

Melotti, P., Ku, D. H., and Calabretta, B. (1994) Regulation of the expression of the hematopoietic stem cell antigen CD34: role of c-*myb*. *J. Exp. Med.* **179**:1023-1028.

Morrow, B., and Kucherlapati, R. (1993) Gene targeting in mammalian cells by homologous recombination. *Curr. Opin. Biotechnol.* **4**:577-582.

Mucenski, M. L., McLain, K., Kier, A. B., Swerdlow, S. H., Schreiner, C. M., Miller, T. A., Pietryga, D. W., Scott, W. J., Jr., and Potter, S. S. (1991) A functional c-*myb* gene is required for normal murine fetal hepatic hematopoiesis. *Cell* **65**:677-689.

Nakayama, K., Yamamoto, R., Ishii, S., and Nakaguchi, H. (1993) Binding of c-*Myb* to the core sequence of the CD4 promoter. *Intl. Immunol.* **5**:817-824.

Ness, S. A., Marknell, A., and Graf, T. (1989) The v-*myb* oncogene product binds to and activates the promyelocyte-specific *mim*-1 gene. *Cell* **59**:1115-11125.

Ness, S. A., Kowenz-Leutz, E., Casini, T., Graf, T., and Leutz, A. (1993) Myb and NF-M: combinatorial activators of myeloid genes in heterologous cell types. *Genes Dev.* **7**:749-759.

Nomura, T., Sakai, N., Sarai, A., Sudo, T., Kanei-Ishii, C., Ramsay, R. G., Favier, D., Gonda, T. J., and Ishii, S. (1993) Negative autoregulation of c-*Myb* activity by homodimer formation through the leucine zipper. *J. Biol. Chem.* **268**:21914-219123.

Osman, F., Tomsett, B., and Strike, P. (1994) Homologous recombination. *Prog. Industr. Microbiol.* **29**:687-732.

Press, R. D., Reddy, E. P., and Ewert, D. L. (1994) Overexpression of C-terminally but not N-terminally truncated Myb induces fibrosarcomas: a novel nonhematopoietic target cell for the *myb* oncogene. *Mol. Cell Biol.* **14**:2278-2290.

Prochownik, E. V., Smith, M. J., Snyder, K., and Emeagwali, D. (1990) Amplified expression of three *jun* family members inhibits erythroleukemia differentiation. *Blood* **76**:1830-1837.

Ratajczak, M. Z., Hijiya, N., Catani, L., DeRiel, K., Luger, S. M., McGlave, P., and Gewirtz, A. M. (1992a) Acute- and chronic-phase chronic myelogenous leukemia colony-forming units are highly sensitive to the growth inhibitory effects of c-*myb* antisense oligodeoxynucleotides. *Blood* **79**:1956-1961.

Ratajczak, M. Z., Kant, J. A., Luger, S. M., Hijiya, N., Zhang, J., Zon, G., and Gewirtz, A. M. (1992b) *In vivo* treatment of human leukemia in a scid mouse model with c-*myb* antisense oligodeoxynucleotides. *Proc. Natl. Acad. Sci. USA* **89**:11823-11827.

Ratajczak, M. Z., Luger, S. M., DeRiel, K., Abrahm, J., Calabretta, B., and Gewirtz, A. M. (1992c) Role of the *kit* proto-oncogene in normal and malignant human hematopoiesis. *Proc. Natl. Acad. Sci. USA* **89**:1710-1714.

Ratajczak, M. Z., Luger, S. M., and Gewirtz, A. M. (1992d) The c-*kit* proto-oncogene in normal and malignant human hematopoiesis. *Intl. J. Cell Cloning* **10**:205-214.

Sakura, H., Kanei-Ishii, C., Nagase, T., Nakagoshi, H., Gonda, T. J., and Ishii, S. (1989) Delineation of three functional domains of the transcriptional activator encoded by the c-*myb* proto-oncogene. *Proc. Natl. Acad. Sci. USA* **86**:5758-5762.

Small, D., Levenstein, M., Kim, E., Carow, C., Amin, S., Rockwell, P., Witte, L., Burrow, C., Ratajczak, M. Z., Gewirtz, A. M., and Civin, C. I. (1994) STK-1, the human homolog of *Flk*-2/*Flt*-3, is selectively expressed in CD34$^+$ human bone marrow cells and is involved in the proliferation of early progenitor/stem cells. *Proc. Natl. Acad. Sci. USA* **91**:459-463.

Szczylik, C., Skorski, T., Ku, D. H., Nicolaides, N. C., Wen, S. C., Rudnicka, L., Bonati, A., Malaguarnera, L., and Calabretta, B. (1993) Regulation of proliferation and cytokine expression of bone marrow fibroblasts: role of c-*myb*. J Exp Med **178**:997-1005.

Takeshita, K., Bollekens, J. A., Hijiya, N., Ratajczak, M., Ruddle, F. H., and Gewirtz, A. M. (1993) A homeobox gene of the *Antennapedia* class is required for human adult erythropoiesis. *Proc. Natl. Acad. Sci. USA* **90**:3535-3538.

Travali, S., Reiss, K., Ferber, A., Petralia, S., Mercer, W. E., Calabretta, B., and Baserga, R. (1991) Constitutively expressed c-*myb* abrogates the requirement for insulinlike growth factor 1 in 3T3 fibroblasts. *Mol. Cell Biol.* **11**:731-736.

Vorbrueggen, G., Kalkbrenner, F., Guehmann, S., and Moelling, K. (1994) The carboxy terminus of human c-Myb protein stimulates activated trans-cription in trans. *Nucleic Acids Res.* **22**:2466-2475.

Weber, B. L., Westin, E. H., and Clarke, M. F. (1990) Differentiation of mouse erythroleukemia cells enhanced by alternatively spliced c-*myb* mRNA. *Science* **249**:1291-1293.

Willnow, T. E., and Herz, J. (1994) Homologous recombination for gene replacement in mouse cell lines. *Methods Cell Biol.* **43**(Pt A):305-314.

Witte, O. N. (1993) Role of the *bcr/abl* oncogene in human leukemia: Fifteenth Richard and Hinda Rosenthal Foundation Award Lecture. *Cancer Res.* **53**:485-489.

Yu, H., Bauer, B., Lipke, G. K., Phillips, R. L., and Van Zant, G. (1993) Apoptosis and hematopoiesis in murine fetal liver. *Blood* **81**:373-384.

8

Clinical Trials with Anti-p53 DNA, OL(1)p53, in Patients with Acute Myelogenous Leukemia and Myelodysplastic Syndrome

Patrick L. Iversen

AntiVirals, Inc., Corvallis, Oregon

I. INTRODUCTION

"A drug may be broadly defined as any chemical agent which affects living protoplasm..." (Goodman and Gilman, 1941). This long held opinion is now truly expanded by the use of synthetic DNA and biological vectors. The pharmacological understanding of oligonucleotides with molecular masses between 5,000 and 10,000 grams/mole is limited to the past 10 to 20 years. This limited history is the basis for controversy regarding the most appropriate therapeutic uses, the selection of appropriate expressed target genes, the mechanism of action, toxicity, and efficient methods of detection. The justification for progression to clinical trials with the antisense p53 phosphorothioate oligonucleotide included reproducible efficacy demonstrated with patient samples, favorable pharmacokinetics, and minimal toxicity in mice, rats, and monkeys. These favorable pharmacological properties were simultaneous with the ability to prepare multiple gram quantities of pure and highly characterized phosphorothioate oligodeoxynucleotides.

Leukemia has been the vanguard for nearly every aspect of the biology, etiology, and novel approaches to therapy of malignant disease.

This is in part due to the ability to recover and study disease cells from a blood sample or bone marrow aspirate. There are approximately 6,400 new cases of AML diagnosed in the United States each year. Current therapy, reviewed in *The Medical Letter on Drugs and Therapeutics* (Anonymous, 1991), includes induction with ara-C and an anthracycline such as daunorubicin, idarubicin or mitoxantrone. Unfortunately, most patients relapse from initial remission (Schiffer and Lee, 1989; Keating *et al.*, 1989; Rees *et al.*, 1986). Nearly one fifth of patients with early relapse experience subsequent treatment related deaths resulting from currently available therapies. In addition, patients with disease that is refractory to ara-C have a very short survival. Hence, a new therapeutic strategy capable of enhanced remission rates, diminished treatment related mortality and salvage for refractory patients is needed.

II. REVIEW OF THE RATIONALE

The AML blast cell expresses p53 protein at higher levels than normal hematopoietic cells (Koeffler *et al.*, 1986; Smith *et al.*, 1986). Further, the p53 expressed in the AML cell is predominantly wild type (Sugimoto *et al.*, 1991; Singerland *et al.*, 1991; Fenaux *et al.*, 1991). These two observations suggest p53 is not acting as a tumor suppressor in AML. The function of p53 as a checkpoint for cells entering S-phase of the cell cycle and the involvement of p53 in apoptosis made this mRNA an attractive antisense target. However, the complexity of p53 function, regulation of expression and myriad of post-translational modifications and chaperone interactions cloud the mechanistic interpretation of the empirical observations of antisense p53.

The selected oligonucleotide has sequence complementary to exon 10 of the *p53* mRNA and is not expected to discriminate between wild type and mutant *p53* (Zambetti *et al.*, 1991). The oligonucleotide sequence, 5'-d(CCCTGCTCCCCCCTGGCTCC)-3' is called OL(1)p53. It was the most active sequence tested, has little or no secondary structure, and is a high melting temperature sequence. Finally, OL(1)p53 has demonstrated a consistent inhibitory effect on the growth and viability of primary cultures of human AML blast cells with no apparent effect on normal hematopoietic cells (Bayever *et al.*, 1994).

III. PHASE I CLINICAL TRIAL WITH OL(1)P53

What are the toxic responses to phosphorothioate oligonucleotides in humans? The goal of initial studies of OL(1)p53 was to evaluate the human responses to phosphorothioate oligonucleotides. Sixteen patients received a 10 day continuous infusion of OL(1)p53 in a dose escalation protocol beginning at 0.05 mg/kg/hr and increasing in 0.05 mg/kg/hr intervals to 0.25 mg/kg/hr. These studies were approved by the University of Nebraska Medical Center Institutional Review Board and the patients gave informed consent to participate. The patient demographics are provided in Table 1. Standard supportive care for leukemia was utilized including prophylactic allopurinol, broad spectrum antibiotics, blood product support, and total parental nutrition as indicated. The patients were diagnosed with either AML (French-American-British [FAB] classification M1-5) or MDS (FAB classifications, refractory anemia with excess blasts [RAEB] and RAEB/T). These patients had either not responded to two cycles of conventional therapy or had relapsed after previous remission. Patients in first relapse were required to have a duration of first remission of <6 months, or to have failed one attempt at re-induction with established regimens. Exclusion criteria were 1) coexpression on leukemic blasts of CD2, 3, 19, and 20, 2) prior chemotherapy within 4-weeks, or treatment with any growth factor within 3 weeks, or 3) patients eligible to immediately proceed with bone marrow transplantation, and in whom chemotherapy prior to bone marrow transplantation was considered unnecessary.

Clinical laboratory monitoring included 1) peripheral blood counts with differential, platelets and reticulocytes determined 1 day prior to and then daily during the OL(1)p53 infusion, 2) a serum chemistry profile with albumin, alkaline phosphatase, aspartate aminotransferase, bilirubin (total), calcium, gamma-glutamyl-transferase, glucose, lactate dehydrogenase (LDH), magnesium, phosphorus, protein (total), and uric acid determined 1 day prior to and then daily during OL(1)p53 treatment, 3) a renal profile with chloride, CO_2, serum creatinine, potassium, sodium, urea nitrogen, and anion gap determined 1 day prior to and then three times a week during OL(1)p53 treatment, 4) determination of serum amyl-

Table 1. Patient Demographic Summary

Dose	Age	Gender		Diagnosis		Ethnic Origin			
mg/kg/hr	years	M	F	AML	MDS	Caucasian	Hispanic	Black	Asian
0.05	53.1	1	2	2	1	3			
0.10	61.8	3	0	1	2	3			
0.15	62.5	2	1	0	3	2	1		
0.20	49.6	2	2	3	1	2	0	1	1
0.25	48.5	2	1	2	1	2	1		

ase, iron, iron-binding protein, ferritin, complement, C3, C4, IgA, IgG, IgM, antinuclear antibody, and anti-neutrophil cytoplasm antibody determined 1 day prior to and then post OL(1)p53 treatment. Bone marrow aspirates for morphology, cytogenetics, and longterm bone marrow culture were collected and determined 1 day prior to initiation, and then on days 7, 14, and 28 post initiation of OL(1)p53 infusion.

Statistical evaluation of the data was conducted by an outside agency to ensure an unbiased perspective. National Cancer Institute (NCI) common toxicity criteria were employed which assigns a grade of 0, 1, 2, 3, or 4 in 22 clinical categories. A grade of 0 means no toxicity or within normal limits and 1, 2, 3 and 4 represent increasing toxicity or severity. Adverse experiences (AEs) are coded according to the COSTART system. Finally the frequency of clinical observations were evaluated for the 10 day intravenous infusion period, and for 28 days post dosing.

Statistical analysis included a paired t-test and the Wilcoxon Signed-Rank test to identify pre- to post-dosing and pre- to post-study changes. Analyses were conducted using SAS, version 6.08. A summary of adverse experiences in the clinical trial is provided in Table 2. The larger values of the risk ratio may be suggestive of an association between dosing and certain adverse experiences.

Table 2. Adverse Experience Summary

Symptom	Frequency During	Frequency After	Risk Ratio
Chills	31.3	46.7	0.7
Fever	56.3	40.0	1.4
Headache	56.3	13.3	4.2*
Inj. site reaction	93.8	26.7	3.5*
Pain	56.3	53.3	1.1
Pallor (CV)	37.5	26.7	1.4
Diarrhea	50.5	26.7	1.9*
Nausea	43.8	26.7	1.6*
Vomiting	37.5	33.3	1.5*
Cough	56.3	26.7	2.1*
Dyspnea	18.8	33.3	0.6
Lung Dis.	50.0	33.3	1.5
Sweat	31.3	33.3	0.9
Scleritis	62.5	13.3	4.7*

Scleritis, headache, and injection site reactions are indicated as the greatest risks associated with the increasing dose regimen. The mechanistic basis of these adverse responses is not obvious at this time. Several symptoms are consistent with a wide variety of antineoplastic agents administered to patients with advanced disease. A detailed summary of common toxicity is provided in Table 3. The fact that all of the patients involved in these studies have pre-existing hematopoietic disease tends to obscure the interpretation of the apparent toxicity towards white blood cells (WBCs), granulocytes and platelets. The patient with toxicity in the cardiac category experienced prior anthracycline-induced cardiac toxicity.

The most significant toxicity that may not be related to preexisting disease or prior chemotherapy are the liver toxicity related to bilirubin, transaminases, hyperglycemia and alterations in calcium homeostasis. A notable absence of toxicity with phosphorothioate oligonucleotides, based upon preclinical studies at several institutions, is the lack of perturbation in fibrinogen, prothrombin time, and partial thromboplastin time (PTT).

The overall interpretation of these findings is that doses of up to 2.65 grams of phosphorothioate oligonucleotide per square meter body surface area (g/m^2) over a period of 10 days failed to produce serious toxicity. This interpretation is significant when compared with a wide variety of more traditional antineoplastic agents currently used in the clinic.

Significant changes from baseline during administration are described in Table 4 and significant changes from baseline to end of 28 days are described in Table 5. Patients with active disease may be responsible for the changes in mean corpuscular hemoglobin (MCH), platelet count, white blood cell count, eosinophils, and red blood cell distribution indicated in these two tables. The changes in albumin are consistent with liver toxicity indicated in Table 3 and deserve careful scrutiny. The decrease in serum uric acid is interesting in that one tends to anticipate an increase in serum uric acid as an indication of diminished renal function. Further, the additional nucleic acid in the oligonucleotide that is added to the blood stream would be expected to increase serum uric acid.

Hence, interpretation of the uric acid decreases may be due to diminished renal tubular reabsorption of uric acid due to enhanced uric acid excretion. We have observed that when probenecid is administered to rats concomitantly with phosphorothioate oligonucleotides, oligonucleotide renal excretion is enhanced, supporting the notion that oligonucleotides are actively reabsorbed in the renal tubules. This may be of significant importance in anticipation of drug interactions, as we reported

Table 3. Common Toxicity Summary

Category	Criterion	0	1	2	3	4	Average
Leukopenia	WBCs	6	1	1	6	2	1.81
	Granulocytes	9	0	1	0	6	1.63
	Lymphocytes	13	0	1	0	2	0.63
Thrombyctopenia	Platelets	1	1	2	1	11	3.25
Anemia	Hemoglobin	0	0	2	13	1	2.94
Hemorrhage	Clinical	10	1	3	0	2	0.94
Infection		12	3	1	0	0	0.31
Fever	No Infection	8	2	5	1	0	0.94
Genitourinary	Creatinine	13	3	0	0	0	0.19
	Proteinuria	13	3	0	0	0	0.19
	Hematuria	12	4	0	0	0	0.25
	BUN	16	0	0	0	0	0.00
Gastrointestinal	Nausea	9	7	0	0	0	0.44
	Vomiting	10	6	0	0	0	0.38
	Diarrhea	8	8	0	0	0	0.50
	Stomatitis	14	2	0	0	0	0.13
Liver	Bilirubin	8	0	6	2	0	1.13
	Transaminases	9	5	1	1	0	0.63
	Alk. Phosphatases	15	0	1	0	0	0.13
	Clinical	16	0	0	0	0	0.00
Pulmonary		8	7	1	0	0	0.56
Cardiac	Dysrhythmias	11	5	0	0	0	0.31
	Cardiac Function	15	0	0	0	1	0.25
	Ischemia	15	0	0	0	1	0.25
	Pericardia	15	0	0	0	1	0.25
Blood Pressure	Hypertension	13	3	0	0	0	0.19
	Hypotension	14	2	0	0	0	0.13
Skin		8	8	0	0	0	0.50
Allergy		16	0	0	0	0	0.00
Phlebitis Local		14	2	0	0	0	0.13
Alopecia		16	0	0	0	0	0.00

Category	Criterion	0	1	2	3	4	Average
Weight Loss/gain		13	2	1	0	0	0.25
Sensory	Neuro-sensory	15	1	0	0	0	0.08
	Neuro-vision	16	0	0	0	0	0.00
	Neuro-hearing	16	0	0	0	0	0.00
Motor	Neuro-motor	13	3	0	0	0	0.19
	Neuro-constip.	15	1	0	0	0	0.08
Psychol.	Neuro-mood	16	0	0	0	0	0.00
	Neuro-cortical	7	9	0	0	0	0.56
	Neuro-cerebellar	13	3	0	0	0	0.19
	Neuro-headache	7	8	1	0	0	0.63
Metabolic	Hyperglycemia	9	4	1	2	0	0.75
	Hypoglycemia	16	0	0	0	0	0.00
	Amylase	16	0	0	0	0	0.00
	Hypercalcemia	15	0	1	0	0	0.69
	Hypocalcemia	14	2	0	0	0	0.13
	Hypomagnesia	14	2	0	0	0	0.13
	Fibrinogen	16	0	0	0	0	0.00
	Prothrombin Time	14	2	0	0	0	0.13
	PTT	16	0	0	0	0	0.00

(Bishop *et al.*, 1996) in one patient that developed non-oligouric renal failure that occurred after initiation of vancomycin.

Finally, the alterations in calcium homeostasis may be due to oligonucleotide induced liver toxicity resulting in altered vitamin D metabolism. There were no corresponding alterations in phosphate homeostasis, reducing concern for interactions with parathyroid function, and limited liver toxicity is suspected based on common toxicity criteria.

IV. REVIEW OF PRECLINICAL PHARMACOKINETICS AND TOXICOLOGY

A pharmacokinetic summary is provided in Table 6. Plasma concentration versus time curves over the period of time during the infusion revealed a rapid rise to steady-state concentration within the initial 24 hours. The plasma concentrations were calculated from material with the

Table 4. Significant Changes from Baseline During Dose

Analyte	Pre-mean	Post-mean	Mean difference	p-value
Albumin	3.75	3.31	-0.44	<0.05
Lymphocytes	43.5	32.5	-11.0	<0.05
MCH	0.34	29.96	-0.38	<0.05
Platelet count	79.5	61.9	-17.5	<0.05
Protein (total)	6.97	6.30	-0.67	<0.05
Uric Acid	5.41	4.33	-1.08	<0.001
WBC count	8.52	16.81	8.29	<0.05

Table 5. Significant Changes from Baseline to End of 28 Days

Analyte	Pre-mean	Post-mean	Mean difference	p-value
Albumin	3.93	3.74	-0.19	<0.05
Calcium	9.32	9.04	-0.28	<0.10
Eosinophils	1.33	0.11	-1.22	<0.10
MCH	30.68	29.93	-0.74	<0.10
RBC distribution	15.94	14.29	-1.66	<0.10
Uric Acid	5.28	4.69	-0.59	<0.10

same retention time as OL(1)p53, which would exclude degradation products which alter the oligonucleotide size:charge ratio. The analysis by high performance electrophoretic chromatography does not exclude the potential loss of sulfur from the phosphate backbone or the presence of abasic sites. The elimination half-life was estimated from the fall in plasma concentration after cessation of the infusion. It is surprising that plasma concentration did not increase as a function of dose over the range of doses administered. A commensurate increase in renal excretion was not observed to account for the minimal changes in plasma concentration and plasma half-life is actually extended at higher doses.

One speculation concerning the nonlinear pharmacokinetics may be an increase in volume of distribution. An accumulation of oligonucleotide in a body compartment that returns the oligonucleotide to the plasma relatively slow providing for the apparent increase in plasma half-life. The calculated steady-state volume of distributions were 34.1, 209.8, 117.7, 356.7, and 454.1 liters for the increasing infusion rates, respectively.

Table 6. Pharmacokinetic Summary of OL(1)p53

Dose (grams)	Plasma Concentration (µg/ml)	Half-life (Hours)	Urinary Excretion (Grams)
0.69 ± 0.23	3.03 ± 0.15	24.4 ± 5.4	0.33 ± 0.09
1.90 ± 0.10	2.08 ± 0.85	42.6 ± 8.27	0.51 ± 0.21
2.79 ± 0.56	3.73 ± 0.88	26.1 ± 4.74	0.88 ± 0.98
4.08 ± 1.00	3.70 ± 1.00	51.1 ± 21.1	0.70 ± 0.32
5.28 ± 1.12	4.15 ± 0.63	62.5 ± 20.7	0.79 ± 0.15

The whole blood-to-plasma ratio was greater than 0.9 indicating little of the compound is associated with circulating red blood cells in whole blood. The OL(1)p53 concentration in the peripheral blood mononuclear cell (PBMC) fraction from one patient that was administered uniformly labeled phosphorothioate oligonucleotide was determined. The average cell volume of PBMCs is 55.1×10^{-11} liters and the average PBMC concentration of OL(1)p53 at steady state was 38.8 ± 9.6 µM. This concentration is well beyond the concentration required to induced cytotoxicity in leukemic blast cells in cell culture. No significant correlation was observed between the change in the number of circulating PBMCs and plasma concentration of OL(1)p53. The same was true for the number of detected blast cells in peripheral blood. However, there is a linear relationship in long-term bone marrow cultures of the mononuclear fraction from bone marrow aspirates in which a decrease in cellularity was observed with dose (Bishop *et al.*, 1996).

V. INSIGHTS INTO FUTURE CLINICAL TRIALS WITH OL(1)P53

The selected phosphorothioate oligonucleotide, OL(1)p53, has demonstrated a consistent inhibitory effect on the growth and viability of human AML blast cells *in vitro* with little or no effect on non-malignant cells (Bayever *et al.*, 1994). The potential utility of this oligonucleotide for the modulation of p53 expression in malignant hematopoietic cells will depend *inter alia* upon the effective concentrations of OL(1)p53 in the target cell. The potential toxicity of this oligonucleotide will depend upon both the role of p53 expression in non-malignant cells and the concentrations of OL(1)p53 in non-target tissues. This study represents the most extensive systemic exposure of humans to a phosphorothioate oligonucleotide to date. The current studies were designed to evaluate the toxicity of OL(1)p53 in patients with AML or MDS and evaluate the pharmacokinetics of OL(1)p53.

Oligonucleotide binding to plasma proteins has been suggested for phosphodiester structures *in vitro* (Emlen and Mannik, 1978,1984) and for phosphorothioate backbones (McCormack *et al.*, 1990). Synthetic oligonucleotides are generally thought of as acidic with isoelectric points of 1.5-4.5 (Levene and Bass, 1931) and albumin binds compounds that are acidic (Kober and Sjöholm, 1980; Sudlow, 1978; Sudlow *et al.*, 1975). Recent studies conducted in our laboratory (Srinivasan *et al.*, 1995) demonstrated that OL(1)p53 and other phosphorothioate oligonucleotides bind to serum albumin *in vitro* with a dissociation constant of $1\text{-}5 \times 10^5$ M, suggesting the majority of OL(1)p53 in the plasma is in the protein bound form. The serum albumin protein binding capacity exceeded the OL(1)p53 plasma concentrations by nearly 3 orders of magnitude. Hence, protein binding may explain the rapid, less than 24 hours, rise in plasma concentration to apparent steady-state levels.

The renal clearance of OL(1)p53 suggests a prominent role for proximal tubular reabsorption. The OL(1)p53 which is bound to albumin would not be filtered but the unbound fraction should increase with the increased dose. Since the amount of OL(1)p53 cleared by the kidney is constant over all dose levels and there is no correlation between renal clearance of OL(1)p53 and creatinine clearance (estimated from serum creatinine after Hull *et al.*, 1985), the role of glomerular filtration is not the determinant of renal clearance. Secretion and absorption mechanisms for OL(1)p53 may also provide insight into the only patient which developed non-oligouric renal failure which occurred only after initiation of vancomycin therapy, an antibiotic whose renal clearance involves secretion.

There was no significant toxicity observed in patients associated with OL(1)p53 administration. Earlier preclinical observations of OL(1)p53 in the non-human primate achieved peak plasma concentrations of up to 4.39 µM for over 6 days also indicated no toxicity (Cornish *et al.*, 1993). These studies evaluated the cardiovascular effects of OL(1)p53 in detail but also examined serum enzymes, pharmacokinetics and overall behavior. Initial observations of patients from dose level 1 of this study indicated no toxicity and were reported earlier (Bayever *et al.*, 1993). Detailed clinical evaluation of patients from all dose levels of this study was also reported (Bishop *et al.*, 1996). Hence, the preclinical observations are consistent with the current human experience and collectively these studies indicate the systemic administration of a phosphorothioate oligonucleotide can achieve potentially effective plasma concentrations without concomitant toxicity.

The active accumulation of OL(1)p53 in PBMCs reported here is 38.8 µM following *in vivo* exposure to steady-state plasma concentrations of 0.53 µM. This is less than anticipated from studies in which 0.41 µM phosphorothioate oligonucleotides were incubated with PBMCs which accumulated between 32 and 182 µM oligonucleotide (Iversen *et al.*, 1992). This may be explained by the relative growth fraction of the cells *in vitro* versus that in peripheral blood as other studies indicate oligonucleotide binding is greater in actively dividing cells (Iversen *et al.*, 1992). In summary, the current human data indicate potentially effective concentrations of OL(1)p53 (Bayever *et al.*, 1994) reached the target cells during the 10 day infusions.

The plasma concentrations and half-life of OL(1)p53 are close to those anticipated from studies in non-human primates. The half-life in this study was 20.6 to 100.6 hours and in the Rhesus monkey we observed half-lives of 7.6 to 43 hours (data not published). The plasma concentrations reached 0.8 µM in the monkey following an infusion of 2.40 g/m^2 over 5 days of continuous infusion and in the current study the plasma concentration of 0.63 µM was observed following an infusion of 2.65 g/m^2 over 10 days. Further, the volume of distribution and clearance parameters are also similar between the human and the non-human primate. These data indicate the non-human primate is an accurate indicator of OL(1)p53 pharmacokinetic behavior.

The apparent disparity between the cellular growth and decreased viability response to OL(1)p53 *in vivo* versus *in vitro* is worth continued investigation. This may be due to the systemic influence of the leukemic blast cell in the patient, which cannot be mimicked *in vitro*. Current studies are focused on the influence of OL(1)p53 on the expression of p53 protein and the role of oxygen tension. Once these studies are complete it is hoped a rationale which will explain the mechanism of action of OL(1)p53 will become obvious thus allowing additional clinical evaluation of OL(1)p53.

VI. REFERENCES

Anonymous (1991) Drugs of choice for cancer chemotherapy. *The Medical Letter on Drugs and Therapeutics* **33**:21-28.

Bayever E., Haines, K.M., Iversen, P.L., Ruddon, R.W., Pirruccello, S.J., Mountjoy, C.P., Arneson, M.A., and Smith, L.J. (1994) Selective cytotoxicity to human leukemic myeloblasts produced by oligodeoxyribonucleotide phosphorothioates complementary to p53 nucleotide sequences. *Leukemia and Lymphoma* **12**:223-231.

Bishop, M.R., Iversen, P.L., Bayever, E., Sharp, J.G., Greiner, T.C., Copple, B.L., Ruddon, R., Zon, G., Spinolo, J., Arneson, M., Armitage, J.O., and Kessinger, A. (1996) Phase I trial of an antisense oligonucleotide OL(1)p53 in hematologic malignancies. *J. Clin. Oncology* **14**:1320-1326.

Emlen, W., and Mannik, M. (1978) Kinetics and mechanisms for removal of circulating single-stranded DNA in mice. *J. Exp. Med.* **147**:684-699.

Emlen, W., and Mannik, M. (1984) Effect of DNA size and strandedness on the *in vivo* clearance and organ localization of DNA. *Clin. Exp. Immunol.* **56**:185-192.

Fenaux, P., Jonveaux, P., and Quiquandon, I. (1991) p53 gene mutations in myeloid leukemia with 17p monosomy. *Blood* **78**:1652-1657.

Goodman, L.A. and Gilman, A.G. (1941) *The Pharmacological Basis of Therapeutics*, 1st Ed., The Macmillan Company, New York, 3.

Hull, J.H. (1981) Influence of range of renal function and liver disease on predictability of creatinine clearance. *Clin. Pharmacol. Ther.* **29**:516-521.

Iversen, P.L., Zhu, S., Meyer, A., and Zon, G. (1992) Cellular uptake and subcellular distribution of phosphorothioate oligonucleotides into cultured cells. *Antisense Res. Dev.* **2**:211-222.

Keating, M.J., Kantarjian, H.M., and Smith, T.L. (1989) Response salvage therapy and survival after relapse in acute myelogenous leukemia. *J. Clin Oncol.* **7**:1071-1080.

Kober, A., and Sjoholm, I. (1980) The binding sites on human serum albumin for some nonsteroidal anti-inflammatory drugs. *Molec. Pharmacol.* **18**:421-426.

Koeffler, H.P., Miller, C., and Nicolson, M.A. (1986) Increased expression of p53 protein in human leukemia cells. *Proc. Natl. Acad. Sci. USA* **83**:4035-4039.

Levene, P.A., and Bass, L.W. (1931) *Nucleic Acids,* American Chemical Society Monograph Series, The Chemical Catalogue Company, Inc., New York.

McCormack, J.J., Bigelow, J.C., Chrin, L.R., and Mathews, L.A. (1990) High-performance liquid chromatographic analysis of phosphorothioate analogues of oligonucleotides in biological fluids. *J. Chrom.* **533**:133-140.

Rees, J.K., Gray, R.G., Swirsky, D., and Hayhoe, F.G. (1986) Principal results of the Medical Research Council's eighth acute myeloid leukemia trial. *Lancet* **2**(8518):1236-1241.

Schiffer, C.A., and Lee, E.J. (1989) Approaches to the therapy of relapsed acute myeloid leukemia. *Oncology* **3**:23-27.

Singerland, J.M., Minden, M.D., and Benchimol, S. (1991) Mutation of the *p53* gene in human acute myelogenous leukemia. *Blood* **77**:1500-1507.

Smith, L.J., McCulloch E.A., and Benchimol, S. (1986) Expression of the *p53* oncogene in acute myelogenous leukemia. *J. Exp. Med.* **164**:751-761.

Srinivasan, S.K., Tewary, H.K., and Iversen, P.L. (1995) Characterization of binding sites, extent of binding and drug interactions of oligonucleotides with albumin. *Antisense Res. Dev.* **5**:131-135.

Sudlow, G. (1978) The specificity of binding sites in serum albumin. In Tillement, J.P., ed., *Advances in Pharmacology and Therapeutics*, Vol. 7, *Biochemical-Clinical Pharmacology.* Pergamon Press, Oxford, 113-123.

Sudlow, G., Birkett, D.J. and Wade, D.N. (1975) The characterization of specific drug binding sites on human serum albumin. *Molec. Pharmacol.* **11**:824-832.

Sugimoto, K., Toyoshima, H., and Sakai, R. (1991) Mutations of the p53 gene in lymphoid leukemia. *Blood* **77**:1153-1156.

Zambetti, G.P., Quartin, R.S., Martinex, J., Georgoff, I., Momand, J., Dittmer, D., Finlay, C.A., and Levine, A.J. (1991) Regulation of transformation and the cell cycle by p53. In *The Cell Cycle, Cold Spring Harbor Symposium on Quantitative Biology* vol. LVI, Cold Spring Harbor Laboratory Press, 219-225.

9

Human *Bcl*-2 Antisense Therapy for Lymphomas

Finbarr E. Cotter
Institute of Child Health, London, United Kingdom

Andrew Webb, Paul Clarke, and David Cunningham
Royal Marsden Hospital, Sutton, United Kingdom

I. INTRODUCTION

Many lymphoma cells have an inherent resistance to chemotherapy due to the altered pattern of gene expression in the tumor cell. It is an attractive proposition to use this difference within the lymphoma cell to eradicate the malignant process. This is the aim of *bcl-2* antisense oligonucleotide (AO) therapy. The new therapeutic approach based on the abnormal biology of the lymphoma cell has been predominantly based on silencing a gene involved in preventing programmed cell death (apoptosis).

Lymphoma associated with the t(14;18) translocation (classically Follicular lymphoma) and deregulated expression of the *bcl-2* gene, conferring chemoresistance and subsequent protection against lymphoma cell death, was an obvious target for AO. Subsequently with our awareness of the increased levels of Bcl-2 protein in a whole range of lymphomas (as well as many solid tumors) extension of *bcl-2* AO therapy to these tumors is worthy of consideration. Interestingly it is the presence of increased Bcl-2 protein levels that confers a worse prognosis, whether or not a cytogenetic abnormality is detected. On this basis, blocking Bcl-2 production should bring about the death of the cell, if the production of

169

that protein is essential for the survival of the lymphoma cell. Non-sequence related effects have caused much skepticism following initial enthusiasm for antisense therapy, however, an increased understanding of these additional oligonucleotide properties has lead to a number of emerging therapeutic AO molecules for clinical use, working through a true antisense downregulation of the target gene and protein. This includes *bcl-2* AO in the lymphoma field, with a phase I studies commenced. There are a number of considerations when using AO to inhibit gene expression with respect to the nature of the target, and type, length and dosage of oligonucleotide. Attempts to regulate gene expression at the level of the mRNA with single stranded DNA oligonucleotides, in relation to lymphoma therapy, is discussed in this chapter.

II. PRECLINICAL STUDIES WITH *BCL-2* ANTISENSE OLIGONUCLEOTIDES

II.A. Choice of Target

The initiating codon of mRNA, the AUG start site, is the point from which protein synthesis is initiated. AO targeted to contain complementary sequences to this area are most frequently used in antisense experiments. Other successful sites include the 5' cap site, the mRNA site that initiates message reading rather than protein synthesis (Daaka and Wickstrom, 1990; Bacon and Wickstrom, 1991) the first splice donor-acceptor site (Daaka and Wickstrom, 1990), the polyadenylation signal, 5' to the polyA tail (Goodchild *et al.*, 1988) and to the junctional sequences of tumor associated chromosomal translocations (Szczylik *et al.*, 1991) Sequences targeted to the RNA loop structures show logarithmically greater hybridization than those in the areas flanking the loop (Lima *et al.*, 1992). The *bcl-2* AO were targeted to the AUG start site initially and proved useful, as determined by their ability to downregulate Bcl-2 protein and induce apoptosis in lymphoma cell lines.

II.B. Oligonucleotide size and internalization

Oligonucleotide length is important. A sequence in the region of 16 to 20 bases is the optimal length of AO, being long enough to provide a unique sequence to the targeted gene and short enough to stop too much nonspecific hybridization (Hélène and Toulmé, 1990; Monia *et al.*, 1992). For the *bcl-2* target two useful AO of 18 (G3139) and 20 bases (G3854) were initially studied both of which showed good intracellular internalization.

II.C. Protection against nuclease degradation

Oligonucleotides based on the normal chemical structure of single-stranded DNA (phosphodiester [PO] backbone) would be sensitive to 3',5'-exonuclease activity in serum and possibly to endonuclease activity inside the cell (Eder *et al.*, 1991) conferring an unrealistically short half-life for therapeutic use. Changing the phosphodiester linkage to a phosphorothioate (PS) (Stein and Cohen, 1989), where sulfur atoms replace oxygen, was accomplished to confer resistance to nuclease enzymes, thereby prolonging the half-life considerably. To date the *in vivo* phase I and II human studies have all been based on PS molecules and is the molecule of choice for current AO anti cancer research.

II.D. Specificity of *bcl-2* antisense effect

The aim for antisense researchers is to show down-regulation of a gene in a sequence-specific manner while control oligonucleotides show little or no down-regulating capability. It follows that all antisense experiments must be interpreted with adequate reference to control parameters. Primarily it is essential to investigate the possibility of the AO having a non-sequence specific effect, which would negate its use for a true "gene silencing" strategy. Whatever the mechanism, the important consideration in assessing antisense effects is to establish sequence specificity of the AO against control oligomers, and to demonstrate a decrease in the amount of protein produced by the gene targeted. This has been demonstrated for G3139 and G3854 using both sense and nonsense controls. Downregulation of the Bcl-2 protein was demonstrated at 72 hours onwards in lymphoma cells with the t(14;18) translocation with antisense oligonucleotides but not with the control sense and nonsense oligonucleotides (Cotter *et al.*, 1994). This result correlates with the Bcl-2 protein having a relatively long half-life. The effect was specific to cell lines dependent on Bcl-2 for survival. No nonspecific antipro-liferative effects were seen (Cotter *et al.*, 1994).

II.E. *In vivo bcl-2* AO pharmacokinetics and toxicities

There is some knowledge about the pharmacokinetics of phosphorothioate oligonucleotides with intravenous, or intraperitoneal infusions (Agrawal *et al.*, 1991; Iversen, 1991). Degradation is not marked (15 to 50%) up to 48 hours, when detectable levels in the tissues are still present. Excretion occurs in the urine with 30% loss within 24 hrs. Plasma clearance by both routes is biphasic with an initial half life of 15-25 minutes and a second half life, related in part to protein binding, of 20 to 40 hours, representing elimination from the body. We have confirmed a similar finding with subcutaneous delivery of *bcl-2* antisense (G3139) in mice (Raynaud *et al.*, 1997). G3139 has a high volume distribution due to high protein binding, particularly to albumin. Organ distribution varies with good uptake into bone marrow, liver, spleen, kidney and lymphatics (hematological sites predominantly), lesser uptake into other tissues and minimal uptake into the central nervous system (Raynaud *et al.*, 1997). Subcutaneous AO infusions prolongs the bioavailability of the molecule and increase steady state levels (Cotter *et al.*, 1996; Raynaud *et al.*, 1997).

Minimal toxicities, at a dose up to 10 mg/kg/day for two weeks, have been observed in primates. This dose is well above the therapeutic range for *bcl-2* antisense (Cotter *et al.*, 1996). Other studies using phosphorothioates have found good tolerance at similar dose ranges (Bishop *et al.*, 1996). The main *in vivo* toxicities break down into two groups, namely those associated with either rapid intravenous bolus administration (easily avoided by a 2 hour infusion) (Calabretta *et al.*, 1996) or the thioate backbone which is less readily avoided. We chose to administer the *bcl-2* antisense subcutaneously, avoiding the potential intravenous bolus toxicities and possibly providing greater bioavailability (Raynaud *et al.*, 1997). Those associated with bolus administration are rapid peripheral vasodilation and a transient anticoagulant effect. Thioate chemistry effects are rapidly reversible and include mild thrombocytopenia (dose dependent) and mild hyperglycemia (not dose dependent) requiring no treatment.

II.F. Preclinical *bcl-2* AO therapy in lymphoma

The systemic use of AO has its attractions where it might be hoped to have a specific antitumor effect while avoiding the many nonspecific toxicities caused by chemotherapeutic substances. It is this approach that has been most beneficial in lymphoma. The *bcl-2* oncogene has been implicated in the oncogenicity of a wide variety of hematological malignancies and cancers including melanoma, breast, lung and bowel carcinomas. Bcl-2 protein directly prolongs cellular survival by blocking programmed cell death. Approaches to downregulate the protein by antisense therapy have been pursued as an antitumor strategy. Initial attempts to manipulate the *bcl-2* gene with antisense DNA consisted of transfecting a *bcl-2* antisense sequence in a plasmid into a human T-cell lymphoma cell line (Reed *et al.*, 1990a). Transfection with a combination of *bcl-2* and c-*myc* sense plasmids markedly enhanced the tumorigenicity of this cell line in a nude mouse model. The *bcl-2* antisense gene reduced survival of the cell line following growth factor deprivation. When *bcl-2* AO was substituted for the antisense gene a similar effect was observed, with reduced expression of Bcl-2 protein (Reed *et al.*, 1990b). In follicular lymphoma cell lines with high and deregulated Bcl-2 protein levels due to the t(14;18) translocation, *bcl-2* AO directed at the open reading frame of *bcl-2*, compared to, control sense and nonsense oligonucleotides gave specific downregulation of Bcl-2 protein with a subsequent induction of apoptosis (Cotter *et al.*, 1994).

The AO has little or no effect on the viability of cell lines not expressing high levels of Bcl-2, suggesting that the AO is specifically targeting cells with high levels of Bcl-2 expression. Down regulating *bcl-2* in a cell that is heavily dependent on Bcl-2 expression for its survival advantage, appears to commit the cell to an apoptotic death, even when *bcl-2* oncogene is subsequently up-regulated again, suggesting an ir-reversible process. It has been demonstrated that the AO is exerting its effect by induction of apoptotic cell death as indicated by DNA fragmentation and characteristic apoptotic bodies. Similar sequence-specific effects are seen with *bcl-2* AO on human leukemia cell line with high Bcl-2 expression (Reed *et al.*, 1990ab).

Using the *in vivo* lymphoma model from the same cell line (Cotter *et al.*, 1994,) an effect against tumor growth *in vivo* (Cotter *et al.*, 1996; Pocock *et al.*, 1995) was also possible with a two week infusion at 100 µg daily achieving a plasma level of approximately 0.1 µM. Both G3139 and G3854 *bcl-2* AO showed similar efficacy. G3139, being shorter by two bases, was chosen to pursue an *in vivo* human study.

Sixty mice with lymphoma were treated with *bcl-2* AO G3139, 5'-TCTCCCAGCGTGCGCCAT-3 (supplied by Genta, San Diego, CA) and showed almost complete abolition of lymphoma in 50 (83%) of the mice, however, disease was still present in the remaining mice (17%). Extension of the treatment to 3 weeks, at the same dose, showed complete eradication of lymphoma in all animals, even at the PCR level. A dose response curve was also established; complete disease eradication was observed at a dose of 300 mg/day for 2 weeks (Cotter *et al.*, 1996). No abolition of lymphoma was seen in any other group of controls, which produced a pattern of disease similar to that of untreated mice.

The experiments were also repeated in NOD/SCID mice that have no NK, B or T cell activity, and similar efficacy was observed. This is of importance as it has been postulated that some AO may work through induction of interferon or tumor necrosis factor released by NK or B cells (Yamamoto *et al.*, 1992). Observation of efficacy in mice incapable of producing these cytokines provided further evidence for a direct *bcl-2* AO effect.

These results suggest that duration of treatment may be as im-portant as the dosage. Considerably less antisense is required to give complete disease eradication with a three week infusion compared to two. Rigorous preclinical evaluation of efficacy, pharmacokinetics, and toxicity were carried out in partnership with Genta, Inc., an antisense research company in San Diego, CA. These studies have now been extended into a phase I study at The Royal Marsden Hospital, Sutton, UK, for lymphoma patients with high Bcl-2 expression, who have failed at least two or more therapy regimens.

II.G. A human phase I *bcl-2* AO trial in lymphoma

Low grade and follicular non-Hodgkin's lymphomas are essentially incurable. They have an indolent course, initially chemosensitive, but eventually relapsing with chemoresistant disease, resulting in a median survival in the vicinity of 10 years from diagnosis. In the later stages of the disease, these tumors develop chemoresistance and often transform to a more aggressive histological appearance. Intermediate/high grade (aggressive, diffuse) non-Hodgkin's lymphoma provide a better chance of long term survival in the region of 40% with combination chemotherapy such as CHOP (cyclophosphamide, adriamycin, vincristine, prednisolone), but 60% will die of their disease having initially responded to therapy.

The use of new drugs, different chemotherapy combinations and high dose chemotherapy with bone marrow or peripheral stem cell rescue have had a minimal impact on patient survival. Bcl-2 expression appears to be a major factor in failure of cure with chemotherapy. As already outlined, overexpression of the *bcl-2* gene results in resistance to programmed cell death (apoptosis) (Hockenbery *et al.*, 1990) leading, to chemoresistance (Miyashita and Reed, 1992). Two recent studies have examined the prognostic significance of Bcl-2 protein overexpression in patients with diffuse, large B cell lymphoma. In both these studies, multivariate analyses confirmed the importance of Bcl-2 expression as an independent prognostic marker (Hermine *et al.*, 1996; Hill *et al.*, 1996).

In low grade, follicular non-Hodgkin's lymphoma, Bcl-2 overexpression is seen in virtually all the tumors. High levels of the Bcl-2 protein expression occur for a number of reasons, one of these being a t(14;18) translocation. However, the prognostic significance in both low grade and more aggressive lymphomas does not lie in the specific translocation but in the overexpression of Bcl-2 (Pezzela *et al.*, 1992; Hill *et al.*, 1996).

Therapies based on the molecular biology of the disease offer an alternative approach that may improve both response and survival. The preclinical data with G3139 *bcl-2* AO suggested that patients with relapsing lymphoma and high Bcl-2 expression were a logical group for a phase I study of the molecule (Cotter *et al.*, 1994).

II.H. Human study design

Patients eligible for the study were required to have B-cell non-Hodgkin's lymphoma of any grade that demonstrated immunohistochemically to overexpress the Bcl-2 protein in a lymph node biopsy. In addition, patients were required to have relapsing disease following completion at least 2 forms of conventional treatment, a life expectancy of more than 12 weeks normal renal and liver function, white blood count greater than $3 \times 10^9/l$ and a platelet count in excess of $100 \times 10^9/l$. Toxicity assessment was scored according to the common toxicity criteria (CTC) (NCI, 1988). This included areas of concern pinpointed in *in vitro* and animal studies.

The G3139 phosphorothioate oligonucleotide, supplied by Genta Inc., was administered as a 2 week subcutaneous continuous infusion using a portable syringe driver, mimicking the *in vivo* animal model. In order to avoid infusion site inflammation, these were changed when early signs of inflammation were observed. Toxicity was monitored for the initial 48 hours of treatment as an inpatient but thereafter as an outpatient. A single 2 week course of treatment was given. If there was evidence of response, a second course was considered.

The starting dose was 4.6 mg/m^2/day, which was equivalent to below the 1/10th LD$_{10}$ in mice. The dose was increased by 100% within each patient cohort unless grade 2 or greater toxicity was observed, according to the EORTC schema (EORTC, 1994). The maximum tolerated dose was defined as the dose that causes grade 3 or 4 toxicity in 50% or more patients. Response was evaluated from CT scan performed pretreatment, at weeks 2 (end of infusion) and 6, and classified using the WHO criteria (Miller *et al.*, 1981).

A complete response (CR) was defined as disappearance of all disease, and a partial response (PR) as 50% or greater reduction in the bidimensional product of measurable disease. Greater than 25% increase in the bidimensional product of measurable disease was defined as progressive disease (PD) and all other states were defined as stable disease (SD). Additional parameters of lymphoma activity were serum LDH and the numbers of circulating lymphoma cells, identified morphologically; changes in excess of 20% were considered relevant.

Samples of blood, bone marrow and fine needle aspirates of lymph nodes were collected at the start of weeks 0, 2, and 6. Mononuclear cells, freshly separated by Ficoll-Isopaque centrifugation (Lymphoprep; Nycomed, Norway), were suspended in 10% DMSO and stored in liquid nitrogen. At the time of analysis, the samples were fixed in 70% ethanol, incubated with an antibody to the Bcl-2 protein (DAKO, clone 124) followed by further incubation with an anti-IgG FITC second antibody. Bcl-2 protein levels were determined by flow cytometric analysis of gated lymphocytes. Within this population, Bcl-2 negative and positive cells were identified. The mean value and standard deviation were determined by gating on the Bcl-2 population. All samples from a single patient were labeled simultaneously under the same conditions. Controls for nonspecific protein expression (patient 3 onwards) were determined by flow cytometry using incubation with FITC conjugated HLA-A,B,C antibody (Scrotec).

III. EFFECTS OF *BCL-2* AO IN HUMANS

Nine patients have been treated over a 6 month period and have now been observed for over a year (Webb *et al.*, 1997). All had been extensively pretreated and had relapsing disease satisfying the entry criteria for the study. Of particular note was that all patients with low grade disease had failed both chlorambucil and fludarabine, which constitute the two most useful agents for this grade of lymphoma. Six patients had low grade tumor and 3 had intermediate or high grade tumor histology. All patients had stage IV disease (Table 1).

III.A. Toxicity

The main aim of the phase I studies was to determine the toxicity related to treating humans with G3139 (Webb *et al.*, 1997). The toxicities seen have essentially been minimal and have not prevented 100% dose increments as planned. The dose reached in the first 9 patients has been 2 mg/kg/day (patients 7,8,9) (Webb *et al.*, 1997). In addition to the results being reported here, a further 5 patients have received the AO at a higher dose (approximately 6 mg/kg/day) but have not yet been evaluated for response. In a few patients at the higher dose, *bcl-2* AO plasma concentration has been measured in the region of 1 μM, although variability is seen and a plateau level has not been reached, unlike the results in mice, where a constant state was reached within 4 days (Raynaud *et al.*, 1997). However, no dose limiting toxicities have been observed and the dose continues to be escalated, until such time as a maximum tolerated dose is reached. Interferon levels were monitored during the treatment and showed no rise. It is of note that the dose is currently well within the therapeutic range observed in the animal models.

Table 1. Baseline Characteristics of Patients Treated with *bcl-2* Antisense (G3139)

Pt. No.	Age yr.	Sex	Dose $\frac{mg}{m^2}$	Classification WF*	REAL*	Previous chemo-therapy regimens	Bone marrow	Circulating lymphoma cells	Palpable peripheral lymphade-nopathy
1.	50	M	4.6	B	Follicular grade 1	2	2	2	2
2.	63	M	9.2	E	Mantle cell	3	+	+	+
3.	64	M	18.4	G	Diffuse large B cell	4	2	2	2
4.	68	M	36.8	B	Follicular grade 1	5	2	2	+
5.	41	F	36.8	B	Follicular grade 1	2	+	2	2
6.	65	F	36.8	B	Follicular grade 1	5	+	+	+
7.	57	F	73.6	A	Small lym-phocytic	4	+	+	+
8.	53	M	73.6	C	Follicular grade 2	4	2	2	2
9.	54	F	73.6	G	Diffuse large B cell	4	2	2	2

*WF: Working Formulation. †REAL: Revised European American Lymphoma classification. (*Reprinted, with permission, from Webb, et al., 1997*).

III.A.1. Hematological toxicity

The evaluation of hematological toxicities in lymphoma with bone marrow disease (commonly present in progressive low grade disease) is not easy, yet it is required to determine the contribution from the disease, versus the contribution of the AO. Taking this into consideration, there was no antisense related hematological toxicity (Table ?) However, patient 8 developed grade 3 leukopenia and grade 2 thrombocytopenia associated with a *Haemophilus influenzae* chest infection at the start of the 2 week treatment infusion. The leukopenia and thrombocytopenia resolved following treatment with intravenous antibiotics, while the antisense infusion continued, suggesting a nonoligonucleotide effect. Patient 9 developed grade 2 thrombocytopenia at the end of the 2 week drug infusion, in addition to marked eosinophilia (up to 31%). Subsequently, at week 6, bone marrow infiltration and progressive disease in lymph nodes was observed.

Table 2. Toxic Effects of Treatment with *bcl-2* Antisense (G3139)

Common toxicity criteria grade	0	1	2	3
Hematological toxicity				
Anemia	6	0	3*	0
Leukopenia	8	2	2	1[†]
Lymphopenia	5	2	2	2*
Thrombocytopenia	7	2	2*[†]	
Clotting	9	2	2	2
Non-hematological toxicity				
Renal	9	2	2	2
Neurological	9	2	2	2
Gastrointestinal	9	2	2	2
Liver	9	2	2	2
Pulmonary	9	2	2	2
Cardiovascular	8	2	2	1[‡]
Hyperglycemia	0	5	4	2
Infection**	5	1	2	1
Local skin reaction	0	6	3	2

*Progressive lymphomatous disease. [†]Secondary to septicemia. [‡]Secondary to obstruction of superior vena cava. [§]Without fasting. **Coincidental secondary to advanced lymphoma. (*Reprinted, with permission, from Webb, et al., 1997*).

The thrombocytopenia and eosinophilia resolved on subsequent treatment with chemotherapy, implying that these effects were more likely to be due to advancing lymphoma than to the antisense oligonucleotide. Lymphopenia was present in 4 patients (patients 3,7,8,9) pretreatment and did not worsen with therapy. The anemia observed (patients 2,5,9) was not dose related and appeared to be equated with advancing bone marrow infiltration. One of the other areas of concern, when evaluating toxicity, was that of potential T cell alterations brought about by *bcl-2* down-regulation in T cell development. The CD4:CD8 ratio was carefully monitored, and no significant alterations were seen. No abnormalities in coagulation or clotting factors were observed.

An additional area of concern was that of stem cell suppression, although a role for Bcl-2 has not been demonstrated in the stem cell. Bone marrow biopsies have been regularly examined in each patient, and no evidence of treatment related aplasia has been detected in any of the patients.

III.A.2. Non-hematological toxicity

The G3139 has been well tolerated and the non-hematological toxicities are shown in Table 2 (Webb *et al.*, 1997). Of those seen, they can be ascribed primarily to the phosphorothioate chemistry and not to the effects of Bcl-2 downregulation. Those of particular note are outlined below. Patient 7 had episodes of transient syncope at rest 2 days following the end of treatment. This was related to progressive mediastinal disease causing superior vena cava (SVC) obstruction. Following chemotherapy to reduce the SVC obstruction, no further episodes have occurred. All 9 patients had a transient rise in non-fasting blood glucose levels, observed within 24 hours of starting the infusion; none exceeded 12 mmol/l. The increases in blood glucose levels were not dose related, and returned to normal after stopping the therapy. No intervention was required.

Similar findings have been documented in a phosphorothioate human trial targeting p53 (Bishop *et al.*, 1996). Four of the 9 patients developed an infection, none of which could be directly attributed to the antisense therapy. The only significant toxicity directly related to antisense therapy was a local skin reaction surrounding the infusion site. In 8 patients this simply required re-siting the line site every 3-4 days. However, one patient (patient 4) suffered a local inflammatory reaction that became unacceptably painful about 12 hours after starting treatment. A skin biopsy from the inflamed area demonstrated increased infiltration of T lymphocytes. Despite several site changes and dilution of drug concentration by 50%, the inflammation persisted and treatment could not to be continued. Two patients treated subsequently at this dose level and 3 with a 100% dose increment did not experience the same degree of reaction.

III.B. Lymphoma response

The response to treatment is shown in Table 3. Unlike chemotherapy responses, which are usually rapid and maximal within a week, the effect of *bcl-2* AO appears to be slower and more prolonged in duration, with continued reduction in lymphoma bulk in excess of 6 weeks after the end of the infusion (Webb *et al.*, 1997). The most impressive response was seen in patient 8, whose largest pretreatment lymph node measured 2.5×2.0 cm. By 6 weeks of therapy, no lymph node masses larger than 1 cm were detected, and a response was achieved in all lymph node sites (Figure 1). This patient remains in remission 8 months after starting treatment, without further therapy. Patient 6 at 6 weeks had near partial responses in right axillary and mediastinal lymph nodes, but in the larger lymph node mass within the abdomen, response was minimal. Two patients (1 and 8) had improvements in the their symptoms. Patient 1 had resolution of sweats and pruritus within 48 hours of starting treatment, which lasted 3 weeks.

Table 3. Tumor Response to Treatment with *bcl-2* Antisense (G3139)

Pt. No.	Tumor response by CT	Circulating lymphoma cells	Serum lactate dehydrogenase	B symptoms	Bcl-2 by flow cytometry
1.	SD	NE	↓	↓	NE
2.	SD	↓	↓	→	PB↓ BM↓ LN→
3.	PD	NE	↑	→	NE
4.	PD	NE	↑	→	LN→
5.	PD	NE	→	→	BM→
6.	SD*	↓	↓	→	LN↓ BM→ PB→
7.	PD	→	↑	→	LN→ BM→ PB-NE
8.	CR	NE	↓	↓	NE
9.	PD	NE	→	→	NE

Abbreviations: CT: computerized tomography; LDH: lactate dehydrogenase; CR: complete response; SD: stable disease; PD: progressive disease; NE: non-evaluable; LN: lymph node; BM: bone marrow; PB: peripheral blood sample.
*Tumor shrinkage but did not satisfy definition of partial response.
(Reprinted, with permission, from Webb, et al., 1997).

On retreatment at the same dose, this patient again had resolution of the sweats and pruritus again. Patient 8's lymphoma related alcohol intolerance resolved such that he was able to drink alcohol for the first time in 2 years. Two out of 3 patients had a reduction in their circulating lymphoma cells (patient 2, patient 6), 4 out of 9 patients had a reduction in their serum LDH levels (patients 1,2,6,8).

III.C. Bcl-2 expression

It was of considerable importance to try and determine if Bcl-2 protein could be downregulated in lymphoma cells, although this could be technically and ethically difficult without ready access to tumor material. It was felt that circulating lymphoma cells could offer this opportunity, as well as fine needle aspirates from lymph nodes when readily accessible, without potentially endangering the patients' well being. Over the treatment period and the subsequent 4 weeks, Bcl-2 levels were measured by flow cytometry in peripheral blood samples containing circulating lymphoma cells (patients 2,6,7), lymphoma infiltrated bone marrow aspirates (patients 2,5,6,7) and FNA of peripheral lymph nodes (patients 2,4,6,7). HLA-A,B,C levels acted as controls and demonstrated little intrapatient variability. Bcl-2 levels were reduced in the peripheral blood and bone marrow aspirates of patient 2 and the lymph node aspirates of patient 6 (Figure 2).

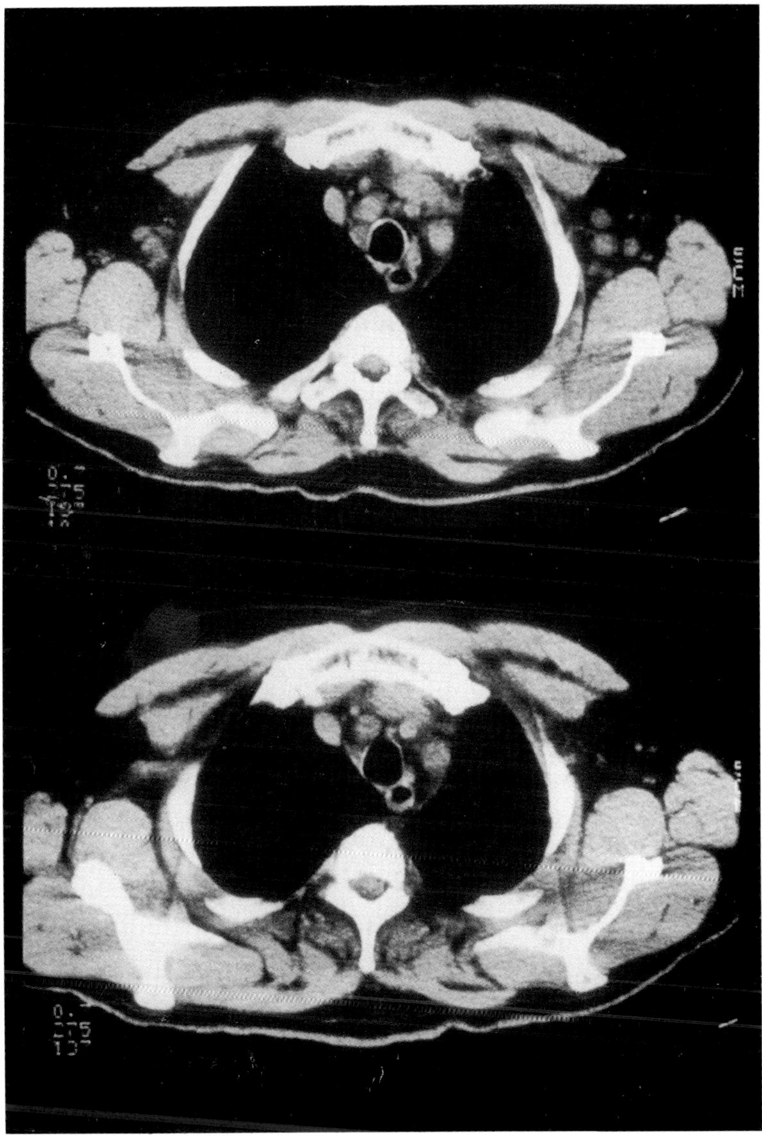

Figure 1. CT scan of the upper chest of patient 8 before (**top**) and after (**bottom**) treatment with *bcl-2* antisense, showing reduction in axillary lymphadenopathy *(reprinted, with permission, from Webb, et al., 1997).*

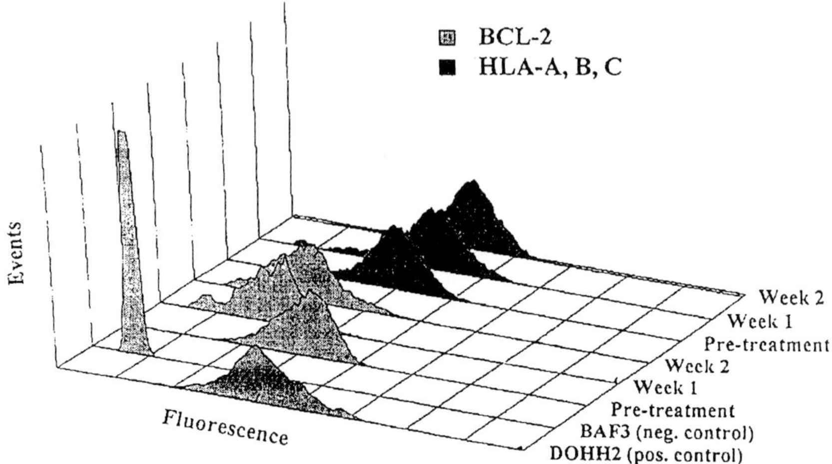

Figure 2. Flow cytometric measurements of Bcl-2 and HLA expression in lymph node aspirates of patient 6 before and after *bcl-2* antisense therapy *(reprinted, with permission, from Webb, et al., 1997).*

The remaining samples that contained lymphoma cells showed little change in Bcl-2 levels. Samples from patients 1, 3, 8, and 9 contained no lymphoma cells and were considered nonevaluable for Bcl-2 analysis.

III.D. Parallel animal model

To further test the specificity of the *bcl-2* AO in an animal model, a portion of the lymph node biopsy from patient 6 was disaggregated, and 10^7 cells were injected intraperitoneally into 3 non-obese diabetic, severe combined immunodeficient (NOD/SCID) mice within 24 hours of the biopsy. Each mouse was fitted with a subcutaneous osmotic Alzet pump which delivers a 2 week infusion. This consisted of either 0.9% saline, a control reverse sense oligonucleotide (0.4 mg/kg/day sequence 5'-TACCGCGTGCGACCCTCT-3) or antisense (0.4 mg/kg/day Anticode G3139 sequence) starting on day 1 for 14 days. The proportion of lymphoma cells in the spleen, liver and bone marrow were determined using flow cytometry to identify cells dual labeling for CD45 (common leukocyte antigen) and CD19 (B cell lymphoma marker). The normal saline and reverse sense treated control mice showed evidence of distress by day 50, and necropsy demonstrated gross lymphoma in spleen and lymph nodes. Both controls had significant infiltration with CD45/19 positive cells in liver (38%, 46%), spleen (62%, 47%) and bone marrow (30%, 29%). The antisense treated mouse remained well with no evidence of lymphoma at day 80.

III.E. Further treatment

Eight out of the 9 patients have gone on to receive further chemotherapy with a variety of regimens. Six of them have achieved partial responses using chemotherapy for which response had been poor previously. Patient 9 was treated with the chemotherapy regimen ChlVPP (chlorambucil, vinblastine, procarbazine, prednisolone) 5 weeks after finishing antisense treatment, and developed complete alopecia, which is unusual with that regimen (Selby *et al.*, 1990).

III.F. Conclusions from the human *bcl-2* AO phase I study

In vitro and *in vivo* studies using antisense oligonucleotides to inhibit expression of Bcl-2 have demonstrated good efficacy with low toxicity (Webb *et al.*, 1997). The results of this first human study investigating antisense therapy targeting *bcl-2* have demonstrated good tolerance, with minimal toxicity and additional tumor regressions, improvement in the laboratory parameters, symptom improvement together with down regulation of target protein expression (Webb *et al.*, 1997). A good chemotherapy agent for a phase I study will have a response rate in the region of 10%. These results, although very preliminary, certainly match those of a good chemotherapeutic drug (Webb *et al.*, 1997).

This is encouraging, as the maximum tolerated dose has not been reached, and the doses used are just above those levels showing efficacy in the mouse model. Two patients had evidence of tumor shrinkage on the CT scanning with one patient at the time of this report (4 months after treatment) maintaining this improvement. Both of these patients had low grade tumors. In rare cases, low grade lymphomas can spontaneously regress, although it would be unlikely that two such events would both occur by chance within a few weeks of completing antisense therapy.

Furthermore, there were reductions in circulating lymphoma cells and serum LDH. In addition, two patients had symptomatic benefit. The combination of this evidence supports the theory that *bcl-2* antisense oligonucleotides have anti-lymphoma activity (Webb *et al.*, 1997). *In vitro* studies demonstrate specific reduction in mRNA, protein expression and induction of apoptosis using *bcl-2* antisense (Cotter *et al.*, 1994). For the first time this study demonstrates evidence of down regulation of the target protein in humans and the parallel animal model provides evidence of a sequence specific anti-lymphoma effect.

The rate of response differs from chemotherapy and it is of note that a steady state has not been reached at the end of two weeks treatment. It is probable that the wide distribution of phosphorothioate AO within the body with strong protein binding leads to the long lag phase before reaching a steady state and there is a slow release of AO for a long period after the infusion has been completed.

Hence, the continued down-regulation of *bcl-2* in the lymphoma cells for some weeks after completing treatment and the continued response seen in some patients. In addition, it may be speculated that the better response to chemotherapy after treatment with *bcl-2* AO in this trial may have been due to a more prolonged downregulation of *bcl-2* gene, making the tumor cells more susceptible to apoptotic cell death. It was observed that 6 out of 8 patients who were treated with chemotherapy following antisense treatment went on to achieve a partial remission. Patient 9 developed complete alopecia with ChlVPP, a treatment regimen with a documented 0.5% incidence of this side effect (Selby *et al.*, 1990).

It is known that Bcl-2 protein levels in the hair follicle are normally high (Stenn *et al.*, 1994) and *bcl-2* gene knockout mice develop abnormalities in hair pigmentation (Veis *et al.*, 1993; Kamada *et al.*, 1995). These results suggest that *bcl-2* antisense, when given prior to chemotherapy may have a sensitizing effect. Potential uses of *bcl-2* antisense in further human trials may extend beyond the field of lymphoma, as overexpression of Bcl-2 protein is seen in breast, lung, melanoma, leukemia, prostate, gastric and colorectal tumors.

An important aim of this study was to establish toxicity of *bcl-2* antisense oligonucleotides. At the dose reached to date, no major dose limiting toxicities have been observed. Therefore, dose escalation in future patient cohorts will be continued to determine the maximum tolerated dose. Concerns about potential toxicity came from a number of sources. Mouse toxicity data using this compound demonstrated myocardial and liver necrosis, splenomegaly and deaths at high dosages (50 mg/kg/day).

The dose levels at present are considerably below the levels at which toxicities were observed in mice. Primate studies revealed no clinical toxicity and autopsy revealed only mild injection site inflammation and nonspecific pathological changes within the kidneys at doses up to 10 mg/kg/day for 2 weeks (Cotter *et al.*, 1996).

Primate studies using high doses of other phosphorothioate compounds demonstrated hypotension and deaths if the drugs were given as a rapid intravenous bolus, but not when given as a slow infusion. Furthermore, the hypotensive events were not sequence related (Galbraith *et al.*, 1994). Another human study, using a 20 base full phosphorothioate oligonucleotide targeted against p53 in patients with acute myeloid leukemia and myelodysplastic syndrome at doses up to 6 mg/kg/day as an intravenous infusion, demonstrated no compound related toxicity (Bishop *et al.*, 1996). There was concern that the *bcl-2* AO might have direct sequence related effects on cells which express high Bcl-2 protein levels under normal conditions, but so far such effects have not been observed.

The memory B cells, mature ($CD4^+$ and $CD8^+$) T cells, neuronal tissue and intestinal mucosa have high Bcl-2 levels. Bone marrow stem cells do not normally overexpress Bcl-2, and another member of the *bcl-2* family, *bcl-*X_L appears to be of greater importance here, so adverse effects against the stem cells were considered unlikely (Park *et al.*, 1995).

It also seems unlikely that neuronal tissue would be damaged as oligonucleotides do not cross the blood brain barrier. In addition, *bcl-2* knockout mice in which both copies of the *bcl-2* gene have been deleted by recombinant DNA techniques are viable without neurological deficit, suggesting that Bcl-2 protein is not of prime importance to the stem cells or neuronal tissue. However, the knockout mice do subsequently, develop lymphopenia, apoptotic involution in the thymus and spleen, polycystic kidneys, gray hair in the second follicle cycle (Veis *et al.*, 1993; Kamada *et al.*, 1995) and distorted small intestine (Kamada *et al.*, 1995). In practice and at the current dose levels major *bcl-2* sequence specific toxicities do not occur.

The only significant treatment related toxicity is almost certainly thioate related. This was an inflammatory response at the injection site. In one patient this was severe enough for the treatment to be stopped. The severity of reaction, however, was not seen in other patients even at higher doses. Transient non-fasting hyperglycemia has also been observed during the treatment period. Elevated blood glucose had no clinical implications, but has also been noted in a recently published report of antisense treatment against p53, suggesting that it may be related to the phosphorothioate backbone, but did not appear to be dose related (Bishop *et al.*, 1996).

The potential uses of *bcl-2* antisense oligonucleotides are many. They include a single agent use, treatment of minimal residual disease or *ex vivo* purging of bone marrow or peripheral stem cell harvests. Potentially, the most important application may be to overcome chemo-resistance. *In vitro* transfection of cells with *bcl-2* gene results in chemoresistance which can be reversed by, antisense mediated reduction of *bcl-2* gene expression (Hockenbery *et al.*, 1990).

The results of this first study investigating antisense therapy targeting *bcl-2* have demonstrated safety at the dose levels used, tumor regressions, improvement in the laboratory parameters, symptom improvements and down regulation of target protein expression.

III.G. The future

Bcl-2 antisense has a definite anti-lymphoma effect *in vitro* and *in vivo* using the SCID lymphoma model, and in addition has been transferred into the clinic with continued evidence for safety and efficacy (Webb *et al.*, 1997). *Bcl-2* AO may well be of benefit in a broad spectrum of tumors with high Bcl-2 expression, in particular to overcome chemoresistance. It may be envisaged that the patient would be treated with a short course of AO to downregulate *bcl-2* as a "priming" procedure and then treated with conventional chemotherapy drugs. The "primed" tumor cells would be considerably more sensitive to the chemotherapy due to the reduced levels of Bcl-2 protein. It is in this manner that we feel the *bcl-2* AO will be of maximum benefit. The use of *bcl-2* AO as an anti-tumor strategy may be a logical extension of the lymphoma work described above.

The potential ability of antisense oligonucleotides to downregulate the expression of oncogenes involved in lymphoma, with minimal toxicity, can be achieved. The possibility of combining *bcl-2* antisense therapy with chemotherapy will provide an interesting means of overcoming tumor cell resistance to chemotherapy in lymphoma and a range of other high Bcl-2 expressing malignancies. As additional antisense molecules targeting oncogenes involved in lymphomas become available (Wickstrom *et al.*, 1988; Monia *et al.*, 1996), it will be possible to combine AO to enhance their efficacy, either targeting the same gene at two sites or more a combination of genes (for example, *bcl-2* and c-*myc* in Burkitt's lymphoma). Of major importance are approaches to improve AO uptake into cells, which is inefficient (Wickstrom *et al.*, 1988; Daaka and Wickstrom, 1990). Methods to improve antisense uptake into cells and tissues *in vivo* are required, and in addition a new generation of oligonucleotides free of the nonspecific phosphorothioate toxicities are required. AO are a dramatic new area of research and as such requires much evaluation if it is to be applied maximally. Both *in vitro* and *in vivo* efficacy has to be established. With care, novel therapies based on the biology of the malignant cell may be determined on a scientific basis and may help improve the treatment of patients with these diseases. As demonstrated in lymphomas, antisense oligonucleotides have a role to play in silencing pathogenic genes.

ACKNOWLEDGEMENTS

We are grateful to Genta, Inc. for supplying the *bcl-2* antisense oligonucleotide Anticode G3139 free of charge. We are indebted to Drs.

Bob Bryan and Zofia Dziewanowska for their continued support, and without whom this work would have been very difficult.

IV. REFERENCES

Agrawal, S, Temsamani, J, and Tang, J. (1991) Pharmacokinetics, biodistribution, and stability of oligodeoxynucleotide phosphorothioates in mice. *Proc. Natl. Acad. Sci. USA* **88**:7595-7599.

Bacon, T.A., and Wickstrom, E. (1991) Walking along human c-*myc* mRNA with antisense oligodeoxynucleotides: maximum efficacy at the 5' cap region. *Oncogene Res.* **6**:13-19.

Bishop, M.R., Iversen, P.L., Bayever, E., Sharp, G., Greiner, T.C., Capple, B.L., Ruddon, R., Zon, G., Spinolo, J., Arenson, M., Armitage, J.O., and Kassinger, A. (1996) Phase I Trial of an antisense oligonucleotide OL(1) p53 in hematological malignancies. *J. Clin. Oncol.* **14**:1320-1326.

Calabretta, B., Skorski, T., Ratajczak, M.Z., and Gewirtz, A.M. (1996) Antisense strategies in the treatment of leukemias. *Semin. Oncol.* **23**:78-87.

Cotter, F., Johnson, P., Hall, P., Pocock, C., Al Mahdi, N., Cowell, J., and Morgan, G. (1994) Antisense oligonucleotides suppress B-cell lymphoma growth in a SCID-hu mouse model. *Oncogene* **9**:3049-3055.

Cotter, F.E., Corbo, M., Raynaud, F., Orr, R., Pocock, C., Bryan, B., Harper, M., Webb, A., Clarke, P., Judson, I., and Cunningham, D. (1996) *Bcl*-2 antisense therapy in lymphoma: *in vitro* and *in vivo* mechanisms, efficacy, pharmacokinetics and toxicity studies. *Annals Oncol.* **7** (suppl 3):32.

Daaka, Y., and Wickstrom, E. (1990) Target dependence of antisense oligodeoxynucleotide inhibition of c-H-*ras* p21 expression and focus formation in T24-transformed NIH3T3 cells. *Oncogene Res.* **5**:267-275.

Eder, P., DeVine, R., Dagle, J., and Walder J. (1991) Substrate specificity and kinetics of degradation of antisense oligonucleotides by a 3' exonuclease in plasma. *Antisense Res. Dev.* **1**:141-151.

European Organization for Research and Treatment of Cancer (EORTC) (1994) A practical guide to EORTC studies.

Galbraith, W.M., Hobson, W.C., Giclas, P.C., Schechter, P.J., and Agrawal, S. (1994) Complement activation and hemodynamic changes following intravenous administration of phosphorothioate oligonucleotides in the monkey. *Antisense Res. Dev.* **4**:201-206.

Goodchild, J., Agrawal, S., Civeira, M.P., Sarin, P.S., Sun, D., and Zamecnik, P. C. (1988) Inhibition of human immunodeficiency virus replication by antisense oligodeoxynucleotides. *Proc. Natl. Acad. Sci. USA* **85**:5507-5511

Hélène, C., and Toulmé J.J. (1990) Specific regulation of gene expression by antisense, sense and antigene nucleic acids. *Biochim. Biophys. Acta* **1046**:99-125.

Hermine, O., Haioun, C., Lepage, E., d'Agay, M.F., Briere, J., Lavignac, C., Fillet, G., Salles, G., Marolleau, J.P., Diebold, J., Reyas, F., and Gaulard, P. (1996) Prognostic significance of *bcl*-2 protein expression in aggressive non-

Hodgkin's lymphoma. Groupe d'Etude des Lymphomes de I'Adulte (GELA) *Blood* **87**:265-272.

Hill, M.E., MacLennan, K.A., Cunningham, D.C., Vaughan-Hudson, B., Burke, M., Clarke, P., Di-Stefano, F., Anderson, L., Anderson, L., Vaughan-Hudson, G., Mason, D., Selby, P., and Linch, D.C. (1996) Prognostic significance of Bcl-2 expression and Bcl-2 major breakpoint region rearrangement in diffuse large cell non- Hodgkin's lymphoma: A British National Lymphoma Investigation study. *Blood* **88**:1046-1051.

Hockenbery, D., Nunez, G., Milliman, C., Schreiber, R.D., and Korsmeyer, S.J. (1990) Bcl-2 is an inner mitochondrial membrane protein that blocks programmed cell death. *Nature* **348**:334-336.

Iversen, P. (1991) *In vivo* studies with phosphorothioate oligonucleotides: pharmacokinetics prologue. *Anti-Cancer Drug Design* **6**:531-538.

Kamada, S., Shimono, A, Shinto, Y., Tsujimura, T., Takahashi, T., Noda, T., Kitamura, Y., Kondoh, H., and Tsujimoto Y. (1995) Bcl-2 deficiency in mice leads to pleiotropic abnormalities: accelerated lymphoid cell death in thymus and spleen, polycystic kidney, hair hypopigmentation, and distorted small intestine. *Cancer Res.* **55**:354-359.

Lima, W.F., Monia, B.P., Ecker, D.J., and Freier, S.M. (1992) Implication of RNA structure on antisense oligonucleotide hybridization kinetics. *Biochemistry* **31**:12055-12061.

Miller, A.B., Hoogstraten, B., Staquet., M., and Winkler, A. (1981) Reporting results of cancer treatment. *Cancer* **47**:207-214.

Miyashita, T., and Reed, J.C. (1992) *Bcl-2* Gene transfer increases relative resistance of S49.1 and WEHI7.2 lymphoid cells to cell death and DNA fragmentation induced by glucocorticoids and multiple chemotherapeutic drugs. *Cancer Res.* **52**:5407-5411.

Monia, B.P., Johnston, J.F., Ecker, D.J., Zounes, M.A., Lima, W.F., and Freier, S.M. (1992) Selective inhibition of mutant Ha-*ras* mRNA expression by antisense oligonucleotides. *J. Biol. Chem.* **267**:19954-19962.

Monia, B.P., Johnston, J.F., Geiger, T., Muller, M., and Fabbro, D. (1996) Antitumor activity of a phosphorothioate antisense oligodeoxynucleotide targeted against C-*raf* kinase. *Nature Med.* **2**:668-675.

National Cancer Institute (1988) Guidelines for reporting of adverse drug reactions. Division of Cancer Treatment, National Cancer Institute, Bethesda, MD.

Park, J.R., Bernstein, I.D., and Hockenbery, D.M. (1995) Primitive human hematopoietic precursors express Bcl-x but not Bcl-2. *Blood* **86**:868-876

Pezzella, F., Jones, M., Ralfkiaer, E., Ersboll, J., Gatter, K.C., and Mason, D.Y. (1992) Evaluation of bcl-2 protein expression and 14;18 translocation as prognostic markers in follicular lymphoma. *Br. J. Cancer* **65**:87-89.

Pocock, C., Malone, M., Booth, M., Evans, M., Morgan, G., Greil, J., and Cotter, F. (1995) BCL-2 expression by leukaemic blasts in a SCID mouse model of biphenotypic leukaemia model associated with the t(4;11)(q21;q23) translocation. *Br. J. Haem.* **90**:885-867.

Raynaud, F.I., Orr, R.M., Goddard, P.M., Lacey, H., Lancashire, H., Judson, I.R.,

Beck, T., Bryan, B., and Cotter, F.E. (1997) Pharmacokinetics of G3139 a phosphorothioate oligodeoxynucleotide antisense to *bcl-2* following intravenous administration or continuous subcutaneous infusion to mice. *J Pharmacol. Exper. Therap.* **281**: 420-427.

Reed J., Cuddy, M., Haldar, S., Croce, C., Nowell, P., Makover, D., and Bradley, K. (1990a) *Bcl2*-mediated tumorigenicity of a human T-lymphoid cell line: synergy with *myc* and inhibition by *bcl2* antisense. *Proc. Natl. Acad. Sci. USA* **87**:3660-3664.

Reed, J., Stein, C., Subasinghe, C., Haldar, S., Croce, C., Yum, S., and Cohen, J. (1990b) Antisense-mediated inhibition of *bcl2* proto-oncogene expression and leukemic cell growth and survival: comparisons of phosphodiester and phosphorothioate oligodeoxynucleotides. *Cancer Res.* **50**:6565-6570.

Selby, P., Patel, P., Milan, S., Meldrum, M., Mansi, J., Mbidde, E., Brada, M., Perren, T., Forgeson, G., and Gore, M. (1990) ChlVPP combination chemotherapy for Hodgkin's disease: long-term results. *Br. J. Cancer.* **62**:279-285.

Stein, C., and Cohen, J. (1989) Phosphorothioate oligodeoxynucleotide analogues. In: Cohen, J., ed., *Oligodeoxynucleotides: Antisense Inhibitors of Gene Expression*. CRC Press, Boca Raton, FL, 97-117.

Stenn, K.S., Lawrence, L., Veis, D., Korsmeyer, S., and Seiberg, M. (1994) Expression of the *bcl*-2 proto-oncogene in the cycling adult mouse hair follicle. *J Invest Dermatol.* **103**:107-111.

Szczylik, C., Skorski, T., Nicolaides, N., Manzella, L., Malaguarnera, L., Venturelli, D., Gewirtz, A., and Calabretta, B. (1991) Selective inhibition of leukemia cell proliferation by BCR-ABL antisense oligodeoxynucleotides. *Science* **253**:562-565.

Veis, D.J., Sorenson, C.M., Shutter, J.R., and Korsmeyer, S.J. (1993) Bcl-2-deficient mice demonstrate fulminant lymphoid apoptosis, polycystic kidneys, and hypopigmented hair. *Cell* **75**:229-240.

Webb, A., Cunningham, D., Cotter, F., Clarke, P., di Stefano, F., Ross, P., Corbo, M., and Dziewanowska, Z. (1997) *Bcl*-2 antisense therapy in patients with non-Hodgkin lymphoma. *Lancet* **349**:1137- 1141.

Wickstrom, E.L., Bacon, T.A., Gonzalez, A., Freeman, D.L., Lyman, G.H., and Wickstrom, E. (1988) Human promyelocytic leukemia HL-60 cell proliferation and *c-myc* protein expression are inhibited by an antisense pentdecadeoxynucleotide targeted against c-*myc* mRNA. *Proc. Natl. Acad. Sci. USA* **85**:1028-1032.

Yamamoto, S., Yamamoto, T., Kataoka, T., Kuramoto, E., Yano, O., and Tokunaga, T. (1992) Unique palindromic sequences in synthetic oligonucleotides are required to induce IFN and augment IFN-mediated natural killer activity. *J. Immunol.* **148**:4072-4076.

10

Retroviral Gene Transfer in Autologous Bone Marrow and Stem Cell Transplantation

Rafat Abonour

Indiana University School of Medicine, Indianapolis, Indiana

Kenneth Cornetta

Indiana Cancer Research Institute, Indianapolis, Indiana

I. INTRODUCTION

Bone marrow has been the target of intense gene therapy research, partially due to existing technology for *ex vivo* manipulation and re-implantation. Retroviral vectors are attractive gene therapy vehicles since they insert stably into the target cell genome and pass the genetic material to all of its progeny. This property is extremely important for cells such as marrow stem cells that may generate millions, possibly billions, of daughter cells.

In this chapter we will briefly review the state of bone marrow transplantation and discuss how retroviral vectors can be used to understand the biology of autologous transplantation. In addition, we will address current gene therapy trials aimed at improving survival after transplantation, highlighting studies utilizing drug resistant gene vectors.

Bone marrow transplantation (BMT) is an increasingly important treatment modality used in patients with a variety of malignant and genetic disorders. The first successful marrow transplants were directed at hematologic disorders and utilized marrow obtained from sibling donors.

This approach combines chemotherapy and radiation (the preparative regimen) in an attempt to eradicate the underlying marrow disorder. The preparative regimen must also eradicate host immunity in order for the new organ, i.e. the donor marrow, to engraft without rejection. Allogeneic BMT, which was limited to the use of HLA match siblings, has now been expanded to include marrow from related, unrelated, and cord blood donors. In contrast, autologous BMT uses the patient's own marrow progenitors and is becoming an increasingly common procedure in the treatment of solid tumors and hematological malignancies.

The International Bone Marrow Transplantation Registry estimates that over 10,000 autologous transplants are performed each year, and the number grows at an annual rate of 20% (Horowitz, 1995). For hematological malignancies, autologous transplantation is an attractive alternative to allogeneic trans-plantation because it does not require a histocompatible donor, lacks the risk of graft-versus-host disease, and can be used in patients up to 70 years of age. In solid tumors, autologous transplantation permits dose intensification while minimizing the dose limiting hematological toxicity associated with high dose chemotherapy.

Initial work in marrow transplantation utilized cells obtained from the bone marrow space. This procedure is most often performed under general anesthesia during which time multiple passes (100-200) are made into the posterior iliac crest of the donor with removal of approximately one liter of bone marrow. As our understanding of hematopoietic stem cell biology advanced, it was recognized that longterm engraftment results from a very small population of marrow cells, referred to as primitive marrow progenitors or stem cells. It was also recognized that these primitive progenitors circulate in the bloodstream and their number can be increased, or mobilized through the use of chemotherapy and/or cytokines (such as GM-CSF or G-CSF).

Peripheral blood progenitor cells (also referred to as peripheral blood stem cells) are obtained through the process of pheresis, in which leukocytes are selectively removed from the bloodstream and the patient's red cells, platelets and plasma are returned to the circulation. This outpatient procedure lasts approximately 4 hours and can be repeated daily. Two to three phereses are usually required to obtain sufficient cells for transplantation. In autologous transplantation, peripheral blood progenitor cells have replaced autologous marrow as the source of transplantable cells in most centers. We and others have also been evaluating the use of mobilized peripheral blood progenitors in the setting of allogeneic transplantation.

In the allogeneic setting, transplanted progenitors cells are most often infused immediately after processing, while autologous cells are usually stored in liquid nitrogen until the patient has completed chemotherapy and/or radiation therapy. Both types of cells are infused intravenously and home to the bone marrow microenvironment, where they restore granulopoiesis in approximately two weeks, with platelet and red cell counts returning within approximately one month.

II. GENE MARKING STUDIES IN AUTOLOGOUS TRANSPLANTATION

II.A. Gene marking in evaluating transplantation biology

Although autologous marrow and peripheral blood progenitor transplants are being performed in large number, several critical issues are poorly understood. One such issue is whether transplanted autologous cells lead to longterm engraftment. Bone marrow contains an array of hematopoietic progenitors at various stages of differentiation and the majority of transplanted cells are at or near terminal differentiation, do not contribute significantly to engraftment, and do not express the CD34 antigen (Strauss *et al.*, 1986; Watt *et al.*, 1987).

Hematopoietic reconstitution appears to involve two populations of cells expressing CD34. The first is responsible for initial engraftment and is composed of committed progenitors with an impressive capacity for expansion and differentiation but with a finite life span (Sutherland *et al.*, 1989; Terstappen *et al.*, 1991). The rapid engraftment of CD34 enriched marrow cells in autologous BMT supports the hypothesis that CD34 cells are responsible for short-term engraftment (Berenson *et al.*, 1991; Broun *et al.*, 1994; Schpall *et al.*, 1994). Longterm hematopoiesis after BMT is believed to arise from a second population of CD34 expressing cells, stem cells, which differentiate into committed progenitors and also capable of self-renewal.

Committed progenitors have a high cycling rate, consistent with their notable sensitivity to chemotherapeutic agents and their relatively high susceptibility to retroviral mediated gene transfer. In contrast, the quiescent stem cells are relatively resistant to chemotherapy and appear difficult to transduce with retroviral vectors. Since many chemotherapy regimens used in autologous transplantation are not believed to be truly myeloablative, it is possible that transplanted committed progenitors are responsible for short-term engraftment while residual stem cells (i.e. those that survived the preparative regimen) comprise the major component of longterm hematopoiesis. Therefore, researchers interpreting gene marking studies and designing gene therapy proposals must consider the types and strength of chemo-radiotherapy administered, since genetic manipulations of the transplanted marrow would only provide transient benefit if longterm hematopoiesis arises from residual stem cells.

The issues of short and longterm engraftment becomes increasingly important as new marrow purification methods are brought to clinical trial. CD34 selection is one such manipulation being implemented in autologous transplantation (Berenson *et al.*, 1991; Broun *et al.*, 1994; Schpall *et al.*, 1994). Since many solid tumors do not express CD34, selection for CD34 expressing cells is being evaluated as a means of decreasing contamination of the graft with malignant cells. Gene-marking of CD34-selected grafts provides a means of assessing the kinetics and longterm engraftment potential of cells exposed to the selection process.

The use of exogenous cytokines in autologous transplantation represents a manipulation capable of improving the rate of engraftment, but its effect on longterm hematopoiesis is still not fully understood and will likely require gene marking. For example, granulocyte-colony stimulating factor (G-CSF) and granulocyte-macrophage colony stimulating factor (GM-CSF) are cytokines which increase the number of committed progenitor cells when administered prior to marrow harvest (Bregni *et al.*, 1991; Hohaus *et al.*, 1993).

These agents are also used to enhance the rate of granulocyte engraftment when administered after transplantation (Masaoka *et al.*, 1989; Nemunaitis *et al.*, 1992). Cytokines have also opened the possibility for *ex vivo* expansion of marrow cells; that is, sufficient cells for transplantation might be expanded from small quantities of blood or bone marrow, eliminating the need for marrow harvest or pheresis (Moore, 1993; Traycoff *et al.*, 1995; Verfaillie and Miller, 1995). Unfortunately cytokine therapy and *ex vivo* manipulation could promote differentiation of hematopoietic stem cells and decrease their number, so that longterm hematopoiesis after transplantation consists solely of recipients' stem cells that survived the preparative regimen. Since an *in vitro* assay of marrow stem cells does not exist, gene marking is the only method currently available to evaluate the effects of marrow manipulation on longterm engraftment.

II.B. Retroviral vectors as markers

Until the precise phenotype and *in vitro* growth characteristics of marrow stem cells are known, their existence can only be evaluated using *in vivo* studies. One approach to determining the kinetics of marrow reconstitution is to mark the transplanted cells.

Radionucleotides have been used extensively as labels, but their *in vivo* use is limited by either the rapid decay of the label or the harm of radiation exposure (Fisher *et al.*, 1989; Griffith *et al.*, 1989). In addition, the loss, reutilization, or sequestration of label is a significant problem for longterm *in vivo* studies. These limitations can be overcome by the use of retroviral-mediated gene transfer (Figure 1).

In addition to stable integration, retroviral vectors are attractive cell markers because very sensitive methods of vector detection, namely the polymerase chain reaction, are available. Also, if the marker gene confers drug resistance, it may be possible to select out marked cells *in vitro*. Lastly, retroviral gene transfer is a simple technical procedure that does not expose the marked cell to toxic compounds or radioactivity that might alter the function of the marked cells. Retroviral vectors have been used extensively *in vitro* to mark human leukemic cell lines with an efficiency of 10-60% using supernate transduction protocols (Hogge and Humphries, 1987; Smith and Benchimol, 1987; Cornetta *et al.*, 1993).

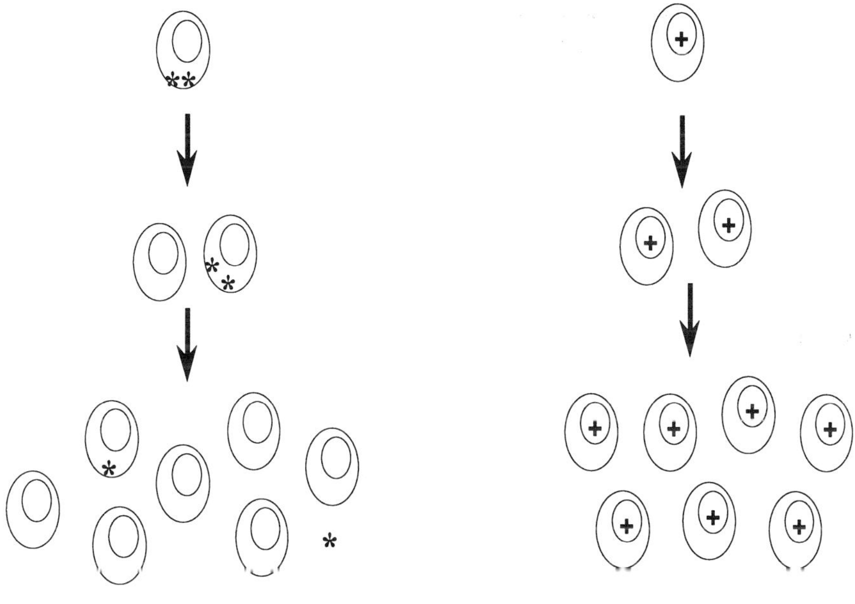

Figure 1. Advantage of gene marking in monitoring autologous cells *in vivo*. Retroviral vectors stably integrate into target cell chromosome such that all daughter cells contain the marker gene. The disadvantage of radiolabeling, and other non-integrated labels, is illustrated on the left showing eventual dilution of the marker (stars) as cell expansion proceeds.

Retroviral vectors in murine syngeneic bone marrow transplantation demonstrate stable integration and expression of exogenous genes (Miller *et al.*, 1984; Eglitis *et al.*, 1985; Keller *et al.*, 1985; Belmont *et al.*, 1986). The first application of retroviral gene transfer in humans utilized retroviral vectors and introduced the neomycin resistance gene into Tumor Infiltrating Lymphocytes, or TIL. The neomycin resistance gene was detected in the circulation for up to two months after infusion (Rosenberg *et al.*, 1990).

II.C. Clinical studies of marrow engraftment

In the setting of bone marrow transplantation, over 20 gene marking studies targeting bone marrow, peripheral blood progenitor cells, and peripheral blood T cells have been proposed. Protocols that have accrued patients, or are currently open, are listed in Table 1. All but one utilize neomycin resistance gene vector(s) as the marker gene. A recent study proposes to mark cells with a vector containing both the neomycin resistance gene and the thymidine kinase gene (Stewart *et al.*, 1995).

Relatively few patients have yet undergone gene marking, and only a few centers have reported their results. Results from four centers will be discussed. Three protocols evaluated patients undergoing autologous transplantation for leukemia (Deisseroth, 1991; Brenner *et al.*, 1993ab; Cornetta *et al.*, 1996), while three studies evaluated patients with neuroblastoma (Brenner, 1991ab), breast cancer, and multiple myeloma (Dunbar, 1993; Dunbar *et al.*, 1995).

Two similar vectors have been utilized in these studies, the G1Na retroviral vector created by Genetic Therapy Incorporated (Gaithersburg, Maryland, USA) and the LNL6 vector (Bender *et al.*, 1987) created by Dr. Dusty Miller (Figure 2). The vectors are based on the Moloney murine leukemia virus, altered by the removal of most of the viral genes and the insertion of the bacterial neomycin resistance gene. The vectors have a number of modifications that decrease the likelihood that transplant recipients will be exposed to replication-competent virus. These include the introduction of a stop codon at the *gag* start codon and substitution or deletion of 5' and 3' sequences to minimize homology between vector and packaging cell line genome (Miller and Buttimore, 1986). Differences in noncoding regions between LNL6 and G1NA permit the design of vector specific PCR primers enabling the vectors to be used in double marking studies.

Table 1. Gene Marking Studies in Autologous Transplantation

Investigator	Institution	Target	S/D	Population
Brenner, M.K. (1991)	St Jude, Memphis	BM	S	Acute leukemia in first remission*
Brenner, M.K. (1991)	St Jude, Memphis	BM	S	Stage D neuroblastoma*
Brenner, M.K. (1991)	St Jude, Memphis	BM	S	Neuroblastoma without marrow involvement*
Deisseroth, A.B. *et al.* (1994)	MD Anderson, Houston	BM	S	Chronic myeloid leukemia*
Cornetta, K. *et al.* (1996)	Indiana University, Indianapolis	BM	S	Adult acute leukemia*
Deisseroth, A.B. *et al.* (1994)	MD Anderson, Houston	BM/PBPC	D	Chronic myeloid leukemia*
Dunbar, C. *et al.* (1995)	NIH, Bethesda	$CD34^+$ BM/PBPC	D	Multiple myeloma*
Dunbar, C. *et al.* (1995)	NIH, Bethesda	$CD34^+$ BM/PBPC	D	Breast cancer*
Schuening, F.G.	University of Wisconsin, Madison	$CD34^+$ PBPC	S	Breast cancer and Hodgkin's disease
Brenner, M.K. (1991)	St Jude, Memphis	BM	D	Purging in Stage D neuroblastoma
Brenner, M.K.	St Jude, Memphis	BM	D	Purging in acute myeloid leukemia
Heslop, H.	St Jude, Memphis	$CD34^+$ BM	D	Cytokine treatment of $CD34^+$ cells
Dunbar, C.	NIH, Bethesda	$CD34^+$ BM/PBPC	D	Chronic myeloid leukemia
Deisseroth, A.B.	MD Anderson, Houston	$CD34^+$ BM/PBPC	D	Purging in indolent lymphomas
Douer, D.	University of S. California	BM/PBPC	D	Lymphoma and breast cancer
Verfaillle, C.	University of Minnesota, Minneapolis	PBPC	-	Chronic myeloid leukemia
Bjorkstrand, B.	Karolinska Institute, Sweden	$CD34^+$ BM/PBPC	-	Multiple myeloma
Stewart, A.K.	University of Toronto, Canada	cultured BM	S	Multiple myeloma

Abbreviations: BM = Bone marrow; PBPC = Peripheral blood progenitor cells; S = Single marker vector; D = double marker vectors.
* indicates findings already published, see bibliography for reference.

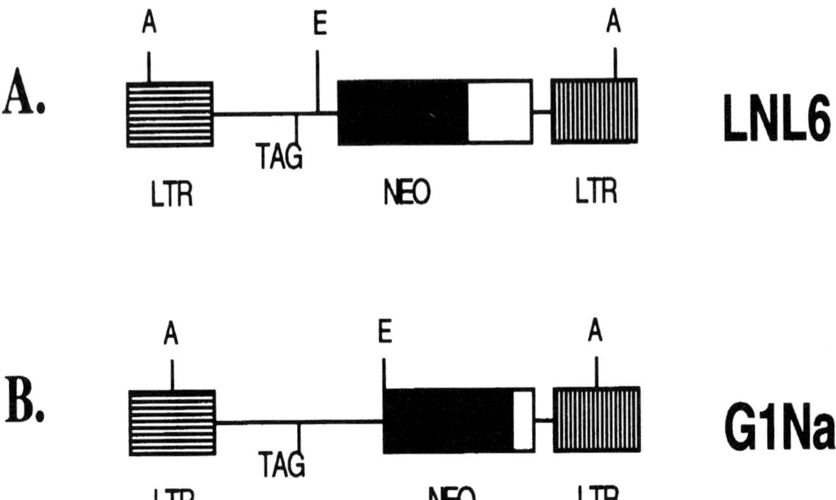

Figure 2. Retroviral vectors utilized in clinical gene marking studies. The LNL6 and G1Na retroviral vector containing the neomycin resistance gene (NEO). Abbreviations used: LTR, long terminal repeat region, Moloney murine sarcoma virus LTR (horizontal striped box), Moloney murine leukemia virus LTR (vertical striped box), neomycin resistance gene (black box), non-coding region of the neomycin resistance gene (empty box), and replacement of the gag start codon with a stop codon (TAG). Restriction enzyme sites; Asp718 (A) and EcoRI (E).

Three centers evaluating leukemic patients utilized short term supernate incubation of target cells and vector supernate (Figure 3). Preclinical studies indicated that gene transfer rates into hematopoietic cells can be increased if cytokines are included in the transduction protocol (Bodine *et al.*, 1989; Nolta and Kohn, 1990; Luskey *et al.*, 1992). However, cytokines were not included in protocols evaluating leukemic patients, since cytokines can stimulate leukemic cell growth and theoretically increase the chance of disease relapse. In these studies, gene transfer at the time of transduction ranged from 1% to 10% of committed progenitors.

In a study of patients with breast cancer, patients' progenitor cells were transduced in the presence of stem cell factor, IL-3, and IL-6 (Dunbar *et al.*, 1995.) Patients with multiple myeloma had progenitor cells transduced in the presence of two cytokines, stem cell factor and IL-3. At the time of transduction, the gene transfer rates were significantly higher in this study (18.4% in breast cancer and 23.7% in myeloma patients) than the less than 10% noted in those studies which did not utilize cytokines.

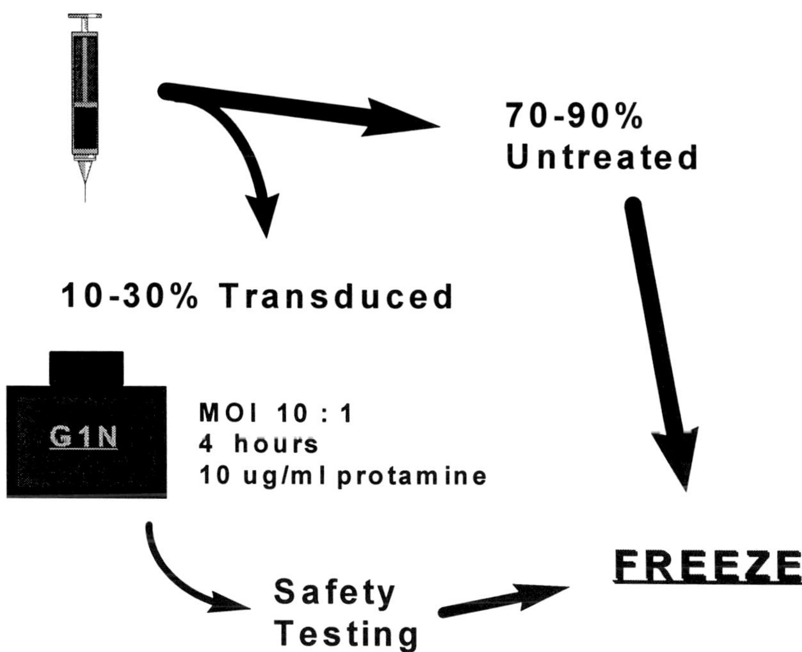

Figure 3. Transduction of bone marrow with marker genes. The figure illustrates the schema used in initial gene marking studies. Bone marrow is aspirated from the iliac crest and a portion is exposed to vector supernate with a multiplicity of infection (MOI) of 10:1. Protamine is used to increased gene transfer. Unmanipulated and vector-exposed marrow are frozen and stored in liquid nitrogen until the day of autologous transplantation.

Other modifications used by Dunbar may also have increased gene transfer, including the use of a multiple transduction regimen over a three day period and *in vivo* chemotherapy priming prior to harvesting and marking the graft cells. The rationale for *in vivo* priming is based on an observation in mice that gene transfer can be increased significantly if performed during recovery from myelosuppressive chemotherapy (Bodine *et al.*, 1991). While cytokines, multiple transduction, and *in vivo* priming are all likely to increase gene transfer into committed progenitors, the relative importance of the individual interventions is currently unknown.

The percentage of gene marked cells at the time of marrow infusion did not always correlate with the longterm engraftment of gene marked cells. Brenner noted a gene marking rate of 5% in the transplanted marrow, which appeared to persist longterm (Brenner *et al.*, 1993). In contrast, our study documented low levels of longterm marking despite having a similar initial gene transfer rate (approximately 5%) in the infused marrow (Cornetta *et al.*, 1996). This discrepancy was even more notable in the study of Dunbar *et al.*. which noted an immediate gene transfer rate of about 20%, but marked cells represented only 0.1-0.01% of marrow elements in those patients with persistent evidence of gene marked cells (Dunbar *et al.*, 1995). There are a number of possible explanations for these findings. First, the highest longterm gene transfer rates were observed in pediatric patients as discussed below (Brenner *et al.*, 1993a). Second, the addition of cytokines, or the three days of *in vitro* culture used by Dunbar, may have decreased the number of primitive stem cells infused (Dunbar *et al.*, 1995).

Differences in the proportion of residual stem cells contributing to longterm hematopoiesis may also be responsible for the variability in longterm engraftment. Our four acute myeloid leukemia (AML) patients were transplanted in second remission (Cornetta *et al.*, 1996) and, as expected, heavy pre-treatment resulted in a considerably longer time to engraftment (a mean of 58 days). Children in the Brenner study were treated in first remission and engrafted in a mean of 35 days (Brenner *et al.*, 1993a). It is possible that very few stem cells contribute to reconstitution after autologous transplantation in heavily pretreated adults, and residual stem cells may comprise a greater component of longterm hematopoiesis.

Gene marking studies have revealed some preliminary findings regarding engraftment of various progenitor cell products. Dunbar used the G1NA and LNL6 vectors to compare the engraftment of bone marrow and peripheral blood cells (Dunbar *et al.*, 1995). While there appeared to be a decreased number of marked progenitors in peripheral blood stem cell product (14.5%) as compared to bone marrow (29.2%), peripheral blood cells appear to have greater rates of longterm engraftment. In regard to marrow manipulations, CD34 selection did not appear to increase the transduction efficacy (Deisseroth *et al.*, 1994), but it does decrease the amount of vector supernate required per patient by at least ten fold.

These early studies demonstrate the usefulness of this technology in understanding the biology of autologous transplantation. All four centers note a portion of patents with gene marked cells detectable one or more years after transplantation. Persistence of gene marked cells, along with the demonstration of marking in multiple lineages (i.e. granulocytes, T cell and B cells), suggests that very primitive multipotential hematopoietic progenitor cells, and possibly stem cells, can be marked with retroviral vectors.

II.D. Gene marking and the biology of disease relapse after autologous transplantation.

Hematologic malignancies and many solid tumors invade the marrow or circulate in the blood. It is therefore possible that transplanted marrow or peripheral blood stem cells contain malignant cells that contribute to disease relapse after transplantation. Relapse may also arise from residual cancer cells that survive the transplantation preparative regimen. Understanding the contribution of each of these potential sources will be important in designing more effective transplant protocols while minimizing the significant toxicity already associated with these procedures.

The first studies to address disease relapse in autologous transplantation were performed on patients with acute leukemia. In this setting, patients are transplanted with marrow obtained in complete remission, i.e. marrow with no visible leukemia when analyzed by pathological examination. Unfortunately, this method may only detect leukemia if it is present at levels greater than 0.1-1%. Since most transplants often utilize over 10^{10} cells, it is possible that up to 10^{8} leukemic cells could inadvertently be infused along with the transplanted marrow progenitors. Because leukemia is a disease of the bone marrow, contaminating leukemic cells in transplanted marrow would intuitively seem to be an important cause of disease relapse. While methods to rid the transplanted marrow of residual acute leukemia cells ("purging") are theoretically attractive, most clinical studies have failed to show a survival advantage in patients transplanted with purged marrow (Ramsay *et al.*, 1985; Chopra *et al.*, 1991; Gulati *et al.*, 1994; Zittoun *et al.*, 1995). The lack of apparent benefit from purging has three potential explanations: there are no leukemic progenitor cells in the transplanted marrow, the purging technique is ineffective, or purging is effective but residual disease also contributes to relapse and negates the benefit of purging. Therefore, a method such as gene marking can help us make educated decisions when assessing the need and efficacy of marrow purging.

Support for marrow purging was first demonstrated by Brenner *et al.* (1993b) who noted gene marked leukemic cells at relapse in two children with AML who had been transplanted. Our institution performed a similar study in adults but did not detect marked cells at relapse. Interestingly, the marked leukemic relapses reported by Brenner *et al.* (1993b) occurred 2 and 6 months after transplantation, while our unmarked relapses occurred relatively late, 11 and 18 months after marrow infusion (Cornetta *et al.*, 1996). It is possible that late relapses may arise from residual disease, but until higher gene transfer rates are attained, the significance of unmarked relapses can not be interpreted with certainty. In contrast to acute leukemia, chronic myeloid leukemia (CML) is a disease with a significant relapse rate even after allogeneic transplantation. Deisseroth and colleagues were successful in demonstrating that infused marrow also plays a significant role in disease relapse, at least for patients with advanced stages of CML undergoing autologous BMT (Deisseroth *et al.*, 1994).

Contamination of the bone marrow graft with tumor cells is also a concern for patients with breast cancer; there are reports suggesting occult bone marrow involvement in up to 50% of patients (Cote *et al.*, 1988). As in acute leukemia, autologous transplant protocols have utilized *ex vivo* drug or monoclonal antibodies for purging (Cheson *et al.*, 1989), without apparent benefit.

In the Dunbar study, of the three breast cancer patients who relapsed at the site of the original disease, biopsies in two evaluable patients failed to detect vector sequences in the tumor mass, suggesting relapse resulted from residual tumor (Dunbar *et al.*, 1995). Since the patient numbers are small and the gene transfer rate relatively low, additional studies will be required to clearly define the need for purging in autologous transplantation for breast cancer.

III. DRUG RESISTANCE GENES IN HEMATOPOIETIC CELL TRANSPLANTATION

III.A. Overview

While gene marking studies help define the role of marrow purging in autologous transplantation, residual tumor is still felt to be a major source of relapse, particularly in solid tumors. In theory, additional chemotherapy given after autologous transplantation could eliminate residual cancer cells, thereby decreasing relapse and increasing survival. Unfortunately, since bone marrow is extremely sensitive to chemotherapeutic agents after transplantation, bone marrow suppression from these agents is severe (Paukovits *et al.*, 1993; Cooper and Einhorn, 1995). Attempts to administer chemotherapy after transplantation are therefore fraught with dose reductions and delayed dosing. In order to overcome this obstacle, several investigators are testing the hypothesis that drug resistance genes expressed in transplanted hematopoietic progenitor cells will abrogate bone marrow suppression and maximize post-transplantation therapy (Figure 4).

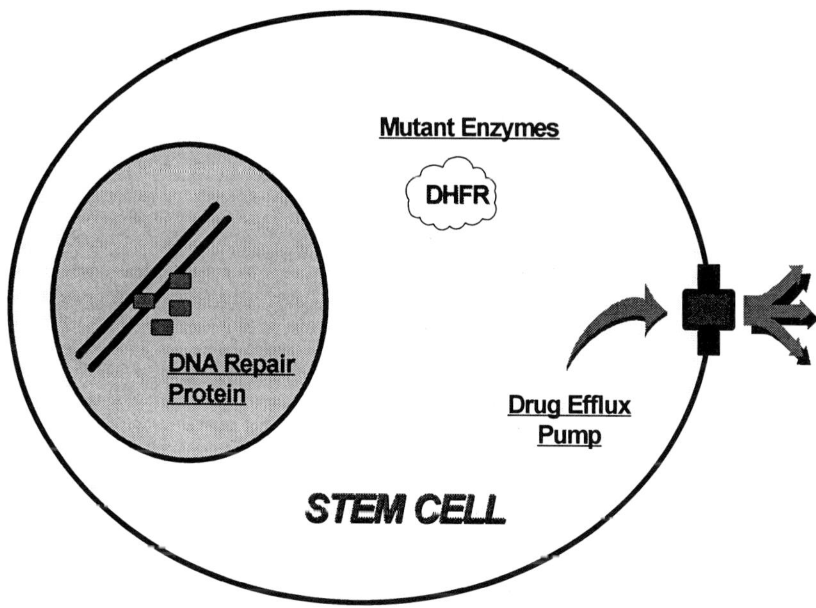

Figure 4. Strategies used to increase the drug resistance of autologous blood and bone marrow progenitor cells. Three methods are currently being used to decrease hematotoxicity of chemotherapy including transduction of progenitor cells with the Multiple Drug Resistance-1 Gene (drug efflux pump); mutant enzyme which have altered affinity for chemotherapy agents (mutant dihydrofolate reductase, DHFR); and increased expression of DNA repair proteins such as O^6-methylguanine DNA methyltransferase.

Table 2. Candidate Chemotherapy Resistance Genes

Gene	Drug
P-glycoprotein Multiple drug resistance gene (MDR-1)	Plant Alkaloids Taxol, anthracyclines
Dihydrofolate reductase	Methotrexate
DNA-repair proteins	Nitrosoureas
	Triazines
	Ionizing radiation
Glutathione-dependent enzymes	Alkylating agents
	Ionizing radiation
Cytosolic aldehyde dehydrogenase	Oxazaphosphorine

Drug resistance genes code for proteins which protect cells by a variety of diverse mechanisms and were discovered by evaluating tumor cells which exhibit drug resistance. A prominent example is the Multiple Drug Resistance Gene-1 (MDR-1), a human gene encoding the p-glycoprotein, a drug efflux pump which is normally expressed in only a select subset of tissues, but is often overexpressed in drug resistant tumor cells.

Another mode of resistance arises from mutant forms of the enzyme dihydrofolate reductase that are less sensitive to inhibition by the chemotherapeutic agent methotrexate than the naturally occurring enzyme. Yet another mechanism conferring drug resistance is the overexpression of DNA-repair proteins to correct damages that result from exposure to ionizing radiation and alkylating agents. The list of enzymes and proteins that prevent the toxic effects of a variety of chemotherapeutic agents is expanding daily (Table 2). Both the cloning of drug resistance genes and our ability to express them in hematopoietic cells have set the stage for several clinical trials.

III.B. The Multiple Drug Resistance Gene-1

The human Multiple Drug Resistance Gene-1 (MDR-1) encodes a 170 kDa p-glycoprotein which functions as an ATP-dependent transmembrane efflux pump for a wide variety of hydrophobic drugs including anthracyclines, *Vinca* alkaloids, epipodophyllotoxins, dactinomycin and taxol (Pastan and Gottesman, 1991). Expression of endogenous MDR-1 normally is limited to a small number of cell types. Interestingly, primitive hematopoietic progenitor cells express MDR-1, which presumably contributes to their known chemoresistance (Chaudhary and Roninson 1991). Its extremely low level of expression in committed marrow precursors correlates with the known sensitivity of this marrow compartment to myelosuppressive chemotherapy.

Introduction of the MDR-1 cDNA into transgenic mice or murine hematopoietic progenitor cells via retroviral gene transfer has lead to chemo-protection *in vitro* and *in vivo* (Galski *et al.* 1989; McLachlin *et al.* 1990; Choi *et al.* 1991; Sorrentino *et al.* 1992). In the setting of murine syngeneic transplantation, Licht *et al.* (1995) detected human MDR-1 DNA in up to 78% of peripheral blood cells following transplantation with vector-transduced marrow. Vector was detectable for up to 6 months after transplantation, but longterm vector expression was variable and short-lived in the absence of *in vivo* selection.

Sorrentino *et al.* (1995) and Podda *et al.* (1992) have demonstrated *in vivo* selection and longterm expression of MDR-1 transduced murine bone marrow cells by administering taxol to mice after transplantation.

Sorrentino *et al.* (1995) detected expression of human MDR-1 in both committed and primitive hematopoietic cells for 8 months after transplantation. Moreover, the expression of human MDR-1 was four times that of murine MDR-1 in primitive bone marrow cells, as judged by quantitative RT-PCR. This finding raises the possibility that MDR-1 expressing vectors can confer added drug resistance to cells constitutively expressing the MDR-1 gene.

Richardson and Bank (1995) addressed longterm engraftment and sustained expression of MDR-1 transduced cells in a series of elegant experiments. They first demonstrated efficiently gene transfer into primitive hematopoietic progenitors by transducing the pHaMDR1/A vector into isolated hematopoietic fetal liver cells, a highly enriched population of very primitive hematopoietic progenitors. The observation of MDR-1 expressing leukocytes one year after transplantation indicated longterm reconstitution and successful retroviral gene transfer into this highly enriched progenitor population.

These investigators also demonstrated the feasibility of *ex vivo* selection by transplanting mice with transduced unselected marrow or transduced marrow expressing the p-glycoprotein purified by flow cytometric cell sorting. At one year, the MDR-1 gene was expressed in 0-5% (median 1.3%) of granulocytes recovered from mice receiving unselected cells while 4-50% (median 20.5%) of circulating granulocytes expressed MDR-1 in mice transplanted with sorted cells. The increased level of MDR-1 gene expression was noted in the absence of any *in vivo* selection. It is possible that the combination of *ex vivo* and *in vivo* selection may lead to better engraftment with drug-resistant cells (Richardson and Bank, 1995; Sorrentino *et al.*, 1995).

The pHaMDR1/A vector has also been used to transduce both low density and CD34$^+$ human bone marrow and cord blood cells with the MDR-1 gene (Bertolini *et al.*, 1994; Ward *et al.*, 1994). Vector expressing cells have been successfully enriched with a multitude of selective agents, including doxorubicin, colchicine and taxol.

Ward *et al.* (1994) demonstrated the enhanced gene transfer of the pHaMDR1/A vector into human CD34$^+$ marrow cells in the presence of IL-3, IL-6, and stem cells factor. Their findings were incorporated into a clinical trial utilizing pHaMDR1/A-modified bone marrow cells in the treatment of women with breast and ovarian cancer (Hesdorffer *et al.*, 1994) and similar studies are underway at the M.D. Anderson Cancer Center in Houston. Indiana University is utilizing this approach for patients with testicular cancer. The general scheme of the clinical trials is shown in Figure 5.

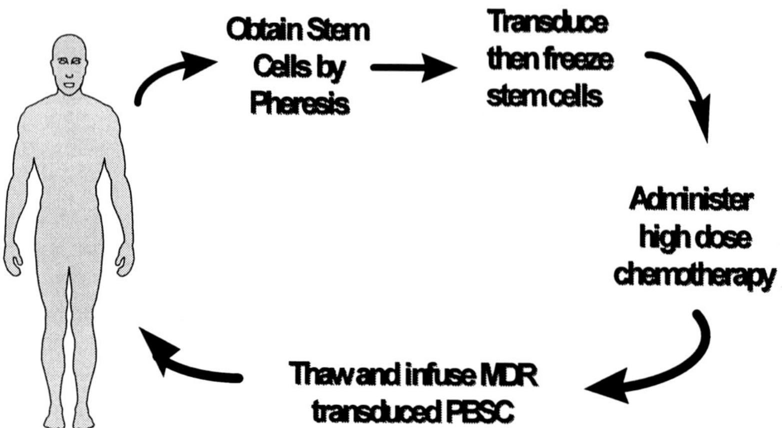

Figure 5. Clinical schema for patients participating in autologous peripheral blood stem transplantation aiming to introduce the Multiple Drug Resistance-1 Gene. Stem cells are obtained by pheresis and transduced with retroviral vector then frozen. After administration of chemotherapy the stored transduced stem cells are infused into the patient. Post-transplantation chemotherapy is administered and patients are monitored for evidence of decreased chemotherapy-related toxicity.

III.C. Dihydrofolate reductase

The 22 kDa dihydrofolate reductase (DHFR) enzyme catalyzes the conversion of dihydrofolate to tetrahydrofolate. Tetrahydrofolate functions as a one carbon group carrier and is required for the synthesis of purine and thymidine nucleotides. Methotrexate is a folate analog that competitively binds to DHFR and inhibits its activity, thus arresting cellular proliferation. Resistance to methotrexate has been reported to occur by a number of mechanisms, including gene amplification, DHFR overexpression, and by mutations in the DHFR gene (McIvor and Simonsen, 1990). DHFR mutations retain catalytic activity but exhibit decreased binding affinity to methotrexate.

Although methotrexate has significant anti-cancer activity, the dose-limiting toxic effects of myelosuppression limits its usefulness. The murine mutant DHFR Arg22, in which arginine replaces leucine in position 22, was one of the first drug resistance genes used to successfully increase the chemoresistance of hematopoietic progenitor cells. Williams *et al.* were able to demonstrate *in vitro* and *in vivo* protection of murine hematopoietic progenitors utilizing a retroviral vector containing the

Arg22 DHFR gene (Williams *et al.*, 1987; Corey *et al.*, 1990). More recently, the Ser31 DHFR mutant has been identified in which serine replaces phenylalanine at position 31. The transduction of murine hematopoietic cells with this mutant was associated with increased resistance to methotrexate (Banerjee *et al.*, 1994; Li *et al.*, 1994). A third mutant, Trp31 DHFR has *in vitro* activity but proved to be inferior to Arg22-transduced marrow in conferring *in vivo* methotrexate resistance in mice (May *et al.*, 1995) Recent work (Vinh and McIvor, 1993) reemphasizes the potential role of *in vivo* selection on sustained gene expression. Murine hematopoietic stem cell transfected with retroviral vectors containing the Arg22 DHFR mutant successfully repopulated lethally irradiated syngeneic mice, as determined by Southern blot analysis, but vector expression was observed only in those mice treated with methotrexate post transplantation.

Drug resistance has been demonstrated in human hematopoietic cells lines using retroviral vectors expressing DHFR mutants, including the newly described Tyr22 DHFR mutant (Braun *et al.*, 1997). Recently, Flasshove *et al.* (1995) successfully transduced human CD34$^+$ cells with the Ser31 DHFR mutant. Umbilical cord or mobilized peripheral blood stem CD34$^+$ cells were incubated with IL-1, IL-3, and c-*kit*-ligand prior to gene transfer. Seventy percent of progenitor colonies were shown to contain vector sequences and transduced cells were resistant to the cytotoxic effects of methotrexate.

III.D. Methylguanine methyltransferase

Chloroethylnitrosoureas such as CCNU (3-cyclohexyl-1-chloroethyl-1-nitrosourea) and BCNU (1,3-bis(2-chloroethyl)-1-nitrosourea) alkylate DNA at the O6 position of guanine which, through intermediary molecular rearrangements, forms an interstrand crosslink between guanine and its corresponding cytosine (Kohn *et al.*, 1981; Tong *et al.*, 1981,1982). The resulting crosslinking leads to disruption in DNA processing and synthesis, and eventually leads to cell death in susceptible tissues (Gralla *et al.*, 1987; Jiang *et al.*, 1989; Pieper *et al.*, 1989). The cellular DNA repair protein O6-methylguanine DNA methyltransferase (MGMT) protects cells from the toxic effects of chloroethylnitrosoureas by preventing interstrand crosslinking (Day *et al.*, 1980ab; Erickson *et al.*, 1980). Elevated MGMT activity has been reported in a variety of cancers, including colon, breast, lung, gliomas, sarcomas and acute myeloid leukemias. The elevated levels of MGMT correlate with drug resistance in *in vitro* assays (Pegg, 1990; Pegg and Byers, 1992) and human tumor xenograft models (Gerson *et al.*, 1993).

Human and murine hematopoietic progenitors express a very low level of MGMT, making these cells vulnerable to the toxic effects of nitrosoureas. Indeed, myeloid suppression is the dose-limiting toxic effects of these agents, which tend to increase with successive cycles of therapy. Moritz *et al.* (1995) have recently demonstrated protection from BCNU-induced hematotoxicity in lethally irradiated syngeneic mice transplanted with marrow containing a retroviral vector expressing the human MGMT protein. Multiple infusions of MGMT-infected marrow cells into non-lethally irradiated animals alleviated the hematological toxicity but to a lesser degree.

BCNU-resistance was not significantly increased in the MGMT-treated group until six weeks after transplantation. Non-ablative radiation given prior to infusion of transduced cells did prolong engraftment of MGMT vector-containing cells and protected animals from a lethal dose of BCNU. This survival advantage was associated with molecular and chemical evidence of MGMT in hematopoietic cells (Maze *et al.*, 1996). Similar results using retroviral vector to confer MGMT to hematopoietic elements and provide *in vivo* protection against BCNU were reported by Gerson and coworkers (Allay *et al.*, 1995).

IV. DISADVANTAGE OF RETROVIRAL MEDIATED GENE TRANSFER

Integration of retroviral vectors into target cells requires cell division. Metabolically active cells (cells residing in the active phases of cell cycle, S and G2+M) integrate retroviral vectors more efficiently than resting cells, such as those in G0 or G1 (Miller *et al.*, 1990). This requirement likely poses the major disadvantage to the use of retroviral mediated gene transfer in bone marrow transplantation, since stem cells are characteristically a very quiescent population. What factors induce a stem cell to undergo self-renewal without irreversibly committing these cells to unilineage differentiation is unknown. Identifying these factors or culture conditions remains a major area of research emphasis, both in regard to gene therapy of hematopoietic disorders and for *ex vivo* expansion of marrow cells.

V. OVERCOMING TRANSDUCTION INEFFICIENCY

Investigators have pursued a number of approaches aimed at increasing gene transfer efficiency. Cell cycling is required for retroviral vector integration, and cytokines have been used to stimulate cell division in hematopoietic cells (Bodine *et al.*, 1989; Nolta and Kohn, 1990; Luskey *et al.*, 1992; Cornetta *et al.*, 1996). Many variations in the transduction protocol have been evaluated, including alterations in temperature, media composition, polycation concentration and the use of centrifugation procedures aimed at increasing vector-cell interactions (Kotani *et al.*, 1994; Cornetta and Anderson, 1989). Unfortunately, most of these modifications are difficult to apply to large scale clinical applications.

Co-culture of target cells and vector producer cell lines has long been known to provide increased gene transfer rates when compared to the use of cell-free vector supernates. Unfortunately, the establishment of stroma required for transducing the large number of cells used in bone marrow transplantation poses significant technical problems. In addition, co-culture raises a number of safety issues, in particular potential contamination with a number of exogenous agents, including replication competent retrovirus (Cornetta, 1992).

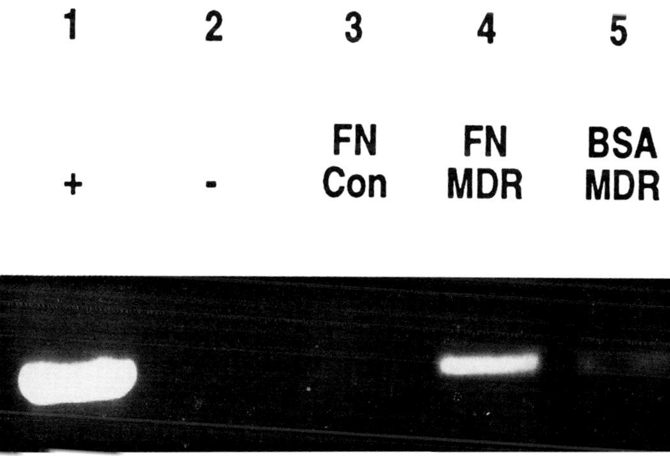

Figure 6. PCR analysis for the presence of the MDR gene of longterm cultured peripheral blood CD34$^+$ cells (PBSC) demonstrating increased gene transfer in the presence of a recombinant fibronectin fragment (FN). PCR were performed on cells incubated for 7 days in non-stromal based longterm culture. Lane 1: MDR-1 containing plasmid; Lane 2: Reagent negative control; Lane 3: Mock transduction of PBSC; Lane 4: PBSC transduced with the AM12M1 retroviral supernate containing the MDR-1 gene in plates coated with recombinant FN; Lane 5: PBSC with the AM12M1 retroviral supernate in plates coated with bovine serum albumin *(reprinted, with permission, from Traycoff et al., 1997).*

Recently, Williams and co-workers at Indiana University described a marked increase in gene transfer efficiency into primitive and committed human and murine progenitors when supernate gene transfer was performed in the presence of the 30/35kDa fragment of human fibronectin (Moritz *et al.*, 1994). They hypothesized that the matrix protein fibronectin may play a role in the increased gene transfer rates associated with co-cultivation procedures. In a subsequent publication they demonstrate that co-localization of marrow cell and vector particles is the mechanism by which increased gene transfer is achieved (Hanenberg *et al.*, 1996). This finding has been confirmed by our group, using normal and leukemic progenitors from patients with chronic myeloid leukemia (Traycoff *et al.*, 1997) and mobilized peripheral blood stem cells obtained from normal donors (Figure 6). Since a recombinant fibronectin fragment is now available (Takara, Japan), this new approach may significantly increase gene transfer rates while maximizing the safety of the procedure.

VI. CONCLUSION

Gene marking studies have begun to make important contributions to our understanding of transplantation biology. The finding of gene marked cells in multiple hematopoietic lineages greater than one year post-transplant confirms the role of transplanted autologous marrow to longterm hematopoiesis. In addition, the finding of marked relapses in children with acute myeloid leukemia and adults with chronic myeloid leukemia emphasizes the need for marrow purging in this population.

Studies currently underway should further our understanding of hematopoiesis following transplantation. Double marking studies will help define the contribution of bone marrow and peripheral blood progenitor cells to short and longterm engraftment. Gene marking has the potential to demonstrate the ability, or inability, of cytokines and *ex vivo* expansion to stimulate proliferation while maintaining the self-renewal potential of marrow stem cells. Although low gene transfer rates remain the major obstacle to retroviral gene transfer into marrow elements, recent manipulations, such as the use of fibronectin fragments to increase the efficiency of supernate transduction, hold promise for providing efficient gene transfer into the stem cell population. Clinical trials introducing drug resistance genes into hematopoietic cells may improve our management of several solid tumors. Finally, knowledge gained in autologous transplantation for malignancies will also provide essential information for investigators interested in treating genetic diseases of the bone marrow using retroviral gene transfer technology.

VII. REFERENCES

Allay, J. A., Dumenco, L. L., Koc, O. N., Liu, L. and Gerson, L. S. (1995) Retroviral transduction and expression of the human alkyltransferase cDNA provides nitrosourea resistance to hematopoietic cells. *Blood* **85**:3342-3351.

Banerjee, D., Schweitzer, B.I., Volkenandt, M., Li, M. X., Waltham, M., Mineishi, S., Zhao, S.C. and Bertino J.R. (1994) Transfection with cDNA encoding Ser 31 or Ser 34 mutant human dihydrofolate reductase into Chinese hamster ovary and mouse marrow progenitor cells confers methotrexate resistance. *Gene* **139**:269-274.

Belmont, J. W., Henkel-Tigges, J., Chang, S. M. W., Wager-Smith, K., Kellems, R. E., Dick, J. E., Magli, M. C., Phillips, R. A., Bernstein, A. and Caskey, C. T. (1986) Expression of human adenosine deaminase in murine haematopoietic progenitor cells following retroviral transfer. *Nature* **322**:385-387.

Bender, M. A., Palmer, T. D., Gelinas, R. E. and Miller, A. D. (1987) Evidence that the packaging signal of Moloney murine leukemia virus extends into gag region. *J. Virol.* **61**:1639-1646.

Berenson, R. J., Bensinger, W. I., Hill, R. S., Andrews, R. G., Garcia-Lopez, J., Kalamasz, D. F., Still, B. J., Spitzer, G., Buckner, C. D., Bernstein, I. D. and Thomas, E. D. (1991) Engraftment after infusion of CD34$^+$ marrow cells in patients with breast cancer or neuroblastoma. *Blood* **77**:1717-1722.

Bertolini, F., de Monte, L., Corsini, C., Lazari, L., Lauri, E., Soligo, D., Ward, M., Bank, A. and Malavasi, F. (1994) Retrovirus-mediated transfer of the multidrug resistance gene into human haemopoietic progenitor cells. *Br. J. Haematol.* **88**:318-324.

Bodine, D. M., Karlsson, S. and Nienhuis, A. W. (1989) Combination of interleukin-3 and 6 preserves stem cell function in culture and enhances retrovirus-mediated gene transfer into hematopoietic stem cells. *Proc. Natl. Acad. Sci. USA* **86**:8897-8901.

Bodine, D. M., McDonagh, K. T., Seidel, N. E. and Nienhuis, A. W. (1991) Survival and retrovirus infection of murine hematopoietic stem cells in vitro: Effects of 5-FU and method of infection. *Exper. Hematol.* **19**:206-212.

Braun, S. E., McIvor, R. S., Davidson, A. S., Hanna, M., Traycoff, C. M., Berebetsky, D. A., Gonin, R., Broxmeyer, H. E. and Cornetta, K. (1997) Retroviral-mediated gene transfer of Arg22 and Tyr22 forms of dihydrofolate reductase into the hematopoietic cell line K562: A comparison of methotrexate resistance. *Cancer Gene Ther.* **4**:26-32.

Bregni, M., Siena, S., Magni, M., Bonadonna, G. and Gianni, A. M. (1991) Circulating hemopoietic progenitors mobilized by cancer chemotherapy and by rhGM-CSF in the treatment of high-grade non-Hodgkin's lymphoma. *Leukemia* **5**:123-127.

Brenner, M. K. (1991a) A phase I trial of high dose carboplatin and etoposide with autologous marrow support for treatment of relapsed/refractory Neuroblastoma without apparent bone marrow involvement. *Hum. Gene Ther.* **2**:273-286.

Brenner, M. K. (1991b) A phase I trial of high dose carboplatin and etoposide with autologous marrow support for treatment of Stage D Neuroblastoma in first remission. *Hum. Gene Ther.* **2**:257-272.

Brenner, M. K., Rill, D. R., Holladay, M. S., Heslop, H. E., Moen, R. C., Buschle, M., Krance, R. A., Santana, V. M., Anderson, F. A. and Ihle, J. N. (1993a) Gene marking to determine whether autologous marrow infusion restores longterm haemopoiesis in cancer patients. *Lancet* **342**:1134-1137.

Brenner, M. K., Rill, D. R., Moen, R. C., Krance, R. A., Mirro, J., Jr., Anderson, W. F. and Ihle, J. N. (1993b) Gene-marking to trace origin of relapse after autologous bone-marrow transplantation. *Lancet* **341**:85-6.

Broun, E. R., Srour, E. F., Traycoff, C., Sledge, G., Schacht, B., Burgess, J., Cornetta, K., Mills, B. and Oldham, F. (1994) Comparison of engraftment between autologous bone marrow and CD34$^+$ cells selected with the Isolex 300. *Exper. Hematol.* **22**:362a.

Chaudhary, P. M. and Roninson, I. B. (1991) Expression and activity of P-glycoprotein, a multidrug efflux pump, in human hematopoietic stem cells. *Cell* **66**:85-94.

Cheson, B. D., Lacerna, L., Leyland-Jones, B., Sarosy, G. and Wittes, R. E. (1989) Autologous bone marrow transplantation. Current status and future directions. *Ann. Int. Med.* **1**:51-65.

Choi, K., Frommel, T. O., Kaplan Stern, R., Perez, C. F., Kriegler, M., Tsuruo, T. and Roninson, I. B. (1991) Multidrug resistance after retroviral transfer of the human MDR1 gene correlates with P-glycoprotein density in the plasma membrane and is not affected by cytotoxic selection. *Proc. Natl. Acad. Sci. USA* **88**:7386-7390.

Chopra, R., Goldstone, A. H. and McMillan, A. K. (1991) Successful treatment of acute myeloid leukemia beyond first remission with autologous bone marrow transplantation using busulfan/cyclophosphamide and unpurged marrow: The British Autograft Group Experience. *J. Clin. Oncol.* **9**:1840-1847.

Cooper, M. A. and Einhorn, L. H. (1995) Maintenance chemotherapy with daily oral VP-16 following salvage therapy in patients with germ cell tumors. *J. Clin. Oncol.* **13**:1167-1169.

Corey, C. A., DeSilva, A. D., Holland, C. A. and Williams, D. A. (1990) Serial transplantation of methotrexate-resistant bone marrow: Protection of murine recipients from drug toxicity by progeny of transduced stem cells. *Blood* **75**:337-343.

Cornetta, K. and Anderson, W. F. (1989) Protamine sulfate as an effective alternative to polybrene in retroviral-mediated gene transfer: Implications for human gene therapy. *J. Virol. Methods* **23**:187-194.

Cornetta, K. (1992) Safety aspects of human gene therapy. *Brit. J. Haematol.* **80**:421-426.

Cornetta, K., Nguyen, N., Morgan, R. A., Muenchau, D. D., Hartley, J. and Anderson, W. F. (1993) Infection of human cells with murine amphotropic replication-competent retroviruses. *Hum. Gene Ther.* **4**:579-588.

Cornetta, K., Moore, A., Leemhuis, T., Moen, R. C., Tricot, G., Leibowitz, D. and Hoffman, R. (1994) Retroviral mediated gene transfer in chronic myelogenous leukemia. *Br. J. Haematol.* **87**:308-316.

Cornetta, K., Srour, E. F., Moore, A., Davidson, A., Broun, E. R., Hromas, R., Moen, R. C., Morgan, R. A., Rubin, L., Anderson, W. F., Hoffman, R. and Tricot, G. (1996) Retroviral gene transfer in autologous bone marrow transplantation for adult acute leukemia. *Hum. Gene Ther.* **7**:1323-1329.

Cote, R. J., Rosen, P. P., Hakes, T. B., Sedira, M., Bazinet, M., Kinne, D. W., Old, L. J. and Osborne, M. P. (1988) Monoclonal antibodies detect occult breast carcinoma metastases in the bone marrow of patients with early stage disease. *Amer. J. Surg. Pathol.* **12**:333-340.

Day, R., Ziolkowski, C., Scudiero, D., Meyer, S., Lubinielki, A., Girardi, A., Galoway, S. and Bynum, G. (1980a) Defective repair of alkylated DNA by human tumor and SV40 transformed human cell strains. *Nature* **288**:724-727.

Day, R., Ziolkowski, C., Scudiero, D., Meyer, S. and Mattern, M. (1980b) Human tumor cell strain defective in the repair of alkylation damage. *Carcinogenesis* **1**:21-32.

Deisseroth, A. B. (1991) Autologous bone marrow transplantation for chronic myelogenous leukemia in which retroviral markers are used to discriminate between relapse which arises from systemic disease remaining after preparative therapy versus relapse due to residual leukemic cells in autologous marrow: A pilot study. *Hum. Gene Ther.* **2**:359-376.

Deisseroth, A. B., Zu, Z., Claxton, D., Hanania, E. G., Fu, S., Ellerson, D., Goldberg, L., Thomas, M., Janicek, K., Anderson, W. F., Korbling, M., Durett, A., Moen, R., Berenson, R., Heimfield, S., Hamer, J., Calvert, L., Tibbits, P., Talpaz, M., Kantarjian, H., Champlin, R., and Reading, C. (1994) Genetic marking shows that Ph$^+$ cells present in autologous transplants of chronic myelogenous leukemia (CML) contribute to relapse after autologous bone marrow in CML. *Blood* **83**:3068-76.

Dunbar, C. E. (1993) Genetic marking with retroviral vectors to study the feasibility of stem cell gene transfer and the biology of hematopoietic reconstitution after autologous transplantation in multiple myeloma, chronic myelogenous leukemia or metastatic breast cancer. *Hum. Gene Ther.* **4**:205-222.

Dunbar, C. E., Cottler-Fox, M., O'Shaughnessy, J. A., Doren, S., Carter, C., Berenson, R., Brown, S., Moen, R. C., Greenblatt, J., Stewart, F. M., Leitman, S. F., Wilson, W. H, Cowan, K., Young, N. S. and Nienhuis, A. W. (1995) Retrovirally marked CD34-enriched peripheral blood and bone marrow cells contribute to longterm engraftment after autologous transplantation. *Blood* **85**:3048-3057.

Eglitis, M. A., Kantoff, P., Gilboa, E. and Anderson, W. F. (1985) Gene expression in mice after high efficiency retroviral-mediated gene transfer. *Science* **230**:1395-1398.

Erickson, L. C., Laurent, G., Sharkey, N. and Kohn, W. (1980) DNA crosslinks and monoadduct repair in nitrosoureas treated human cells. *Nature* **288**:727-729.

Fisher, B., Packard, B. S. and Read, E. J. (1989) Tumor localization of adoptively transferred Indium-111 labeled tumor infiltrating lymphocytes in patients with metastatic melanoma. *J. Clin. Oncol.* **7**:250-261.

Flasshove, M., Banerjee, D., Bertino, J. R. and Moore, M. A. S. (1995) Increased resistance to methotrexate in human hematopoietic cells after gene transfer of the Ser 31 DHFR mutant. *Leukemia* **9** (Suppl 1):S34-S37.

Galski, H., Sullivan, M., Willingham, M. C., Chin, K., Gottesman, M. M., Pastan, I. and Merlino, G. T. (1989) Expression of a human multidrug resistance cDNA (MDR1) in the bone marrow of transgenic mice: Resistance to daunomycin-induced leukopenia. *Mol. Cell. Biol.* **9**:435-4363.

Gerson, S. L., Zborowska, E., Norton, K., Gordon, N. H. and Willson, J. K. W. (1993) Synergistic efficacy of O6-benzylguanine and 1,3-bis(2-chloroethyl)-1-nitrosourea (BCNU) in a human colon cancer xenograft completely resistant to BCNU alone. *Biochem. Pharmacol.* **45**:483-491.

Gralla, J., Sasse-Dwight, S. and Poljak, L. (1987) Formation of blocking lesion at identical sequences by the nitrosourea and platinum class of anticancer drugs. *Cancer Res.* **47**:5092-5096.

Griffith, K. D., Read, E. J. and Carrasquillo, J. A. (1989) *in vivo* distribution of adoptively transferred Indium-111-labeled tumor infiltrating lymphocytes and peripheral blood lymphocytes in patients with metastatic melanoma. *J. Natl. Cancer Inst.* **81**:1709-1717.

Gulati, S. C., Romero, C. E. and Ciavarella, D. (1994) Is bone marrow purging proving to be of value? *Oncology* **8**:19-24.

Hanenberg, H., Xiao, X. L., Dilloo, D., Hashino, K., Kato, I. and Williams, D. A. (1996) Colocalization of retrovirus and target cells on specific fibronectin fragments increases genetic transduction of mammalian cells. *Nature Med.* **2**:1-6.

Hesdorffer, C., Antman, K., Bank, A., Fetell, M., Mears, G. and Begg, M. (1994) Human MDR gene transfer in patients with advanced cancer. *Hum. Gene Ther.* **5**:1151-1160.

Hogge, D. E. and Humphries, R. K. (1987) Gene transfer to primary normal and malignant human hematopoietic progenitors using recombinant retroviruses. *Blood* **69**:611-617.

Hohaus, S., Goldschmidt, H., Ehrnardt, R. and Haas, R. (1993) Successful autografting following myeloablative conditioning therapy with blood stem cells mobilized by chemotherapy plus rhG-CSF. *Exper. Hematol.* **21**:508-514.

Horowitz, M. M. (1995) New IBMTR/ABMTR slides summarize current use and outcome of allogeneic and autologous transplants. *IBMTR Newsletter* **2**:1-8.

Jiang, B., Bawr, B. and Hsiang, Y. (1989) Lack of drug induced DNA crosslinking in chlorambucil resistant Chinese hamster ovary cells. *Cancer Res.* **44**:5514-5517.

Keller, G., Paige, P., Gilboa, E. and Wagner, E. F. (1985) Expression of a foreign gene in myeloid and lymphoid cells derived from multipotent haematopoietic precursors. *Nature* **318**:149-154.

Kohn, K. W., Erickson, L. C. and Laurent, G. L. (1981) DNA alkylation, crosslinking and repair. In *Nitrosoureas in cancer treatment, INSERM Symposia No. 19*. New York: Elsevier Press, 33-48.

Kotani, H., Newton, P. B., Zhang, S., Chiang, Y. L., Weaver, L., Blaese, R. M., Anderson, W. F., and McGarrity, G. J. (1994) Improved methods of retroviral vector transduction and production for gene therapy. *Hum. Gene Ther.* **5**:19-28.

Li, M. X., Banerjee, D., Zhao, S. C., Schweitzer, B. I., Mineishi, S., Gilboa, E. and Bertino, J.R. (1994) Development of a retroviral construct containing a human mutated dihydrofolate reductase cDNA for hematopoietic stem cell transduction. *Blood* **83**:3403-3408.

Licht, T., Aksentijevich, I., Gottesman, M. M. and Pastan, I. (1995) Efficient expression of functional human MDR1 gene in murine bone marrow after retroviral transduction of purified hematopoietic stem cells. *Blood* **86**:111-121.

Luskey, B. D., Rosenblatt, M., Zsebo, K. and Williams, D. A. (1992) Stem cell factor, IL-3 and IL-6 promote retroviral-mediated gene transfer into murine hematopoietic stem cells. *Blood* **80**:396-402.

Masaoka, T., Takaku, F., Kato, S., Moriyama, Y., Kodera, Y., Kanamaru, A., Shimosaka, A., Shibata, H. and Nakamura, H. (1989) Recombinant human granulocyte colony-stimulating factor in allogeneic bone marrow transplantation. *Exper. Hematol.* **17**:1047-1050.

May, C., Gunther, R. and McIvor, R. S. (1995) Protection of mice from lethal doses of methotrexate by transplantation with transgenic marrow expressing drug-resistant dihydrofolate reductase activity. *Blood* **86**:2439-2448.

Maze, R., Carney, J. P., Kelley, M. R., Glassner, B.J., Williams, D.A. and Samson, L. (1996) Increasing DNA repair methyltransferase levels via bone marrow stem cell transduction rescues mice from the toxic effects of 1,3-bis(2-chloroethyl)-1-nitrosourea, a chemotherapeutic alkylating agent. *Proc. Natl. Acad. Sci. USA* **93**:206-210.

McIvor, R. S. and Simonsen, C. C. (1990) Isolation and characterization of an altered dihydrofolate cDNA from methotrexate-resistant mouse L5178Y cells. *Nucleic Acids Res.* **18**:7025-7032.

McLachlin, J. R., Eglitis, M. A., Ueda, K., Kantoff, P. W., Anderson, W. F. and Gottesman, M. M. (1990) Expression of a human complementary DNA for the multidrug resistance gene in murine hematopoietic precursor cells with the use of retroviral gene transfer. *J. Natl. Cancer Inst.* **82**:1260-1263.

Miller, A. D., J., E. R. and Jolly, D. J. (1984) Expression of a retrovirus encoding human HPRT in mice. *Science* **230**:1395-1398.

Miller, A. D. and Buttimore, C. (1986) Redesign of retrovirus packaging cell lines to avoid recombination leading to helper virus production. *Molec. Cell. Biol.* **6**:2895-2902.

Miller, D. G., Mohammed, A. D. and Miller, A. D. (1990) Gene transfer by retrovirus vector occurs only in cells that are actively replicating at the time of infection. *Molec. Cell. Biol.* **8**:4239-4242.

Moore, M. A. S. (1993) *Ex vivo* expansion and gene therapy using cord blood CD34+ cells. *J. Hematotherapy* **2**:221-224.

Moritz, T., Mackay, W. and Glassner, B. J. (1995) Retrovirus-mediated expression of a DNA repair protein in bone marrow protects hematopoietic cells from nitrosourea-induced toxicity *in vitro* and *in vivo*. *Cancer Res.* **55**:2608-2614.

Moritz, T., Patel, V. P. and Williams, D. A. (1994) Bone marrow extracellular matrix molecules improved gene transfer into human hematopoietic cells via retroviral vectors. *J. Clin. Invest.* **93**:1451-1457.

Nemunaitis, J., Anasetti, C., Storb, R., Bianco, J. A., Buckner, C. D., Onetto, N., Martin, P., Sanders, J., Sullivan, K., Mori, M., Shannon-Dorcy, K., Bowden, R., Appelbaum, R., Hansen, J. and Singer, J. W. (1992) Phase II Trial of recombinant human granulocyte-macrophage colony-stimulating factor in patients undergoing allogeneic bone marrow transplantation from unrelated donors. *Blood* **79**:2572-2577.

Nolta, J. A. and Kohn, D. B. (1990) Comparison of the effects of growth factors on retroviral vector-mediated gene transfer and the proliferative status of human hematopoietic progenitor cells. *Hum. Gene Ther.* **1**:257-268.

Pastan, I. and Gottesman, M. M. (1991) Multidrug resistance. *Ann. Rev. Med.* **42**:277-286.

Paukovits, W. R., Moser, M. H. and Paukovits, J. B. (1993) Pre-CFU-S quiescence and stem cell exhaustion after cytostatic drug treatment: protective effects of the inhibitory peptide. *Blood* **81**:1755-1761.

Pegg, A. E. (1990) Mammalian O6-alkylguanine-DNA-alkyltransferase: regulation and importance in response to alkylating carcinogenic and therapeutic agents. *Cancer Res.* **50**:6119-6129.

Pegg, A. E. and Byers, T. L. (1992) Repair of DNA containing O6-alkylguanine. *FASEB J.* **6**:2302.

Pieper, R. O., Futscher, B. W. and Erickson, L. C. (1989) Transcription termination lesions induced by alkylating agents *in vitro*. *Carcinogenesis* **10**:1307-1314.

Podda, S., Ward, M., Himelstein, A., Richardson, C., de la Flor-Weiss, E., Smith, L., Gottesman, Pastan, I. and Bank, A. (1992) Transfer and expression of the human multiple drug resistance gene into live mice. *Proc. Natl. Acad. Sci. USA* **89**:9676-9680.

Ramsay, N., LeBien, T. and Nesbit, M. (1985) Autologous bone marrow transplantation for patients with acute lymphoblastic leukemia in second or subsequent remission: Results of bone marrow treated with BA-1, BA-2, and BA-3 with complement. *Blood* **66**:508-513.

Richardson, C. and Bank, A. (1995) Preselection of transduced murine hematopoietic stem cell populations leads to increased longterm stability and expression of the human drug resistance gene. *Blood* **86**:2579-2590.

Rosenberg, S. A., Aebersold, P. M., Cornetta, K., Kasid, A., Morgan, R. A., Moen, R., Karson, E. M., Lotze, M. T., Yang, J. C., Topalian, S. L., Merino, M. H., Culver, K., Miller, A. D., Blaese, M. D. and Anderson, W. F. (1990) Gene transfer into humans-immunotherapy of patients with advanced melanoma, using tumor infiltrating lymphocytes modified by retroviral gene transduction. *New Engl. J. Med.* **323**:570-578.

Schpall, E. J., Jones, R. B., Franklin, W. A., Archer, P. G., Curiel, T., Bitter, M., Claman, H., Bearman, S., Stemmer, S., Purdy, M., Myers, S., Hami, L., Taffs, S.,

Heimfeld, S., Hallagan, J. and Berenson, R. J. (1994) Transplantation of enriched CD34-positive autologous marrow into breast cancer patients following high-dose chemotherapy: Influence of CD34-positive peripheral blood progenitors and growth factors on engraftment. *J. Clin. Oncol.* **12**:28-36.

Smith, L. J. and Benchimol, S. (1987) Introduction of new genetic material into human myeloid leukemic blast stem cells by retroviral infection. *Molec. Cell. Biol.* **8**:974-977.

Sorrentino, B. P., Brandt, S. J., Bodine, D., Gottesman, M., Pastan, I., Cline, A. and Nienhuis, A. W. (1992) Selection of drug-resistant bone marrow cells *in vivo* after retroviral transfer of human MDR1. *Science* **257**:99-103.

Sorrentino, B. P., McDonagh, K. T., Woods, D. and Orlic, D. (1995) Expression of retroviral vectors containing the human multidrug resistance 1 cDNA in hematopoietic cells of transplanted mice. *Blood* **86**:491-501.

Stewart, A. K., Dube, I. D., Kamel-Reid, S., and Keating, A. (1995) A phase I study of autologous bone marrow transplantation with stem cell gene marking in multiple myeloma. *Hum. Gene Ther.* **6**:107-119.

Strauss, L. C., Rowley, S. D., LaRussa, V. F., Sharkis, S. J., Stuart, R. K. and Civin, C. I. (1986) Antigenic analysis of hematopoiesis. V. Characterization of My-10 antigen expression by normal lymphohematopoietic progenitor cells. *Exper. Hematol.* **14**:878-886.

Sutherland, H. J., Eaves, C. J., Eaves, A. C., Dragowska, W. and Lansdorp, P. M. (1989) Characterization and partial purification of human marrow cells capable of initiating longterm hematopoiesis in vitro. *Blood* **74**:1563-1570.

Terstappen, L. W. M. M., Huang, S., Safford, D. M., Lansdrop, P. M. and Loken, M. R. (1991) Sequential generations of hematopoietic colonies derived from single nonlineage committed CD34$^+$/CD38$^-$ progenitor cells. *Blood* **77**:1218-1227.

Tong, W. P., Kirk, M. C. and Ludlum, D. B. (1981) Molecular pharmacology of haloethylnitrosoureas: formation of 6-hydroxyethylguaninine in DNA treated with N,N'-bis(2-chloroethyl)-N-nitrosourea. *Biochem. Biophys. Res. Commun.* **100**:351-357.

Tong, W. P., Kirk, M. C. and Ludlum, D. B. (1982) Formation of crosslink 1-[N3-deoxycytidyl], 2-[N1-deoxyguanosinyl] ethane in DNA treated with N,N'-bis(2-chloroethyl)-N-nitrosourea. *Cancer Res.* **42**:3102-3105.

Traycoff, C. M., Kosak, S. T., Grigsby, S. and Srour, E. F. (1995) Evaluation of *ex vivo* expansion potential of cord blood and bone marrow hematopoietic progenitor cells using cell tracking and limiting dilution analysis. *Blood* **85**:2059-2068.

Traycoff, C., Srour, E. F., Dutt, P. and Cornetta, K. (1997) The 30/35 kD chymotrophic fragment of fibronectin enhances retroviral-mediated gene transfer in chronic myelogenous leukemia CD34$^+$ HLA-DR$^+$ and CD34$^+$ HLA-DR$^-$ bone marrow progenitors. *Leukemia* **11**:159-167.

Verfaillie, C. M. and Miller, J. S. (1995) A novel single-cell proliferation assay shows that longterm culture-initiating cell (LTC-IC) maintenance over time results from the extensive proliferation of a small fraction of LTC-IC. *Blood* **86**:2137-2145.

Vinh, D. B. N. and McIvor, R. S. (1993) Selective expression of methotrexate-resistant DHFR activity in mice transduced with DHFR retrovirus and administered methotrexate. *J. Pharm. Exper. Therap.* **267**:989-996.

Ward, M., Richardson, C., Pioli, P., Smith, L., Podda, S., Goff, S., Hesdorffer, C. and Bank, A. (1994) Transfer and expression of the human multiple drug resistance gene in human CD34$^+$ cells. *Blood* **84**:1408-1414.

Watt, S. M., Karhi, K., Gatter, K., Furley A.J., Katz, F. E., Healy, L. E., Altass L.J., Bradley N.J., Sutherland D.R., Levinsky, R., and Melvyn G.F. (1987) Distribution and epitope analysis of the cell membrane glycoprotein (HPCA-1) associated with human hemopoietic progenitor cells. *Leukemia* **1**:417-421.

Williams, D. A., Hsieh, K., DeSilva, A. and Mulligan, R. C. (1987) Protection of bone marrow transplant recipients from lethal doses of methotrexate by the generation of methotrexate-resistant bone marrow. *J. Exper. Med.* **166**:210-218.

Zittoun, R. A., Mandelli, F., Willemze, R., de Witte, T., Labar, B., Resegotti, L., Leoni, F., Damasio, E., Vasani, G., Papa, G., Caronia, F., Hayat, M., Stryckmans, P., Rotoli, B., Leoni, P., Peetermans, M. E., Dardenne, M., Vegna, M. L., Petti, M. C., Solbu, G., Suciu, S., EORTC and GIMEMA (1995) Autologous or allogeneic bone marrow transplantation compared with intensive chemotherapy in acute myelogenous leukemia. *New Engl. J. Med.* **332**:217-223.

11

Adenoviral Gene Transfer of the Herpes Virus Thymidine Kinase Gene for Treating Gliomas

Jane B. Alavi, Jason G. Smith, and Stephen L. Eck
University of Pennsylvania School of Medicine, Philadelphia, Pennsylvania

I. INTRODUCTION

Virally mediated gene transfer strategies are being widely explored for the treatment of many cancers (Jolly, 1994; Eck and Wilson, 1995; Weitzman *et al.*, 1995). Indeed such strategies are extremely flexible with respect to the function of the transferred gene and provide a seemingly limitless number of therapeutic avenues. Nonetheless, gene transfer for the treatment of cancer (or any other disease) has yet to be reduced to practice, in large part because of the limitations of the vectors employed. Several major vector design issues have been identified in preclinical work, including targeting of the vector to the tumor, obtaining efficient gene transfer, and limiting the host immune response to the vector. While significant advances have been made in addressing these problems, no vector system has overcome all of them satisfactorily.

Adenoviral vectors are among the most extensively studied vectors and offer tremendous potential because of their plasticity and safety. They are, therefore, a useful paradigm for understanding the problems associated with this new therapeutic strategy. Their application to brain tumors is a logical starting point for their clinical development because of the relatively localized nature of these tumors (which makes targeting easier) and the prior clinical experience with adenoviral infections of the

CNS. In addition, adenoviral vectors are among the most efficient vectors in *in vivo* gene transfer applications (Figure 1). Their *in vivo* gene transfer efficiency stems in part from the ability to produce very high titer stocks of recombinant adenovirus (Berkner, 1988). This has important practical considerations in that the volume occupied by a therapeutic dose of virions approximates the mass of tumor cells that are to be treated.

For brain tumors this is of obvious importance in that one is limited by the volume that one can deliver into the CNS without inducing pressure related problems. For other vector systems, the volume of the delivered vector is usually much larger than that of the target tumor. Although this can be potentially overcome by either introducing vector producer cells (Culver *et al.*, 1992) or by using a replicating virus (Valyi-Nagy *et al.*, 1994), each of these techniques imposes additional problems. For example, viral producer cells elicit a host immune response, and replicating viral vectors pose the risk of disseminated infection. For these reasons, several clinical trials using recombinant adenoviruses are now underway.

Among the therapeutic genes being studied for brain tumor applications, the prodrug activating enzymes (Eck and Wilson, 1995) have received the most attention. This approach is attractive because of its simplicity and potential applicability to other tumors. Other approaches, which may be useful in the treatment of other malignancies, such as insertion of tumor suppressor genes or immunomodulatory agents, are less attractive for the treatment of brain tumors given our current understanding of their mechanisms.

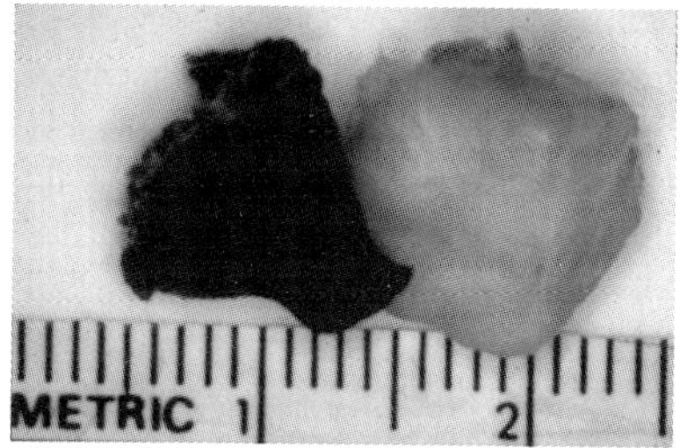
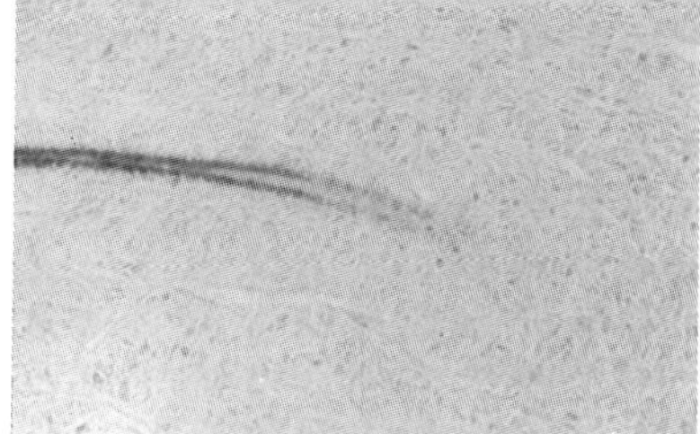

Figure 1. Adenoviral-mediated gene transfer into a solid tumor. T2994 cells were grown as a solid tumor mass in Balb/c mice and then injected with either saline or saline containing 10^9 plaque forming units (pfu) of adenovirus encoding β-galactosidase (Ad.lacZ). Three days later the tumors were removed and stained *en bloc* with X-gal (**left**). The extensive dark staining seen on microscopic examination is indicative of gene transfer throughout the entire tumor mass (**right**).

Genes that alter the cell cycle processes (e.g., *p53*) are difficult to introduce because they require the tranduction of all target cells. Immunomodulatory agents (e.g., cytokines and costimulatory molecules) by design elicit an immune response which may be associated with increasing edema and intracranial pressure. For example, intratumoral delivery of interleukin-2 can enhance the immune response to CNS tumors but also induces severe vasogenic edema (Tjuvajev *et al.*, 1995). Of the prodrug activating genes available, the Herpes simplex virus thymidine kinase gene has received the most attention (Moolten, 1986; Moolten and Wells, 1990; Ezzeddine *et al.*, 1991; Golumbek *et al.*, 1992; Takamiya *et al.*, 1992; Hasegawa *et al.*, 1993; Smythe *et al.*, 1995), and is currently in use in clinical trials (Eck *et al.*, 1996). The application of this strategy to human brain tumors is best understood by appreciating the difficulties associated with the current management of this disease.

II. CLINICAL ASPECTS OF MALIGNANT BRAIN TUMORS

Malignant brain tumors are diagnosed in approximately 8,000 adult Americans yearly, and account for about 2.5 percent of deaths from cancer (Boring *et al.*, 1993). In adults, the primary brain tumors are most often malignant gliomas. The gliomas are classified as astrocytoma (generally benign, usually found in children), anaplastic astrocytoma, and glioblastoma multiforme. Occasionally, mixed tumors of intermediate grade are seen, such as the oligodendroglioma with anaplastic astrocytoma. The cell of origin for these malignancies is the astrocyte, a supporting cell in the brain. Therefore, the tumors arise in white matter and can be found in any location, although about 90 percent are in the cerebrum. The gliomas spread locally, compressing and infiltrating normal brain. Growth into ventricles or meninges is uncommon, and metastases via blood stream or lymphatics are extremely rare (Burger *et al.*, 1991). Frontal lobe tumors may cross to the opposite side of the brain through the corpus callosum. However, the spread is contiguous and multifocal disease is not often seen. Autopsy studies show that the tumors usually remain localized throughout their course (Hochberg and Pruitt, 1980; Burger *et al.*, 1988). After surgery or radiation therapy, the recurrences are close to the original tumor bed.

The population affected by gliomas consists of all ages and races, and there is only a slight male predominance. Most of the patients are older than 50 years (Mahaley *et al.*, 1989). In the younger age group anaplastic astrocytomas are more common, but in patients over age 50 most of the tumors are glioblastomas. There is no known cause or means of prevention. The disease is devastating, due to its debilitating neurologic consequences and rapidly fatal termination. No one survives glioblastoma; only a few patients are longterm survivors of anaplastic astrocytoma. Depending on the regions of the brain affected, the patients may suffer from seizures, paralysis, incoordination, aphasia, confusion, memory loss, sensory deficits or visual loss. In addition, they usually require large doses of corticosteroids early, and late in their illness may experience the disabling side effects of this treatment such as edema, proximal myopathy, diabetes, fungal infections or deep vein thrombosis. Few patients in the older age group are able to work after the diagnosis. Most of the patients are incapable of self-care for several months before death. Survival depends on the histologic type of the tumor as well as the location in the brain and the functional ability of the patient (measured as the Karnofsky performance status). The median survival varies from 58 months for those with anaplastic astrocytoma who are less than 50 years old with normal mental status, to 18 months for those with glioblastoma less than 50 years old and normal Karnofsky status, to 9 months for those with glioblastoma older than 50 years and fairly good Karnofsky status. Unfortunately, the majority of patients fall into the last group, so that the median survival for all adults with malignant gliomas is 9-12 months (Curran *et al.*, 1993; Leibel *et al.*, 1994).

As can be surmised from the survival data, current treatment for this disease is largely ineffective. Surgery and radiation therapy are the primary modalities of treatment, but certain characteristics of the glioma render them quite resistant to cure. Surgical resection is a goal in treatment of any localized malignancy. Indeed, these tumors are often small compared to cancers elsewhere in the body (less than 5 cm in diameter at diagnosis), but they cannot be completely removed for several reasons: they lack a defined capsule or edge, they infiltrate into normal-appearing brain, and they may be located within or adjacent to critical brain regions which cannot be removed without permanently disabling the patient. Therefore, partial resection is performed in most cases. In some patients only a biopsy can be done safely due to the site of the tumor (Broca's area or the thalamus). Surgery has become safer and it is generally believed that partial resection is beneficial, because it creates some space into which the residual tumor can grow before causing pressure symptoms. Patients who are able to undergo resection appear to live longer than those whose tumors are only biopsied. When possible, second and third resections are performed for recurrent tumors in patients with good performance status.

Radiation therapy is applied to all patients with malignant glioma, except when the neurologic condition is too poor to permit it. The tumors do show response to radiation in many cases, but high doses are necessary to achieve control. The normal brain around the tumor can generally tolerate no more than 60 cGy, a dose that is below the curative level for glioma. Although a number of techniques have been devised to maximize the tumor dose (e.g., hyperfractionation, radiation-sensitizing drugs and interstitial radiation (Walker *et al.*, 1980; Scharfen *et al.*, 1992), these techniques are neither uniformly applicable nor very effective. Other approaches to glioma treatment have included chemotherapy, either as an adjuvant to radiation (Walker *et al.*, 1980) or at the time of relapse (Kornblith and Walker, 1988), and immunotherapy with interferons (Yung *et al.*, 1992), intratumoral LAK cell instillation (Merchant *et al.*, 1988), radiolabeled monoclonal antibodies, or other techniques (Jaekle, 1994). None of these methods is curative. Adjuvant chemotherapy is only helpful in a minority of patients but probably does prolong median survival at least for anaplastic astrocytoma. For the bulk of patients (mostly glioblastoma), adjuvant chemotherapy increases median survival only from 9.4 months to 12 months (Fine *et al.*, 1993). The poor results from chemotherapy may be partly due to the inability of most chemotherapy drugs to cross the intact blood-brain barrier at the edges of the tumor (Ausman *et al.*, 1977). Areas of necrosis and areas with poor blood supply within the tumor are less accessible to chemotherapy. Additionally, the tremendous heterogeneity of these tumors results in chemotherapy resistance in many of the cells. Attempts have been made to improve delivery of chemotherapy into brain tumors by giving the drugs directly into the carotid artery, by temporarily opening the blood brain barrier with hypertonic agents such as mannitol (Neuwelt *et al.*, 1986), and by implantation of chemotherapy containing polymers directly into the brain tumors (Brem *et al.*, 1991). There have been significant complications of some of these methods without consistent improvement in outcome. High dose chemotherapy with bone marrow rescue has also been utilized, but again the toxicity has been high and the results poor (Wolff *et al.*, 1987).

Despite a large body of clinical research in the treatment of malignant gliomas, there has been little overall improvement in the outcome for these patients. Gene therapy with the transfer of the drug susceptibility gene Herpes virus thymidine kinase (HSVTK) has shown promise in a number of animal models including CNS tumors. Given the extremely poor prognosis of patients with recurrent brain tumors and the development of new gene transfer technology, we developed a clinical study of HSVTK gene transfer for the treatment of adult patients with glioma. This study is designed to evaluate the use of adenovirus-mediated transfer of the HSVTK gene into primary brain tumors followed by systemic treatment with ganciclovir (GCV). While gene transfer strategies have been proposed as therapy for many malignancies, CNS tumors are particular well suited to the early application of this technology because of their relatively localized distribution, their high mitotic rates compared to the surrounding normal brain parenchyma, and their poor response to conventional therapy. Of the several genetic vectors available (Jolly, 1994), we selected the adenovirus vector system because it is the most efficient genetic vector for *in vivo* delivery and it is amenable to extensive modifications that may improve its function (Weitzman *et al.*, 1995).

III. PRECLINICAL STUDIES: THE USE OF HSVTK WITH GCV IN CANCER MODELS

The failure of conventional brain tumor therapies stems from their inability to functionally distinguish neoplastic from normal cells. Despite considerable efforts, tumor specific molecular attributes that are suitable therapeutic targets have not been identified. Creation of "artificial" differences in biochemical function is an attractive option. Insertion of the HSVTK gene into malignant cells in conjunction with the systemic administration of GCV has become a prototypic gene therapy system for the selective destruction of cancer cells (Moolten, 1986), although several other prodrug activating genes have been subsequently developed for antitumor therapy (Eck and Wilson, 1995).

GCV is an acyclic nucleoside normally not metabolized by mammalian cells; however, it is converted to the 5' monophosphate form by HSVTK. This allows it to be subsequently metabolized to the triphosphate form by mammalian cellular kinases. GCV-triphosphate is a toxic nucleotide analog competing with normal nucleotides in DNA replication. This toxicity, which consists of inhibition of cell growth as well as cell killing, is thought to be due to inhibition of somatic mammalian cell replication mechanisms. Unlike conventional chemotherapy, a large therapeutic index separates HSV-TK$^+$ cells from HSVTK$^-$ cells in the presence of GCV.

Many investigators have shown that the expression of the HSVTK gene confers a negative selectable phenotype to cancer cells *in vitro*. Moolten (1986) demonstrated acquired GCV sensitivity in a murine sarcoma cell line transduced with a retroviral vector that produces HSVTK. The transduced sarcoma tumor cells were 200-1000 times more sensitive to GCV than control tumor cells. This finding has been reproduced in several rodent and human cancer model systems including lung cancer (Hasegawa *et al.*, 1993), mesothelioma (Smythe *et al.*, 1994,1995), hepatocellular carcinoma (Caruso *et al.*, 1993), leukemia (Abe *et al.*, 1993), melanoma (Golumbek *et al.*, 1992), and CNS tumor models (Culver *et al.*, 1992; Takamiya *et al.*, 1992; Barba *et al.*, 1993,1994; Takamiya *et al.*, 1993; Chen *et al.*, 1994; Myklebust *et al.*, 1994). The efficacy of this approach varies significantly and may be due to a variety of factors including promoter function, target cells studied, and efficiency of transduction.

Retroviral vectors were employed for many of the early experiments with HSVTK but were limited by the inability to produce high titer retroviruses. Several modifications have been introduced to overcome this difficulty. Takamiya and colleagues demonstrated that rat glioma cells co-infected with a retroviral TK vector and a wild type (replication competent) vector were 300 fold more sensitive to the toxic effect of GCV than those cells infected with TK vector alone (Takamiya *et al.*, 1992). The co-infection permits the continued production of TK bearing virus and subsequent infection of neighboring tumor cells not transfected by the initial inoculation. This process in effect creates a retrovirus packaging cell line within the tumor. This system was improved by introducing a murine retrovirus packaging cell line directly into the tumor (Takamiya *et al.*, 1993). This approach has been studied in other CNS tumor models (Culver *et al.*, 1992), in experimental hepatic metastases (Caruso *et al.*, 1993), and has become the basis of an ongoing clinical trial for the treatment of brain tumors (Culver *et al.*, 1992).

The tumoricidal activity of the HSVTK/GCV system is due to several factors. In dividing cells, the phosphorylated GCV inhibits DNA synthesis. This effect is not confined to cells that are directly transduced with HSVTK, as neighboring cells are also affected (Wu *et al.*, 1994). This phenomenon, which likely occurs as a result of several mechanisms, has been termed the "bystander effect," and has been observed in several tumor types including CNS tumors. Transfer of the phosphorylated GCV between cells, ("metabolic cooperation") via gap junctions has been proposed as a possible mechanism. Phagocytosis by neighboring cells of GCV phosphate containing apoptotic vesicles (from dying transduced cells) has also been proposed (Freeman *et al.*, 1993). Immune-mediated processes may also account for significant killing of non-transduced cells. In one report antitumor immunity was observed following TK mediated killing of experimental brain tumors. This protective immunity led to eradication of subsequent tumor cell inoculi when the animal was rechallenged at a remote site (Barba *et al.*, 1994). Whether the tumor immunity is TK dependent or merely a manifestation of inherent tumor cell immunogenicity has yet to be established in this rodent model.

More recently, adenovirus vectors have been used for gene therapy of brain tumors. Chen *et al.* (1994) demonstrated regression of experimental gliomas following *in vivo* adenovirus-mediated gene transfer and GCV treatment. The tumor deposits were not completely eliminated by this treatment, however. Tumor cells close to the injection site were more readily transduced than those distant, as judged by parallel marker gene transfer experiments. Furthermore, these more distant cells escaped GCV toxicity because of a diminished bystander effect attributed to a paucity of gap junctions in the rodent brain tumor cell line employed (Chen *et al.*, 1994). This might be overcome in the clinical setting by more precise stereotactic treatment planning (aided by MRI and PET studies) and by multiple tumor injections.

These important findings have served to point out the limitations of this approach. Golumbek and colleagues demonstrated that HSVTK was unable to completely eliminate nonimmunogenic tumor cells even when 100% of the introduced cells exhibited TK expression. They observed a delayed outgrowth of TK$^+$ tumor cells occurring many days after cessation of GCV administration (Golumbek *et al.*, 1992). These cells were likely in G0 arrest during the interval of GCV therapy and were, therefore, able to escape the toxic effects of GCV metabolites. Alternatively, they were sequestered in an area of poor GCV penetration and thus received sublethal doses of the drug.

IV. TOXICITY OF ADENOVIRUS IN THE CNS

IV.A. Wild type adenovirus infections of the CNS

Adenovirus CNS infection is sporadic and uncommon, although its true incidence can not be estimated from case reports. Undoubtedly many cases are neither diagnosed (aseptic meningitis) nor reported. From the reports available, adenovirus meningoencephalitis is usually self-limited and the clinical outcomes governed by the presence of underlying disease in the affected patients. We anticipate that the use of a replication defective adenovirus will substantially limit the spread of the virus and attenuate the pathologic consequences. This will be established more definitively in our ongoing toxicity studies. In the course of developing an adenoviral vaccine, the CNS toxicities of wild type adenovirus strains were studied. Rorke and colleagues systematically studied the neurovirulence of several adenovirus serotypes (1, 2, 3, 4, 5, 7, 14, and 21) following direct injection of the virus into the CNS of rhesus monkeys (Rorke *et al.* 1973; Rubin and Rorke, 1994). Focal injection of the virus into the CNS (up to 0.5 ml into each thalamus, and up to 10^7 TCID$_{50}$/ml) elicited a spectrum of responses depending on the serotype of the virus. They demonstrated that, independent of serotype, viral antigen could be detected in the ependyma, choroid plexus, and leptomeninges within 72 hours of injection into the thalamus. This indicates that the wild type virus is potentially capable of spreading within the substance of the brain. The histologic findings ranged from no significant pathologic findings to lymphocytic infiltration of the leptomeninges and choroid plexus with ependymitis and focal neuronal destruction. Interestingly, this resulted in no overt symptoms of CNS disease. The authors note that "clinical observation disclosed no signs or symptoms among the monkeys with evidence of inflammatory lesions that distinguished them from those without such lesions" (Rubin and Rorke, 1994). Serotype 5 adenovirus produced lesions of mild to moderate intensity. The results were very consistent within each serotype and appeared to be "an all-or-none phenomenon." CNS virus titers decline over a 21 day period suggesting that viral infection is a self-limited process. Furthermore, intranuclear inclusion bodies that are typically associated with adenoviral infections in permissive tissues are specifically absent in CNS. This finding is consistent with the relative lack of CNS tropism of adenovirus (Rubin and Rorke, 1994).

IV.B. Recombinant adenovirus-mediated gene transfer in the rodent CNS

Adenovirus-mediated gene transfer, unlike retrovirus gene transfer, has the potential advantage of transducing nondividing cells. Brain tumor cells are heterogeneous with respect to the fraction in active cell cycle, yet this is unlikely to alter their transducability. Those cells undergoing gene transfer while in G0 will be rendered sensitive to GCV provided that transgene expression persists and that GCV is being administered when these cells eventually enter active cell cycle. Experimental evidence suggests that the adenoviral-mediated gene expression will persist in the absence of an immune response to the virus. Since brain tumors are not an immune privileged site (especially following surgery and radiation therapy) it is likely that an anti-adenoviral immune response will eradicate transduced cells that are not directly killed by the effects of GCV (Yang *et al.*, 1994ab).

Several reports have demonstrated the efficiency of adenoviral gene transfer in the CNS (Akli *et al.*, 1993; Bajocchi *et al.*, 1993; Davidson *et al.*, 1993; Le Gal La Salle *et al.*, 1993). Direct injection of adenovirus into the brain parenchyma of mice results in efficient gene transfer followed by gradual decline in transgene expression, but could be detected up to 30 days post injection (Akli *et al.*, 1993; Davidson *et al.*, 1993). Newer adenoviral vectors that have more limited synthesis of endogenous viral proteins have prolonged gene expression because of the diminished immune response to virally transduced cells (Engelhardt *et al.*, 1994ab; Yang *et al.*, 1994c).

Significant replication and distant spread of the virus does not take place following intracranial administration of recombinant adenovirus. The liver and spleen are readily transduced by intravenously administered adenovirus. However, when recombinant adenovirus was injected into the brains of rats, virus could not be found in the liver or spleen (Davidson *et al.*, 1993; Smith *et al.*, 1997). We have directly administered recombinant adenovirus into the cerebrospinal fluid (CSF) in order to assess its potential for spread to distant sites. We observed no histologic evidence of viral dissemination following intrathecal administration of 10^9 pfu of virus, nor could the virus be cultured from the CSF (Smith *et al.*, 1997).

IV.C. Rodent and primate toxicity studies of intracranial recombinant adenovirus

We have performed additional studies with adenovirus expressing HSVTK (H5.010RSVtk) in the normal rat and nonhuman primate brain in order to assess the toxicity of this vector. Wistar rats were injected in the frontal lobe with 15 ml of vector or a control solution at doses up to 3.8×10^{11} pfu/ml (5.7×10^9 pfu total dose). Rats then received intraperitoneal GCV (10 mg/kg/day) for a period of up to 14 days. The rats showed no clinical evidence of toxicity over a followup period of 3 months. Necropsy evaluations were performed on groups of rats at intervals of 3, 7, 18, 28, 60, and 90 days post injection of virus.

Systematic evaluation of all major organs showed no histologic evidence of toxicity. Blood samples obtained before virus injection and at the time of necropsy show no significant virus related changes in serum chemistries, PT, PTT, white blood cell count (including differential), hemoglobin, hematocrit, or platelet count. Analysis of the CSF and urine at the time of necropsy showed no culturable adenovirus. Pathologic evaluation of the brain specimens showed a multifocal inflammatory process with areas of hemorrhage (Figure 2).

These lesions are not apparent on gross evaluation. Unilateral astrocytosis and necrosis was the predominant finding at the injection site. Focal necrosis and hemorrhage were the most significant findings with occasional areas of mild meningeal and perivascular inflammation. These findings resolved with time (except for necrosis) and were not uniformly present in all animals (Smith *et al.*, 1997). Representative histological findings are shown in Figure 2.

The clinical and histological findings following administration of the virus to rhesus monkeys were quite similar. Doses of up to 1.5×10^{11} pfu were well tolerated and resulted in magnetic resonance image (MRI) changes near the injection site consistent with focal brain edema (Figure 3). The edema was readily apparent three days following the injection and gradually resolved over a period of several weeks.

There were focal histologic changes near the injection site (focal necrosis, inflammation, and meningeal infiltrates) similar to that which we observed in the rats (Smith *et al.*, 1997). We could find no evidence of viral dissemination by histologic methods. Shine and colleagues have performed similar studies using virtually the same vector design (Goodman *et al.*, 1996).

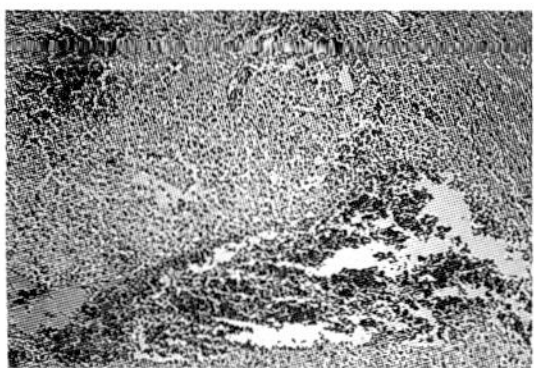 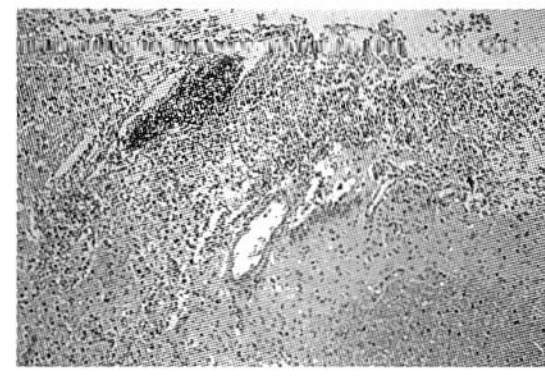 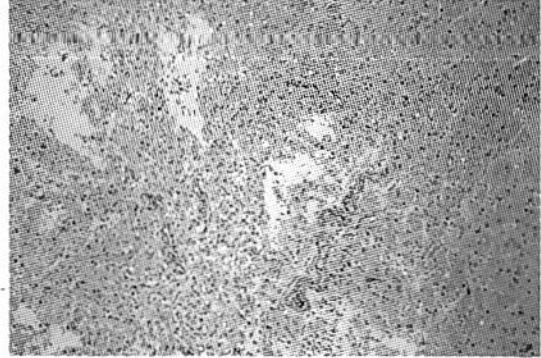

Figure 2. Histologic findings in the brain of rats treated with H5.010RSVtk. Wistar rats were stereotactically injected with a total of 5.7×10^9 pfu of H5.010RSVtk on day 1. Animals were sacrificed on days 3 (**left**), 7 (**center**), and 28 (**right**). The predominant findings were hemorrhage, necrosis, and astrocytosis at the injection site *(reprinted, with permission, from Smith et al., 1997)*.

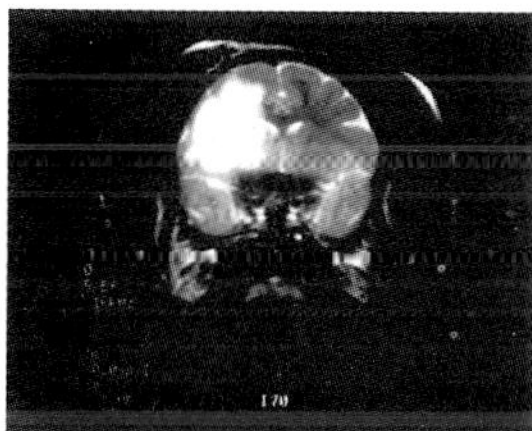  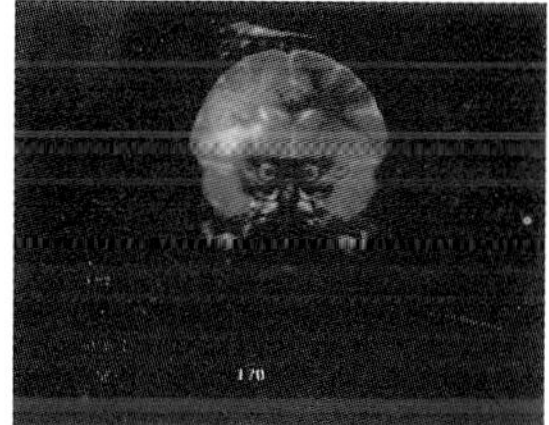

Figure 3. MRI following intracranial injection of H5.010RSVtk in a rhesus monkey. A total of 1.5×10^{11} pfu of virus (the highest dose tested) was injected on day 1. By day 4 there was substantial edema as seen on T2 weighted images (**left**). This gradually declined over from day 8 (**center**) to day 14 (**right**). Long term followup of a similarly treated high dose animal by MRI showed only a small abnormality at the injection site after one year.

They were able to detect evidence of viral spread by PCR; however, these findings were not associated with histologic changes or clinical toxicity. Interestingly, the degree of CNS toxicity observed in their experiments was much more severe than what we observed (Smith *et al.*, 1997) and resulted in the death of two monkeys at a dose that was approximately 100 fold lower than our high dose which was not fatal (Goodman *et al.*, 1996). It is not clear whether the differences in toxicity observed are explained by differences in experimental design (e.g., species of primate or means of drug administration). However, our findings were consistently observed in three different animal models and have been confirmed by our early clinical work (see below).

V. CLINICAL TRIAL OF H5.010RSVTK IN PATIENTS WITH RECURRENT GLIOMAS

Having established that adenoviral mediated delivery of HSVTK was potentially useful in the treatment of brain tumors and that it was not likely to be exceedingly toxic, we began a phase I study of H5.010RSVtk in patients with newly diagnosed or relapsed malignant gliomas (Eck *et al.*, 1996) (Figure 4). Currently, we have treated ten patients with this vector in conjunction with intravenous GCV.

All of the patients with malignant gliomas had progression of their tumors following standard therapy (i.e., surgery and radiation therapy). Patients underwent stereotactic injection of the virus into the tumor (from 10^8 to 10^{10} pfu) followed by systemic GCV. One week after their initial treatment the tumors were partially resected and a second dose of adenovirus was injected into the residual tumor bed. GCV therapy was continued for two more weeks.

We have observed no clinical signs of virus mediated toxicity or viral dissemination in the followup period. Histologic examination of the resected material showed focal areas of necrosis (within the tumor) surrounded by an area of inflammation (Figure 5). These preliminary findings are remarkably similar to that which we described in our animal models (Smith *et al.*, 1997). Studies to assess the nature of the immune response and the extent of gene transfer are currently underway. We plan to enroll additional patients in this trial at doses up to 1×10^{11} pfu.

VI. FUTURE DIRECTIONS

This treatment strategy is only in its early stages of development, with many unanswered questions. A primary concern is whether suffic-

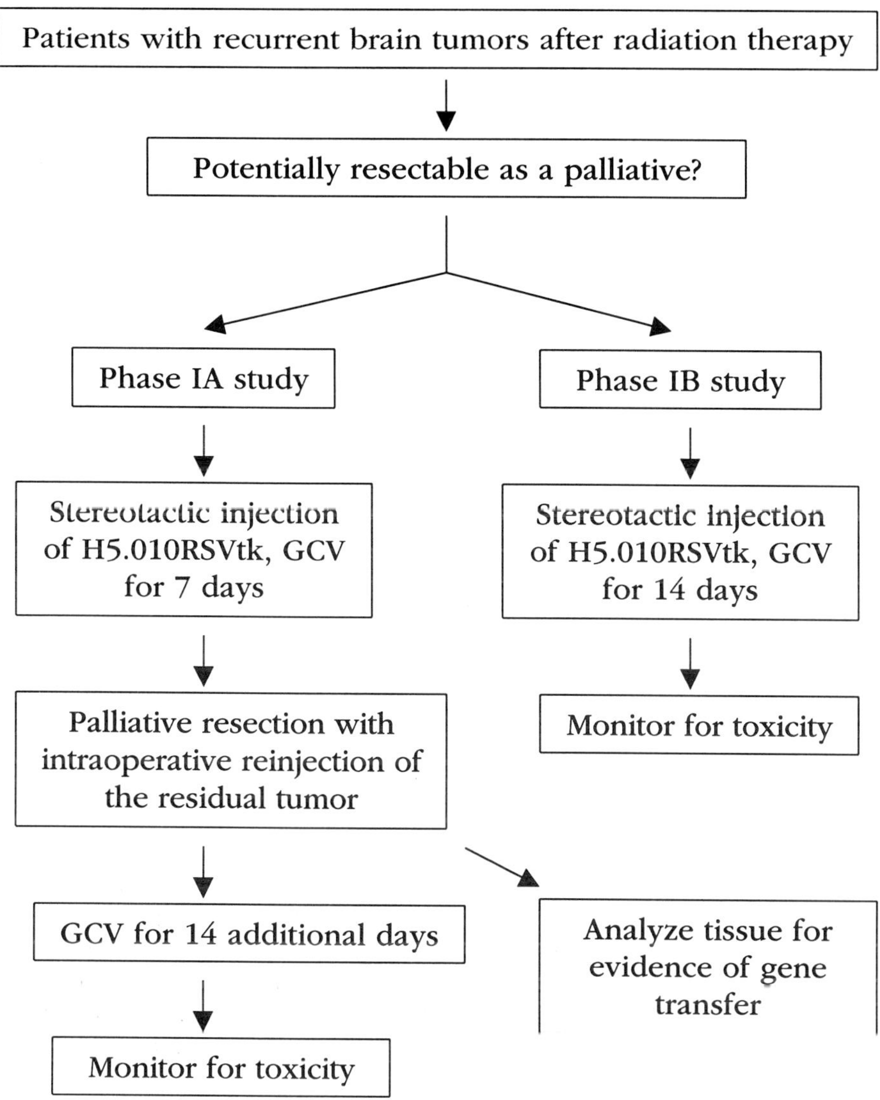

Figure 4. Phase I clinical trial design. Patients are recruited to one of two studies based on the ability to perform a palliative resection of the tumor. Those amenable to palliative resection undergo stereotactic injection of the virus followed by re-treatment at the time of resection. Those patients who are not able to undergo palliative resection receive only a single stereotactic injection of the virus.

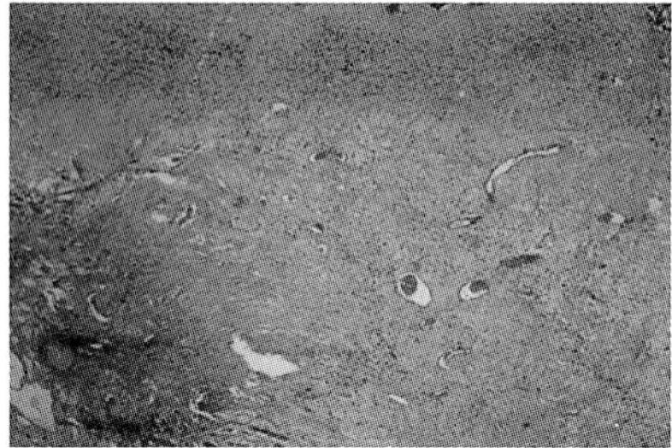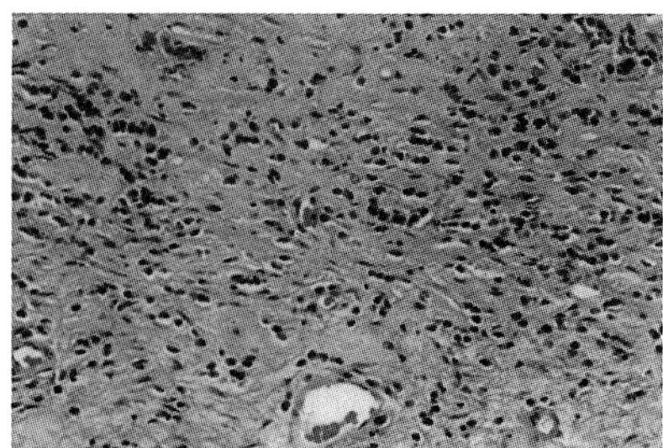

Figure 5. Histologic findings of resected human glioma following gene therapy. Histologic examination of the tumor tissue removed one week after initial treatment showed an area of necrosis (**left**) surrounded by an infiltrate of mononuclear cells (**right**). The infiltrate contains an abundance of T cells (CD3$^+$) and plasma cells *(reprinted, with permission, from Eck et al., 1996)*.

ient vector can be delivered into the tumors, and whether the vector can be distributed throughout the tumor mass and into tumor cells that have infiltrated into the surrounding normal brain. Additionally, it remains to be determined what is the optimal dose and scheduling of GCV administration so as to adequately treat more slowly dividing cells.

Given the flexibility of this system one can be reasonably optimistic that substantial improvements can be made in both the vector design and in its clinical application. One possibility is the use of this approach in conjunction with radiation therapy. We have recently found that external beam radiation enhances the uptake of adenovirus (S. Eck and C. Stevens, unpublished observations). In addition, Freytag and colleagues have shown that antiviral agents (e.g., acyclovir, GCV) can enhance the radiation induced cell killing of tumor cells transduced with HSVTK (Kim *et al.*, 1994,1995). These findings suggest that the earlier application of this strategy to newly diagnosed tumors in a multimodal therapeutic plan (gene transfer, GCV, and radiation therapy) may be more effective.

VII. REFERENCES

Abe, A., Takeo, T., Emi, N., Tanimoto, M., Ueda, R., Yee, J.-K., Friedmann, T. and Saito, H. (1993) Transduction of a drug-sensitive toxic gene into human leukemia cell lines with a novel retroviral vector. *Proc. Soc. Exp. Biol. Med.* **203**:354-359.

Akli, S., Caillaud, C., Vigne, E., Stratford-Perricaudet, L.D., Perricaudet, M., Kahn, A. and Peschanski, M.R. (1993) Transfer of a foreign gene into the brain using adenovirus vectors. *Nature Genet.* **3**:224-228.

Ausman, J. I., Levin, V. A., Brown, W. E., Rall, D. P. and Fenstermacher, J. D. (1977) Brain-tumor chemotherapy. Pharmacological principles derived from a monkey brain-tumor model. *J. Neurosurg.* **46**:155-164.

Bajocchi, G., Feldman, S. H. and Crystal, R. G. (1993) Direct *in vivo* gene transfer to ependymal cells in the central nervous system using recombinant adenovirus vectors. *Nature Genet.* **3**:229-234.

Barba, D., Hardin, J., Ray, J. and Gage, F. H. (1993) Thymidine kinase-mediated killing of rat brain tumors. *J. Neurosurg.* **79**:729-735.

Barba, D., Hardin, J., Sadelain, M. and Gage, F. H. (1994) Development of anti-tumor immunity following thymidine kinase-mediated killing of experimental brain tumors. *Proc. Natl. Acad. Sci. USA* **91**:4348-4352.

Berkner, K. (1988) Development of adenovirus vectors for the expression of heterologous genes. *BioTechniques* **6**:616-629.

Boring, C. C., Squires, T. S. and Tong, T. (1993) Cancer statistics. *CA, Cancer J. Clinicians* **43**:7-26.

Brem, H., Mahaley, M. S. J. and Vick, N. A. (1991) Interstitial chemotherapy with drug polymer implants for the treatment of recurrent gliomas. *J. Neurosurg.* **74**:441-446.

Burger, P. C., Heinz, E. R. and Shibata, T. (1988) Topographic anatomy and CT correlations in the untreated glioblastoma multiforme. *J. Neurosurgery* **68**:698-704.

Burger, P. C., Scheithauer, B. W. and Vogel, F. S. (1991) *Surgical Pathology of the Nervous System and its Coverings*, Churchill Livingston, New York, NY.

Caruso, M., Panis, Y., Gagandeep, S., Didier, H., Salzmann, J.-L. and Klatzmann, D. (1993) Regression of established macroscopic liver metastases after in situ tranduction of a suicide gene. *Proc. Natl. Acad. Sci. USA* **90**:7024-7028.

Chen, S. H., Shine, H. D., Goodman, J. C., Grossman, R. G., and Woo, S. L. (1994) Gene Therapy for brain tumors: Regression of experimental gliomas by adenovirus-mediated gene transfer *in vivo. Proc. Natl. Acad. Sci. USA* **91**:3054-3057.

Culver, K. W., Ram, Z., Wallbridge, S., Ishii, H., Oldfield, E. H., and Blaese, R. M. (1992) *In vivo* gene transfer with retroviral vector-producer cells for treatment of experimental brain tumors. *Science* **256**:1550-1552.

Curran, W. J., Jr., Scott, C. B., Horton, J., Nelson, J. S., Weinstein, A. S., Fischbach, A. J., Chang, C. H., Rotman, M., Asbell, S. O., Krisch, R. E., and Nelson, D. F. (1993) Recursive partitioning analysis of prognostic factors in three radiation therapy oncology group malignant glioma trials. *J. Natl. Cancer Inst.* **85**:704-710.

Davidson, B. L., Allen, E. D., Kozarsky, K. F., Wilson, J. M. and Roessler, B. L. (1993) A model system for *in vivo* gene transfer into the central nervous system using an adenoviral vector. *Nature Genet.* **3**:219-223.

Eck, S. L., Alavi, J. B., Alavi, A., Davis, A., Hackney, D., Judy, K., Mollman, J., Phillips, P. C., Wheeldon, E. B. and Wilson, J. M. (1996) Treatment of advanced CNS malignancies with the recombinant adenovirus H5.010RSVTK: a phase I trial. *Hum. Gene Ther.*7:1465-1482.

Eck, S. L. and Wilson, J. M. (1995) Gene-based therapy. In Hardman, J.G., Limbird, L. E., Molinoff, P. B., and Ruddon, R. W., eds., *Goodman & Gilman: The Pharmacological Basis of Therapeutics,* McGraw-Hill, New York, NY, 77-102.

Engelhardt, J. F., Litzky, L. and Wilson, J.M. (1994a) Prolonged transgene expression in cotton rat lung with recombinant adenovirus defective in E2a. *Hum. Gene Ther.* 5:1217-1229.

Engelhardt, J. F., Ye, X., Doranz, B. and Wilson, J. M. (1994b) Ablation of E2A in recombinant adenoviruses improves transgene persistence and decreases inflammatory response in mouse liver. *Proc. Natl. Acad. Sci. USA* 91:6196-6200.

Ezzeddine, Z. D., Martuza, R. L., Platika, D., Short, P. M., Malick, A., Choi, B. and Breakfield, X. O. (1991) Selective killing of glioma cells in culture and *in vivo* by retrovirus transfer of the herpes simplex virus thymidine kinase gene. *New Biol.* 3:608-614.

Fine, H. A., Dear, K. B., Loeffler, J. S., Black, P. M., and Canellos, G. P. (1993) Meta-analysis of radiation therapy with and without adjuvant chemotherapy for malignant gliomas in adults. *Cancer* 71:2585-2597.

Freeman, S., Abboud, C. N., Whartenby, K. A., Pakman, C. H., Koeplin, D. S., Moolten, F. L. and Abraham, G. N. (1993) The bystander effect: tumor regression when a fraction of the tumor mass is genetically modified. *Cancer Res.* 53:5274-5283.

Golumbek, P. T., Hamzeh, F. M., Jaffee, E. M., Levitsky, H., Lietman, P. S. and Pardoll, D. M. (1992) Herpes simplex-1 virus thymidine kinase gene is unable to completely eliminate live, nonimmunogenic tumor cell vaccines. *J. Immunother.* 12:224-230.

Goodman, J. C., Yrask, T. W., Chen, S.-H., Woo, S. L., Grossman, R. G., Carey, K. D., Hubbard, G. B., Carrier, D. A., Rajagopalan, S., Aguilar-Cordova, E. and Shine, H. D. (1996) Adenoviral-mediated thymidine kinase gene transfer into the primate brain followed by systemic ganciclovir: pathologic, radiologic, and molecular studies. *Hum. Gene Ther.* 7:1241-1250.

Hasegawa, Y., Emi, N., Shimokata, K., Abe, A., Kawabe, T., Hasegawa, T., Kirioka, T. and Saito, H. (1993) Gene transfer of herpes simplex virus type I thymidine kinase gene as a drug sensitivity gene into human lung cancer cell lines using retroviral vectors. *Amer. J. Respir. Cell Mol. Biol.* 8:655-661.

Hochberg, F. H. and Pruitt, A. (1980) Assumptions in the radiotherapy of glioblastoma. *Neurology* 30:907-911.

Jaekle, K. A. (1994) Immunotherapy of malignant gliomas. *Sem. Oncol.* 21:249-259.

Jolly, D. (1994) Viral vector systems for gene therapy. *Cancer Gene Ther.* 1:51-64.

Kim, J. H., Kim, S. H., Brown, S. L. and Freytag, S. O. (1994) Selective enhancement by an antiviral agent of the radiation-induced cell killing of human glioma cells transduced with HSV-tk gene. *Cancer Res.* 54:6053-6056.

Kim, J. H., Kim, S. H., Kolozsvary, A., Brown, S. L., Kim, O. B. and Freytag, S. O. (1995) Selective enhancement of radiation response of herpes simplex virus thymidine kinase transduced 9L gliosarcoma cells *in vitro* and *in vivo* by antiviral agents. *Int. J. Rad. Oncol. Biol. Phys.* **33**:861-868.

Kornblith, P. L. and Walker, M. (1988) Chemotherapy for malignant gliomas. *J. Neurosurg.* **68**:1-17.

Le Gal La Salle, G., Robert, J. J., Berrard, S., Ridoux, V., Stratford-Perricaudet, L. D., Perricaudet, M. and Mallet, J. (1993) An adenovirus vector for gene transfer into neurons and glia in the brain. *Science* **259**:988-990.

Leibel, S. A., Scott, C. B. and Loeffler, J. S. (1994) Contemporary approaches to the treatment of malignant gliomas with radiation therapy. *Sem. Oncol.* **21**:198-219.

Mahaley, M. S. J., Mettlin, C., Natarajan, N., Laws, E. R. J. and Peace, B. B. (1989) National survey of patterns of care for brain-tumor patients. *J. Neurosurg.* **71**:826-836.

Merchant, R. E., Merchant, L. H., Cook, S. H., McVicar, D. W. and Young, H. F. (1988) Intralesional infusion of lymphokine-activated killer (LAK) cells and recombinant interleukin-2 (rIL-2) for the treatment of patients with malignant brain tumor. *Neurosurgery* **23**:725-732.

Moolten, F. (1986) Tumor chemosensitivity conferred by inserted herpes thymidine kinase genes: paradigm for a prospective cancer control strategy. *Cancer Res.* **46**:5276-5281.

Moolten, F. and Wells, J. M. (1990) Curability of tumors bearing herpes thymidine kinase genes transferred by retroviral vectors. *J. Natl. Cancer Inst.* **82**:297-300.

Myklebust, A. T., Godal, A. and Fodstad, O. (1994) Targeted therapy with immunotoxins in a nude rat model for leptomeningeal growth of human small cell lung cancer. *Cancer Res.* **54**:2146-2150.

Neuwelt, E. A., Howieson, J., Frenkel, E. P., Specht, H. D., Weigel, R., Buchan, C. G. and Hill, S. A. (1986) Therapeutic efficacy of multiagent chemotherapy with drug delivery enhancement by blood-brain barrier modification in glioblastoma. *Neurosurgery* **19**:573-582.

Rorke, L. B., Rubin, B. A. and Myers, J. (1973) Neurovirulence of adenoviruses. *J. Neuropathol. Exp. Neurol.* **32**:161-162.

Rubin, D. A. and Rorke, L. B. (1994) Adenovirus vaccines. In Plotkin, S. A., and Mortimer, E. A., eds., *Vaccines,* W.B. Saunders, Philadelphia, 474-502.

Scharfen, C. O., Sneed, P. K., Wara, W. M., Larson, D. A., Phillips, T. L., Prados, M. D., Weaver, K. A., Malec, M., Acord, P., Lamborn, K. R., Lamb, S. A., Ham, B., and Gutin, P. H. (1992) High activity I-125 interstitial implant for gliomas. *Int. J. Radiat. Oncol. Biol. Phys.* **24**:583-591.

Smith, J. G., Raper, S. E., Wheeldon, E., Hackney, D., Judy, K., Wilson, J. M., and Eck, S. L. (1997) Intracranial administration of adenovirus expressing HSV-TK in combination with ganciclovir produces a dose-dependent, self-limiting inflammatory response. *Hum. Gene Ther.* **8**:943-954.

Smythe, W. R., Hwang, H. C., Amin, K. M., Eck, S. L., Davidson, D. L., Wilson, J. M., Kaiser, L. R. and Albelda, S. M. (1994) Use of recombinant adenovirus to transfer the HSV-thymidine kinase gene to thoracic neoplasms: an effective *in vitro* drug sensitization system. *Cancer Res.* **54**:2055-2059.

Smythe, W. R., Hwang, H. C., Elshami, A. A., Amin, K. M., Eck, S. L., Davidson, B. L., Wilson, J. M., Kaiser, L. R., and Albelda, S. M. (1995) Treatment of experimental human mesothelioma using adenovirus transfer of the herpes simplex-thymidine kinase gene. *Ann. Surg.* **222**:78-86.

Takamiya, Y., Short, M. P., Ezzeddine, Z. D., Moolten, F. L., Breakefield, X. O. and Martuza, R. L. (1992) Gene therapy of malignant brain tumors: a rat glioma line bearing the herpes simplex virus type 1-thymidine kinase gene and wild type retrovirus kills other tumor cells. *J. Neuroscience Res.* **33**:493-503.

Takamiya, Y., Short, M. P., Moolten, F. L., Fleet, C., Mineta, T., Breakefield, X. O. and Martuza, R. L. (1993) An experimental model of retrovirus gene therapy for malignant brain tumors. *J. Neurosurg.* **79**:104-110.

Tjuvajev, J., Gansbacher, B., Desai, R., Beattie, B., Kaplitt, M., Matei, C., Koutcher, J., Gilboa, E. and Blasberg, R. (1995) RG-2 glioma growth attenuation and severe brain edema caused by local production of interleukin-2 and interferon gamma. *Cancer Res.* **55**:1902-1910.

Valyi-Nagy, T., Fareed, M. U., O'Keefe, J. S., Gesser, R. M., MacLean, A. R., Brown, S. M., Spivack, J. G. and Fraser, N.W. (1994) The herpes simplex virus type 1 strain 17+ gamma 34.5 deletion mutant 1716 is avirulent in SCID mice. *J. Gen. Virol.* **75**:2059-2063.

Walker, M. D., Green, S. B., Byar, D. P., Alexander, E., Jr., Batzdorf, U., Brooks, W. H., Hunt, W. E., MacCarty, C. S., Mahaley, M. S., Jr., Mealey, J., Jr., Owens, G., Ransohoff, J. d., Robertson, J. T., Shapiro, W. R., Smith, K. R., Jr., Wilson, C. B., and Strike, T. A. (1980) Randomized comparisons of radiotherapy and nitrosoureas for the treatment of malignant glioma after surgery. *New Engl. J. Med.* **303**:1323-1329.

Weitzman, M. D., Wilson, J. M. and Eck, S. L. (1995) Adenovirus vectors in cancer gene therapy. In Sobol, R. E., and Scanlon, K. J., eds., *The Internet Book of Gene Therapy: Cancer Therapeutics,* Appleton and Lange, Stamford, CT, 17-25.

Wolff, S. N., Phillips, G. L. and Herzig, G. P. (1987) High dose carmustine with autologous bone marrow transplant for the adjuvant treatment of high grade gliomas of the central nervous system. *Cancer Treat. Rep.* **71**:183-185.

Wu, J. K., Cano, W. G., Meylaerts, S. A. G., Qi, P., Vrionis, F. and Cherington, V. (1994) Bystander tumoricidal effect in the treatment of experimental brain tumors. *Neurosurgery* **35**:1094-1102.

Yang, Y., Ertl, H. C. and Wilson, J. M. (1994a) MHC class I-restricted cytotoxic T lymphocytes to viral antigens destroy hepatocytes in mice infected with E1-deleted recombinant adenoviruses. *Immunity* **1**:433-442.

Yang, Y., Nunes, F. A., Berencsi, K., Furth, E. E., Gonczol, E. and Wilson, J. M. (1994b) Cellular immunity to viral antigens limits E1-deleted adenoviruses for gene expression. *Proc. Natl. Acad. Sci. USA* **91**:4407-4411.

Yang, Y., Nunes, F. A., Berencsi, K., Gonczol, E., Engelhardt, J. F. and Wilson, J. M. (1994c) Inactivation of E2a in recombinant adenoviruses improves the prospect for gene therapy in cystic fibrosis. *Nature Genet.* **7**:362-369.

Yung, W. K., Prados, M., Levin, V. A., Fetell, M. R., Bennett, J., Mahaley, M. S.,. Salcman, M and Etcubanas, E. (1992) Intravenous recombinant interferon beta in patients with recurrent malignant gliomas: a phase I/II study. *J. Clin. Oncol.* **9**:1945-1949.

12

Distribution and Toxicity of Retroviral Vectors after Intracavitary Delivery in Mouse and Man

Patrice S. Obermiller, Carlos L. Arteaga, and Jeffrey T. Holt
Vanderbilt University, Nashville, Tennessee

Anne M. Pilaro
Food and Drug Administration, Rockville, Maryland

I. INTRODUCTION

Safe and effective viral gene therapy requires an understanding of the distribution and metabolism of vectors *in vivo*. Although the factors which influence vector distribution are largely unknown, the vector titer and local factors such as inflammation could influence the pharmacokinetics of viral vectors (Watt *et al.*, 1979; Sanui *et al.*, 1982). Vector within cells should be readily detectable by PCR assays, which can be designed to have exquisite sensitivity. Expression of vector-transferred genes within transduced tissues can be analyzed by quantitative assays for mRNA expression, such as nuclease protection assays. Prior studies indicated that local injection of viral vectors or vector producer lines did not result in sustained systemic spread of the vector to other tissues (Sajjadi *et al.*, 1994). Because our phase I clinical trial employs direct peritoneal injection of the XM6:anti-*fos* retroviral vector into malignant ascites, we studied the distribution of this amphotropic vector in immunocompetent Balb/c mice following intraperitoneal (i.p.) injection.

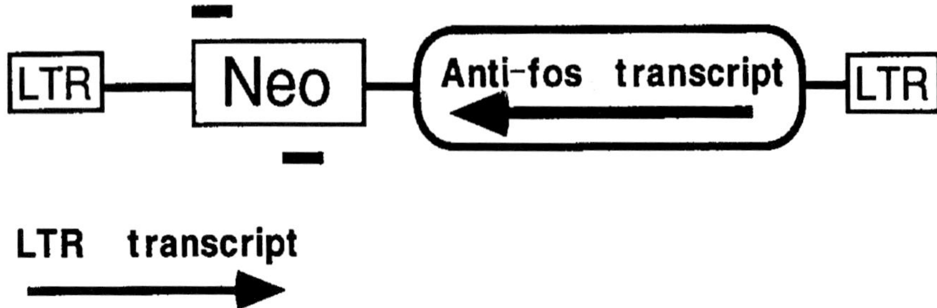

Figure 1. XM6:anti-*fos* retroviral vector. Solid bars show locations of 5' and 3' primers.

The XM6:anti-*fos* retroviral vector (Figure 1) contains a Neo-selectable marker, regulated by transcription from the LTR, and an MMTV-based antisense c-*fos* transcription unit, with breast-targeted or hormone regulated expression (Cato *et al.*, 1989; Choi *et al.*, 1987; Holt *et al.*, 1986; Matsui *et al.*, 1990). This vector inhibits the growth of established human breast xenografts in nude mice (Arteaga and Holt, 1996). Because the MMTV-regulated transcript is transcribed from the non-coding strand of the vector, we can differentiate between LTR-based transcription and breast-targeted MMTV-based transcription, thereby permitting analysis of tissue-specific expression of the vector during pharmacological studies.

Using the XM6:anti-*fos* retroviral vector as a model, we analyzed the dissemination, tissue uptake and expression of intraperitoneal vector in mice. We demonstrated that peritonitis markedly increased detection of vector in the serum during the first 24 hours, resulting in a greater delivery of vector to distant tissues, such as the kidney. Although vector was circulating, no positive tissues were found after the 4 hour time point. Antisense genes from the transcribed vector were not expressed in distant tissues, presumably because of the effects of the breast-targeted MMTV promoter.

II. MATERIALS AND METHODS

II.A. Cell Culture

The retroviral vector producer cell line PA317 was grown in DMEM high glucose (Life Technologies) with L-glutamine, 10% fetal bovine serum (Hyclone) and antibiotic/antimycotic (Sigma) added. When in the artificial capillary system for mass production, the antibiotic/antimycotic was changed to a penicillin/streptomycin solution (Sigma), and the temperature was reduced to 34°C. Before injection into mice, cells were scraped (not trypsinized) off the plates, pelleted and resuspended in warm medium to an appropriate density.

II.B. Production of Retroviral Vector

The retroviral vector was prepared by transfecting PA317 cells with the XM6:anti-*fos* retroviral vector DNA, which had been purified by alkaline lysis and ultracentrifugation twice on CsCl gradients. Following transfection, the PA317 cells were split and then treated with G418 until individual clones could be identified and expanded. Each clone was then assayed for vector titer by analyzing the supernatant's ability to transfer G418 resistance to MCF-7 cells. The clones which had the highest titer of vector production were then frozen in numerous aliquots and tested for sterility, presence of replication-competent retroviral vector, and presence of mycoplasma (Cornetta *et al.*, 1993). The media from the XM6:anti-*fos* retroviral producer clone passed the following quality control tests: MAP test (mouse antibody production), *in vitro* test for adventitious viruses, bacterial and fungal sterility testing, mycoplasma testing, southern blot analysis of the retroviral producer clone, ability to transfer vector DNA to other cells (analyzed by transfer of G418 resistance and by southern blotting of target cells to demonstrate efficient gene transfer). Because distribution is only meaningful if replication competent vector is not present, we performed a PG-4 S+L- assay and 3T3 amplification followed by a PG-4 S+L- assay. Cells were tested for replicative retroviral contamination both by co-culture of producer cells and by amplification of retroviruses in supernatants by culture with *Mus dunni* cells. These tests, which included appropriate controls, showed no detection of replication competent viruses.

The CELLMAX-QUAD Artificial Capillary Cell Culture System (Cellco, Inc., Germantown, MD) using the 400-008 Cellulose cartridge was used to produce high titer vector supernatants. Culture conditions were as specified by the manufacturer. Briefly, vector producer cells were inoculated into the system at 34°C and the glucose and lactate production level of the circulating medium was monitored on a daily basis. Starting with an initial circulating medium volume of 50 ml, the same volume of media was added once the lactate level reached 0.6-0.7 mg/ml, thereby doubling the volume, until it reached one liter. Thereafter, the circulating medium was completely changed once appropriate lactate levels occurred. When vigorously producing vector, each cartridge required one liter of fresh medium per day. The extracapillary space (ECS) volume into which retroviral vector was concentrated was 7 ml. We were able to harvest 10 ml of retroviral vector from each ECS, by washing with 3 ml of medium, which was kept in a reservoir syringe attached to the system. We simultaneously maintained three cartridges, harvesting 30 ml of high titer retroviral vector four times daily. Aliquots were removed for titering and testing purposes. Prior to injection, cellular debris was removed by centrifugation, and polybrene was added to 8 mg/ml final concentration.

II.C. Distribution Study

Groups of three immunocompetent Balb/c mice (12 weeks or older, female, non-lactating, Harlan) were injected intraperitoneally with either 0.5 ml of sterile water, or sterile oyster glycogen 0.5% (type II, Sigma). Oyster glycogen is a standard model for peritonitis (Sanui *et al.*, 1982; Watt *et al.*, 1979). After two days vector was injected intraperitoneally into the mice at three dose levels: 0 (control), 4.5×10^6 (low) and 4.5×10^7 (high) particles for a series of 4 daily injections of 0.5 ml. Some injections also included ß-estradiol (Sigma) at 36 µg/25g mouse (+Est) or dexamethasone (American Reagent Laboratories, Shirley, NY) 80 µg/25g mouse (+Dex) (Table 1).

Table 1. Schedule of XM6:anti-*fos* Treatment in Mice

Mouse	Time	Anti-*fos*
Group 1	4 hours	No Vector+Gly+Dex
Group 2	"	Low Vector+Gly+Dex
Group 3	"	High Vector+Est
Group 4	"	High Vector+Dex
Group 5	"	High Vector+Gly+Est
Group 6	"	High Vector+Gly+Dex
Group 7	24 hours	No Vector+Gly+Dex
Group 8	"	Low Vector+Gly+Dex
Group 9	"	High Vector+Est
Group 10	"	High Vector+Dex
Group 11	"	High Vector+Gly+Est
Group 12	"	High Vector+Gly+Dex

Low Vector = 4.5×10^6 particles, High Vector = 4.5×10^7 particles
Gly = Type II Oyster Glycogen, Est= Estrogen, Dex = Dexamethasone
(Reprinted, with permission, from Arteaga and Holt, 1996).

The dose of dexamethasone administered is the human equivalent that suppresses the pituitary-adrenal axis. Mice were terminated at 4, 24 and 48 hours, one week, and two weeks after injections. The following tissues were isolated: blood; peritoneum; spleen; mammary gland; bone marrow (vertebrae); lung; liver; thymus; ovary; kidney; pancreas; salivary gland; brain and jejunum (not shown). These tissues were immediately frozen after removal from the animal, and stored at -70°C.

A fraction of each tissue (approximately 20 mg) was taken and used to make DNA. The blood was allowed to clot for 30 minutes and then serum was collected following centrifugation. Approximately 300 µl of serum were harvested from each animal, 5 µl of which were used for the PCR detection. PCR was performed on one µl of DNA with oligonucleotides constructed for the Neo region. Additional PCR primers were also designed to test for a proviral DNA with double LTRs.

III. RESULTS

III.A. Development of Retroviral Vector Detection Assay

In order to monitor the distribution of the retroviral vector within the animal and its tissues, we utilized a sensitive PCR assay. Primers were designed to recognize the Neo region (Figure 1) of the retroviral construct. Because PCR cannot prime RNA, the vector itself cannot be detected unless reverse transcription and DNA replication occurs, presumably within a transduced cell. A portion of the DNA from the tissues was tested and the results were easily visible on an agarose gel. The vector plasmid DNA was used as the positive control and diluted sequentially to determine the sensitivity of the assay (Figure 2). The figure displays dilutions of the positive control down to 0.003 picograms (pre-amplification), with and without actual sample DNA. This also shows that the tissue samples do not contain nucleases, which degrade DNA, or other components that might interfere with the PCR assay.

III.B. Hematogenous Spread of Retroviral Vector

The PCR assay was first performed on the sera of mice taken at 4 hours and 24 hours post injection (Table 1, groups 1-12). These data provided an initial measurement of the amount and persistence of the retroviral vector. Figure 3 shows serum from samples of mouse numbers 1-72. No signal is observed in the first samples from animals which received no vector, group 1 (3 mice/group).

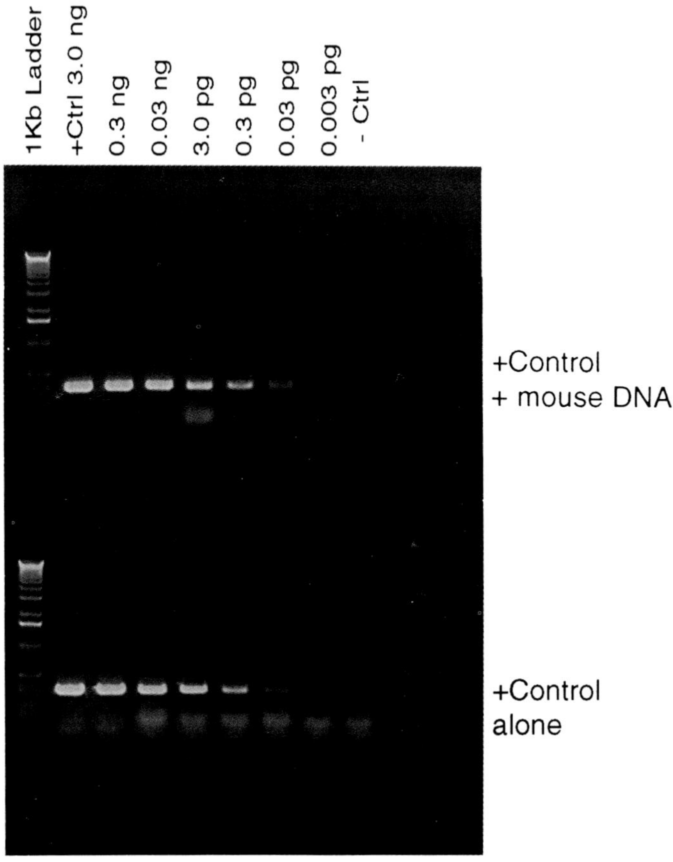

Figure 2. Retroviral detection assay strategy and sensitivity. Primers were designed for the detection of the retrovirus in mouse tissues. The 5' primer was 5'-dCCGGCCGCTTGGGTGGAGA-3' (19-mer, Tm = 66°C), and the 3' primer was 5'-dCAGGTAGCCGGATCAAGCGTATGC-3' (24-mer, Tm = 76°C). Initial denaturation occurred at 94°C for 2 min., followed by 40 cycles of 94°C for 30 sec., 50°C for 1 min., and 72°C for 1 min. were performed. The reaction volume was 50 µl; 10 µl were analyzed on a 1.5% agarose gel stained with ethidium bromide. The positive control (359 bp) was diluted 10-fold each time, starting at 3.0 nanograms and ending at 0.003 picograms. The last lane in each set is the negative control. Sample #2 (4 hours, no vector+Gly) was used as a negative sample added to the positive control serial dilution (top row). The positive control alone, serially diluted, is shown in the bottom half of the gel. The positive control titration is the same with or without sample DNA present. This shows the sensitivity of the PCR assay, and that the mouse sample did not interfere with the determination of the presence of the vector *(reprinted, with permission, from Arteaga and Holt, 1996)*.

Trace amounts are visible in groups 2, 3 and 4 which were inoculated with either low vector+dex+gly (group 2) or high vector+est (group 3) or high vector+dex (group 4). Positive amounts are evident in groups 5 and 6 which received high vector+gly in the presence of either estrogen or dexamethasone. The glycogen was inoculated prior to the retroviral vector, to induce the inflammation observed in malignant peritoneal effusions and determine its influence on vector distribution.

III.C. Vector Tissue Distribution is Influenced by Dose and Inflammation

Of the 4 hour groups which showed only trace amounts of vector present in the sera, using PCR, one mouse out of three was chosen and all 13 tissues were analyzed for vector. All of these tissues were negative (Table 2). Thus, mice receiving low vector, even in the presence of glycogen induced inflammation, or mice receiving high vector without glycogen, failed to show evidence of vector in their tissues. Of those groups showing positive results in their sera, the tissues from all six mice were analyzed.

In the group that received high vector+Gly+Est there were six tissues that were positive among the three mice: three kidneys; one lung; one brain and one thymus. In the group that received high vector+Gly+Dex, however, only two tissues were positive among the three mice: one kidney and one ovary (Figure 4). Some of the sera from 24 hours post injection showed only trace amounts of vector. One mouse from each group had all 13 tissues analyzed at this later time point. All of these tissues were negative, indicating that the mice had already cleared any of the vector that had been present.

The estrogen was employed because of its ability to induce the MMTV promoter, but has no anti-inflammatory properties. The dexamethasone induces the MMTV promoter and has potent anti-inflammatory properties in oyster glycogen induced peritonitis (Berenkopf and Weichman, 1987). The fact that fewer tissues were positive for vector when the inflammation was diminished by dexamethasone, or simply not induced, suggests that the inflammation did indeed influence the distribution of the vector.

III.D. Reverse Transcription and Expression

To determine whether the vector present had entered cells and replicated, a PCR assay was developed to detect only the DNA proviral

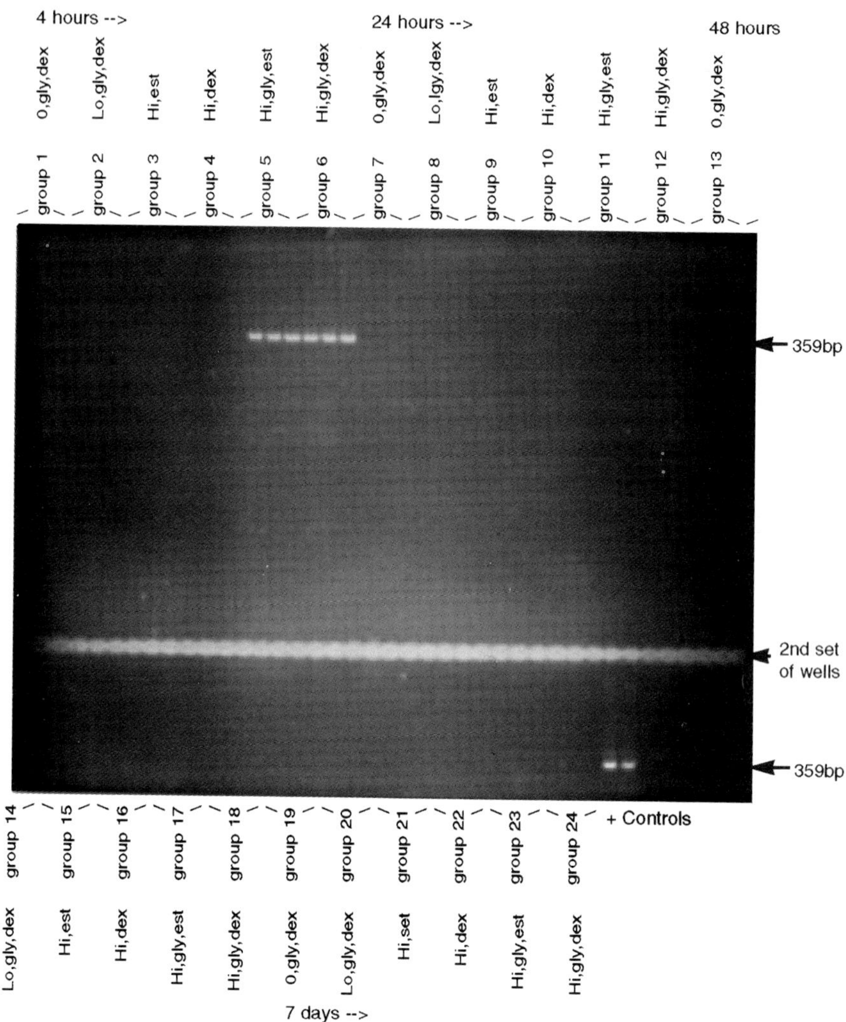

Figure 3. Distribution of vector. Five ml of serum from each mouse were analyzed for presence of retroviral vector by PCR. Analyses of 72 murine sera are seen on this two-tiered gel; the positive controls are visible in the lower half. Groups of 3 mice are as shown. Lo = vector of 4.5×10^6 particles, Hi = vector of 4.5×10^7 particles, Gly = oyster glycogen, Dex = dexamethasone, est = estrogen *(reprinted, with permission, from Arteaga and Holt, 1996).*

Table 2. Distribution Summary of High Vector* at 4 hr. Post Injection

Conditions	Serum Positives	Tissue Positives
-Gly +Est	3/3 trace	0/39
+Gly +Est	3/3	6/39
-Gly +Dex	3/3 trace	0/39
+Gly +Dex	3/3	2/39

*High Vector = 4.5×10^7 particles; Gly = Type II Oyster Glycogen; Est = Estrogen; Dex = Dexamethasone; trace = faint signal; positive = firm positive
(Reprinted, with permission, from Arteaga and Holt, 1996).

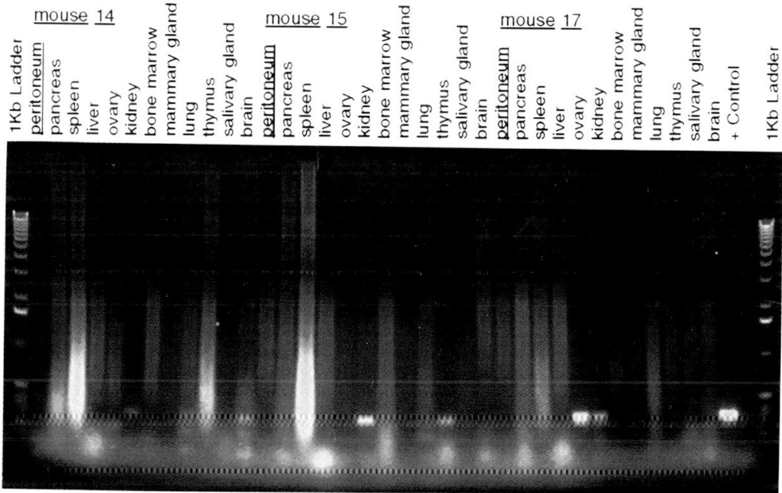

Figure 4. Examples of tissues assayed by PCR for presence of retroviral vector. Positive samples are in lanes 6: 14-kidney; 9: 14-lung; 12: 14-brain; 18: 15-kidney; 22: 15-thymus; 29: 17-ovary; 30: 17-kidney. Positive control is seen in the last lane before the 1 kbp Ladder. The expected fragment and positive control are 359 bp. Note that the gel shown has two sets of lanes (upper and lower) *(reprinted, with permission, from Arteaga and Holt, 1996).*

form of the retroviral vector. The PCR primers for this assay make use of the fact that viral vector regions are rearranged from the sequence of R-U5-space-U3, to U3-R-U5, when reverse transcribed following complete transcription (Figure 5A).

The tissues that contain proviral DNA would give a gel fragment of 402 base pairs long. Of the eight tissues that were positive for the vector by PCR, all four kidneys were positive, as well as one thymus. The lung, brain and ovary samples were negative for proviral DNA (Figure 5B).

All eight tissues were tested again, for gene expression. A nuclease protection assay was performed with an internal control of glyceraldehyde-3-phosphate dehydrogenase (GADPH) (Figure 6). These data demonstrated that none of the tissues expressed MMTV-specific transcript from the vector (Jensen *et al.*, 1994) even in the presence of estrogen or dexamethasone induction.

Following the above preclinical studies we have initiated Phase I trials in patients using two different vectors and have noted some divergences with the results obtained in animals. Most notably, although peritoneal injection into nude mice resulting in frequent systemic vector detected in the blood we have not observed any systemic vector in patients treated intraperitoneally with similar doses of the retroviral vector LXSN-BRCA1sv (Tait *et al.*, 1997).

This may be influenced by the extent of inflammation as well as the site of vector injection, since we did detect systemic vector during the first week following intrapleural injection of XM6:anti-*fos* in one of two treated patients.

Although intraperitoneal injection of the XM6:anti-*fos* vector produced no apparent toxicity in the immunocompetent mouse model, injection of a similar dose of LXSN-BRCA1sv retroviral vector produced a severe transient peritonitis in animals pretreated with oyster glycogen.

Fifteen of the 92 mice in the LXSN-BRCA1sv toxicity study received both high dose vector and oyster glycogen and all animals examined at early time points showed peritonitis. Two of the 15 mice died of peritonitis, whereas none of the animals in the other treatment groups died during this 2 week study. No other toxicity was observed.

Slight traces of intraperitoneally injected LXSN-BRCA1 vector were found by PCR in the following tissues at the 4 hour time point: pancreas, spleen, liver, ovary and colon. Because intraperitoneal injection of LXSN-BRCA1sv produced peritonitis in the above preclinical studies, patients were carefully evaluated for clinical and laboratory signs of acute peritonitis. Three of the 12 patients developed peritonitis as evidenced by patient discomfort, fever, peritoneal fluid counts, and

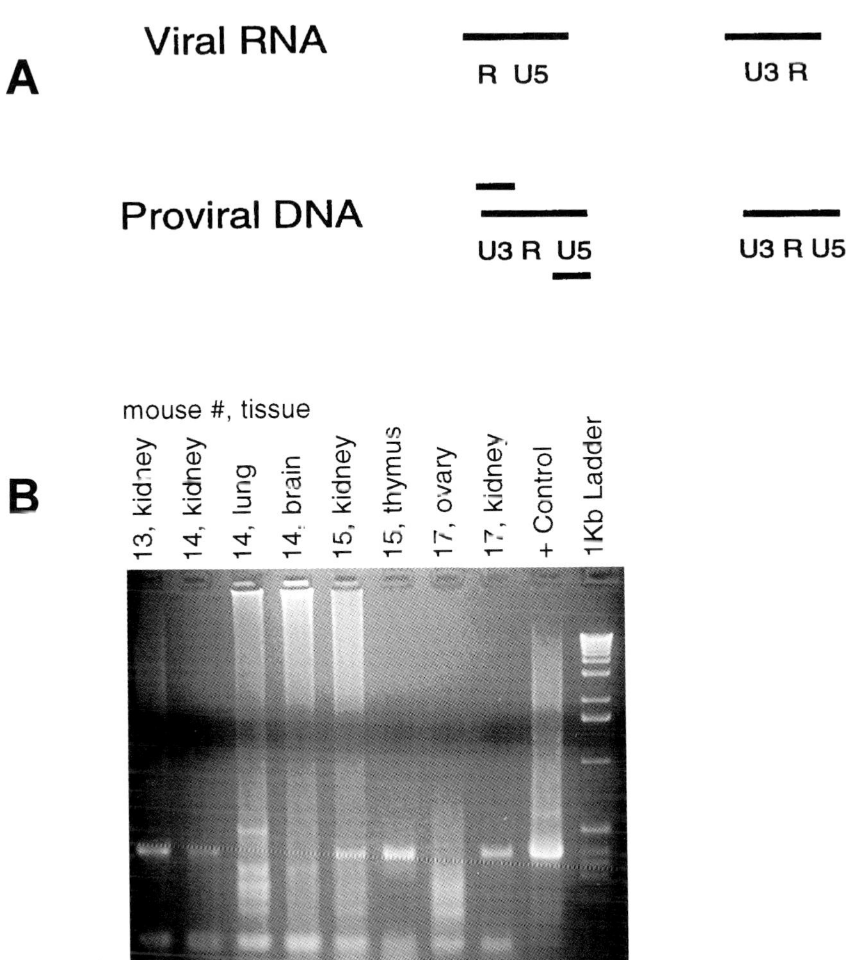

Figure 5. Assay for proviral DNA. **A.** The 5' primer was 5'-GCAAGGCATGGAAAAATACA-3' (20-mer, Tm = 56°C), and the 3' primer was 5'-TGGCGCCCCGAGTGAGG-3' (17-mer, Tm = 60°C). The PCR conditions were the same as for the vector detection test, Figure 3A. The transcribed fragment produced is 402 base pairs long. When not transcribed, no band is produced because the U3 and U5 regions are separated. **B.** All eight tissues that tested positive for the retrovirus were tested for transcription. The samples are as follows: #13 kidney; #14 kidney; #14 lung; #14 brain; #15 kidney; #15 thymus; #17 ovary; #17 kidney; + Control; and 1 kb ladder *(reprinted, with permission, from Arteaga and Holt, 1996).*

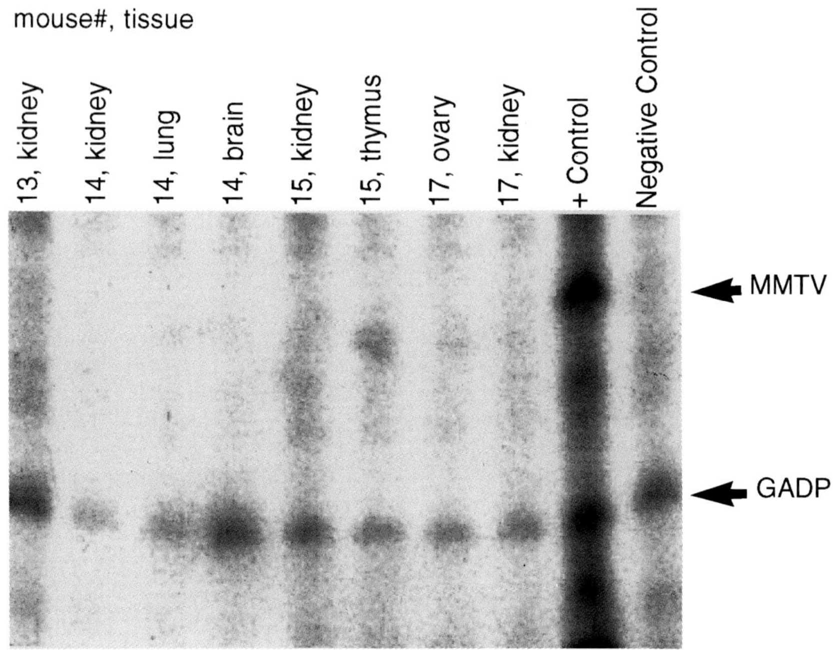

Figure 6. Nuclease protection assay for expression of viral vector. Tissue extracts were probed for presence of MMTV-regulated antisense transcripts (MMTV, 220 bp), and a constitutive control gene, glyceraldehyde-3-phosphate dehydrogenase (GADP, 140 bp) *(reprinted, with permission, from Arteaga and Holt, 1996)*.

negative bacterial cultures, which resolved within 24-48 hours after treatment was stopped. No recurrence of peritonitis was observed, even after two additional dose escalations. Other toxicities included fever in four patients and nausea in two patients, presumably from abdominal distension produced by the intraperitoneal infusion of vector plus 1 liter of saline.

IV. DISCUSSION

We have tested the serum and tissue distribution of XM6:anti-*fos*, a retroviral vector expressing an antisense RNA complementary to the c-*fos* oncogene, in a murine model of peritonitis. Previous data from our laboratories have shown that this vector can selectively transduce established MCF-7 human breast xenografts in estrogen-supplemented athymic mice, decrease the expression of the endogenous c-*fos* mRNA in target cells, inhibit and/or eliminate established tumors and tumor invasiveness, and prolong the survival of tumor-bearing animals (Arteaga and Holt, 1996).

These studies provided the basis for a phase I clinical trial of XM6:anti-*fos*, in which patients with metastatic breast cancer are receiving intrapleural or intraperitoneal inoculations of this vector (Holt and Arteaga, 1996; Tait *et al.*, 1997). In addition to the transduction selectivity in MCF-7 cells (above), there are two other theoretical arguments to expect selective integration and expression of XM6:anti-*fos* in breast tumor but not in normal cells present in the above mentioned transcellular spaces. First, integration may occur preferentially in the more mitotically active tumor cells (Miller *et al.*, 1990). Second, expression of the antisense *fos* is dependent on activation of the upstream MMTV long terminal repeat steroid-responsive promoter enhancer (Fig 1). The specificity in mammary cells and the steroid responsiveness of this promoter have been confirmed in studies with cultured cells and transgenic mice (Cato *et al.*, 1989; Choi *et al.*, 1987; Holt *et al.*, 1986; Matsui *et al.*, 1990). Despite these molecular and practical arguments, gene knockout reagents like XM6:anti-*fos* require a more rigorous investigation of tissue (cell lineage) specificity and preclinical toxicity, which we now report.

PCR amplifications of whole blood DNA, utilizing primers from the Neo gene of XM6:anti-*fos* (Fig. 3), demonstrated that circulating retroviral vector was detectable 4 hours post low vector and 4 and 24 hours after high vector administration (Table 2). All serum samples were negative for XM6:anti-*fos* vector sequences at 48 hours, one- and two-weeks post-treatment. Lack of detectability in PCR amplified mouse mammary gland DNA, even in the presence of a large dose of exogenous steroids (see below), would suggest that the levels of circulating XM6:anti-*fos* at the administered titer were too low to infect distant cells capable of MMTV regulation. Since drugs delivered intraperitoneally are frequently detected in the systemic circulation, our finding illustrates that retroviral vectors share this feature with other intracavitary treatments (Benet, 1972; Wadler *et al.*, 1986; Howell *et al.*, 1987).

Is it possible that the vector detected in sera and tissue samples integrated into cells? Although this would be the simplest explanation for our data, there is no way to guarantee integration by employing a PCR-based method, and the level of DNA present precludes the use of Southern blotting. Although PCR cannot detect the vector RNAs present in vector stocks, there could be some cDNA present within the vector that is produced by the reverse transcriptase within the particle. We think this is unlikely because we have been unable to detect PCR products when these quantities of vector particles have been tested in our assay. We cannot exclude the possibility that the 4 positive kidney samples, and the positive thymus samples constitute double LTR proviral DNAs which are present in the cytoplasm and have not integrated. This might explain why tissue samples were no longer positive for the vector at later time points, since non-integrated vectors may be unstable and disappear from cells at later time points.

Analyses of vector distribution in tissues at only 4 hours after vector administration i.p. revealed the presence of vector DNA in six tissues in group 5, which were treated with high vector, oyster glycogen and estradiol, but in only two tissues in group 6, which were treated with the same vector dose, oyster glycogen and dexamethasone (Table 2). Vector DNA was not detected, however, in any subsequent time point tested. Peritoneum, pancreas, spleen, mammary gland, bone marrow, liver, brain, jejunum and salivary gland specimens were all negative for XM6:anti-*fos*. Four of the 8 positive samples were kidneys. In 4/4 kidneys and one thymus a 402 base pair band was identified with primers which require the U5 and U3 regions to be in continuity (Figure 5), indicating complete reverse transcription of the vector sequence.

However, when a nuclease protection assay was performed, an MMTV transcript was not present (Figure 6); and additionally, the levels of endogenous c-*fos* were not different than those in untreated controls. It is also of interest that the gastrointestinal tract was not positive even though it is a very highly proliferating tissue. Taken together, these data support the absence of antisense *fos* expression in peripheral tissues despite complete reverse transcription. There were no histological abnormalities in any of the kidney sections (data not shown). In addition, BUN and creatinine were universally normal, indicating lack of clinical nephrotoxicity. The apparent selectivity of XM6:anti-*fos* for proviral DNA in kidney is unclear. Whether this is due to the rich vascular supply of the kidneys is not known; however, the fact that there was no instance of proviral DNA in liver (another highly vascular tissue), cannot support such a hypothesis. It is possible, however, that XM6:anti-*fos* is filtered by the kidney tubular system and cleared by tubular cell resorption and metabolism in less than 24 hours. In all cases of positive tissues, XM6:anti-*fos* was present in serum at the same time. Hence, tissue vector levels may well reflect those of contaminating blood in these instances. Histologically these and all other organs were normal. Blood counts, electrolytes, liver enzymes, and other laboratory tests were normal (data not shown).

These data indicate that local inflammation can modulate systemic absorption of XM6:anti-*fos*, since serum and tissue detectability of vector DNA was consistently enhanced by local pretreatment with oyster glycogen (Table 2). Supporting this notion is the fact that the anti-inflammatory agent, dexamethasone, decreased the tissue distribution of vector DNA when co-administered with oyster glycogen prior to antisense *fos* administration. There were no animal deaths due to toxicity in the present study, providing evidence that replication-incompetent retroviral vectors can be administered safely even under conditions in which systemic dissemination occurs. In addition, this is the first demonstration of systemic absorption and complete reverse transcription of a retroviral vector into host cell-DNA at a distant anatomical site.

These findings have implications regarding the use of vectors as anticancer therapies, especially in metastatic tumors requiring systemic drug delivery. Current studies are investigating whether higher doses and/or more prolonged systemic administrations of XM6:anti-*fos* may result in systemic transduction of widespread cancer metastases.

Gene therapy has often been promoted as disease-specific therapy with few side effects, but discovery of toxicities specifically associated with gene therapy should not be surprising. The LXSN-BRCA1sv peritonitis observed in mice and in some patients is rapidly reversible and appears to resolve without sequelae. The peritonitis was not clearly dose-related in patients (unlike the preclinical mouse studies where it was dose related in the presence of chemical peritonitis), although this will be easier to evaluate in the larger phase II and subsequent studies. The preclinical animal studies (in retrospect) heralded the peritonitis observed in patients, although the incidence was lower and the consequences less severe. However, the distribution studies in mice did not correlate well with the patient studies since mice showed systemic vector and patients generally did not. Animal studies are clearly a useful adjunct to evaluation of biotherapies but there will be differences between observed results in patients and animals.

V. SUMMARY

Injection of breast specific retroviral vector encoding XM6:anti-*fos* into the peritoneal cavity of immunocompetent mice poses questions concerning the effectiveness of gene transfer as well as of the dissemination of the vector to peripheral sites. We describe the distribution of our retroviral vector after 4 daily intraperitoneal injections in a murine model of peritonitis. To induce inflammation representative of breast cancer effusions, immunocompetent mice were injected i.p. with oyster glycogen. Others were given sterile water. Forty-eight hours later all mice received 4 daily injections of 0 (control), 4.5×10^6 (low), or 4×10^7 (high) particles of XM6:anti-*fos* retroviral vector. Estrogen was added to some groups of animals to induce specificity of gene expression, while other groups were treated with dexamethasone to induce gene expression via the promoter, as well as provide an anti-inflammatory effect. The sera and tissue samples that tested positive for the vector by PCR were obtained predominantly from mice with the glycogen-induced inflammation. A dose-related increase in the number of vector positive tissues was observed by PCR in samples from mice treated with glycogen and estrogen.

By contrast, the number of positive tissues obtained was decreased when the same dose of vector was administered to glycogen-primed mice in the presence of dexamethasone. PCR testing demonstrated that some of the tissues contained vector; however, no tissues were positive for XM6:anti-*fos* gene expression. These results show that distribution of retroviral vector from the peritoneal cavity is enhanced in the presence of an inflammatory response in mice and may have practical implications with regard to the treatment of metastatic cancer.

ACKNOWLEDGMENTS

We would like to acknowledge Martha Bass and Victor Stephen Koo for expert technical assistance, and Drs. Harold Moses and Mitch Steiner for critical review of the manuscript. This work was funded by the Frances Williams Preston Laboratory of the T.J. Martell Foundation, the Vanderbilt Cancer Center, the Elsa U. Pardee Foundation (CLA) and NIH Grant CA62212-01A1 (CLA).

VI. REFERENCES

Arteaga, C.L., and Holt, J.T. (1996) Tissue-targeted antisense *fos* retroviral vector inhibits established breast cancer xenografts in athymic mice. *Cancer Res.* **56**:1098-1103.

Benet, L.Z. (1972) General treatment of linear mammillary models with elimination from any compartment as used in pharmacokinetics. *J. Pharm. Sci.* **61**:536-541.

Berenkopf, J.W., and Weichman, B.M. (1987) Differential effects of anti-inflammatory drugs on fluid accumulation and cellular infiltration in reverse passive arthus pleurisy and carrageenan pleurisy in rats. *Pharmacology* **34**:309-325.

Cato, A.C., Weinmann, J., Mink, S., Ponta, H., Henderson, D., and Sonnenberg, A. (1989) The regulation of expression of mouse mammary tumor virus DNA by steroid hormones and growth factors. *J. Steroid Biochem.* **34**:139-144.

Choi, Y., Henrard, D., Lee, I., and Ross, S.R. (1987) The mouse mammary tumor virus long terminal repeat directs expression in epithelial and lymphoid cells of different tissues in transgenic mice. *J. Virol.* **61**:3013-3030.

Cornetta, K., Nguyen, N., Morgan, R.A., Muenchau, D.D., Hartley, J.W., Blaese, R.M., and Anderson, W.F. (1993) Infection of human cells with murine amphotropic replication-competent retroviruses. *Hum. Gene Ther.* **4**:579-588.

Holt, J.T., and Arteaga, C.L. (1996) Clinical protocol: gene therapy for the treatment of metastatic breast cancer by *in vivo* transduction with breast-targeted retroviral vector expressing antisense c-*fos* RNA. *Hum. Gene Ther.* **7**:1367-1380.

Holt, J.T., Gopal, T.V., Moulton, A.D., and Nienhuis, A.W. (1986) Inducible production of c-*fos* antisense RNA inhibits 3T3 cell proliferation. *Proc. Natl. Acad. Sci. USA* **83**:4794-4798.

Howell, S.B., Zimm, S., and Markman, M. (1987) Long term survival of advanced ovarian carcinoma patients with small-volume disease treated with intraperitoneal chemotherapy. *J. Clin. Oncol.* **5**:1607-1612.

Jensen, R.A., Page, D.L., and Holt, J.T. (1994) Identification of genes expressed in premalignant breast disease by microscopy-directed cloning. *Proc. Natl. Acad. Sci. USA* **91**:9237-9261.

Matsui, Y., Halter, S.A., Holt, J.T., Hogan, B.L.M., and Coffey, R.J. Jr. (1990) Development of mammary hyperplasia and neoplasia in MMTV-TGF alpha transgenic mice. *Cell* **61**:1147-1155.

Sajjadi, N., Kamantigue, E., Edwards, W., Howard, T., Jolly, D., Mento, S. and Chada, S. (1994) Recombinant retroviral vector delivered intramuscularly localizes to the site of injection in mice. *Hum. Gene Ther.* **5**:693-699.

Sanui, H., Yoshida, S.-I., Nomoto, K., Ohhara, R., and Adachi, Y. (1982) Peritoneal macrophages which phagocytose autologous polymorphonuclear leucocytes in guinea pigs. I: Induction by irritants and microorganisms and inhibition by colchicine. *Br. J. Exp. Path.* **63**:278-284.

Tait, D. L., Obermiller, P. S., Redlin-Frazier, S., Jensen, R. A., Welcsh, P., Dann, J., King, M-C., Johnson, D.H., and Holt, J.T. (1997) A phase I trial of retroviral BRCA1sv gene therapy in ovarian cancer *Clin. Can. Res.,* in press.

Wadler, S., Faks, J.Z., and Wiernik, P.H. (1986) Phase I and phase II agents in cancer. *J. Clin. Pharmacol.* **26**:491-509.

Watt, S.M., Burgers, A.W., and Metcalf, D. (1979) Isolation and surface labeling of murine polymorphonuclear neutrophils. *J. Cell. Physiol.* **100**:1-22.

Yoshinaga, M., Nakamura, S., and Hayashi, H. (1975) Interaction between lymphocytes and inflammatory exudate cells I. Enhancement of thymocyte response to PHA by product(s) of polymorphonuclear leukocytes and macrophages. *J. Immunol.* **115**:533-538.

13

Clinical and Immunologic Responses to Gene Transfer of an Allogeneic Major Histocompatibility Complex Antigen

Alison T. Stopeck and Evan M. Hersh
Arizona Cancer Center, Tucson, Arizona

I. INTRODUCTION

A major goal of cancer immunotherapy is to generate specific and long-lasting antitumor immune responses capable of destroying existing tumor cells and preventing subsequent disease recurrences. Immunotherapics can be generally divided into two approaches. In the first approach, agents that activate host defense cells, such as interleukin-2 (IL-2) or interferons, are used to augment host effector cell function. The second approach attempts to increase tumor immunogenicity sufficiently to induce specific host antitumor immune responses.

Recent advances in immunology, particularly in our understanding of the cellular pathways required for antigen processing (Lanzavecchia, 1993; Germain *et al.*, 1993), the role of cytokine or helper molecules in T cell recognition (June *et al.*, 1994; Clark *et al.*, 1994; Schwartz, 1992; Guinan *et al.*, 1994), and the discovery of tumor-associated antigens capable of inducing host immune activation (Pardoll, 1993; Kawakami *et al.*, 1996) promise to revolutionize immunotherapy. Specificity in the antitumor immune response is desirable to limit toxicities observed with systemic, nonspecific immunotherapies such as IL-2 (Rosenberg *et al.*, 1987,1989). Thus, several immunotherapy approaches attempt to target host responses against tumor-associated antigens. Many tumor-specific antigens, such as mutated oncogenes, are intracellular proteins. Without surface expression of the antigen, antibody or humoral responses are largely worthless. Cellular or T cell immune responses are dependent on surface presentation of the antigenic peptide in the context of self MHC. Intracellular proteins can be processed into peptide fragments that are then transported to the cell surface in association with MHC molecules. In this manner, abnormal intracellular proteins can be presented appropriately for T cell immune recognition. Cellular immunity is both potent and long lasting. Consequently, the generation of cytotoxic T cell responses is the primary goal of several antitumor immunization strategies (Pardoll, 1993).

Several clinical trials have focused on developing tumor cell vaccines that generate antitumor cytotoxic T lymphocytes (CTLs). These efforts have been largely disappointing with low response rates and few longterm remissions (Morton *et al.*, 1992; Dalgleish, 1994). Using DNA transfer technology, specific active antitumor immunity is now achievable in select animal models using genetically modified tumor cell vaccines (reviewed in Pardoll, 1993; Dalgleish, 1994; Gilboa *et al.*, 1994). In murine models, tumor cells genetically modified to produce immuno-stimulatory cytokines (Colombo *et al.*, 1994; Fearon *et al.*, 1990; Golumbek *et al.*, 1991), express immunogenic antigens or co-stimulatory molecules (Townsend *et al.*, 1993; Chen *et al.*, 1992), or block tumor produced inhibitors of immune function (Fakhrai *et al.*, 1996) are usually superior to unmodified tumor cell vaccines. Encouraged by these successes in murine tumor models, clinical trials have been initiated in cancer patients using gene transfer technologies.

Transfer of an allogeneic or foreign major histocompatibility complex (MHC) gene directly into tumor sites is one approach that has been studied in clinical trials involving multiple tumor types. The data generated from these trials illustrate both the promise and problems facing gene therapy and immunotherapy clinical trials in cancer patients today. The clinical, molecular, and immunologic results from these trials are the focus of this chapter.

II. TUMOR IMMUNOLOGY AND MHC

Despite documentation of numerous tumor-associated antigens and mutated oncogenes capable of inducing CTL and antibody responses, most human tumors do not generate effective host antitumor immune responses. Recognition of foreign antigens by the cellular immune system requires presentation of antigen peptide fragments in the context of self MHC class I or class II (Lanzavecchia, 1993; Germain *et al.*, 1993; Ruiter *et al.*, 1991). Multiple tumor types express decreased amounts of MHC class I on their cell surface (Plaksin *et al.*, 1988; Hammerling *et al.*, 1987; Ruiter *et al.*, 1991; Goodenow *et al.*, 1985). Tumor cells from metastatic sites have been found to express decreased levels of MHC protein compared to cells from primary tumors (Kaklamanis *et al.*, 1995; Bruntsch *et al.*, 1990). Decreased MHC expression has also been associated with more aggressive clinical courses in patients with melanoma (van Duinen *et al.*, 1988).

Gamma interferon has many immunomodulatory effects, including markedly increasing surface expression of MHC class I and II molecules (Steimle *et al.*, 1994; Restifo *et al.*, 1992; Ogasawara *et al.*, 1993; Boyer *et al.*, 1989). In several murine models, transfer of the gamma interferon gene into tumor cells has resulted in decreased tumor growth and protected the mice from tumor growth when rechallenged with unmodified, or wild-type tumor cells (Gansbacher *et al.*, 1990; Porgador *et al.*, 1993; Panelli *et al.*, 1996). However, vaccination protocols using tumor cells genetically modified to produce gamma interferon have not been universally effective in murine models (Dranoff *et al.*, 1993; Saito *et al.*, 1994). Additionally, there is data to suggest that the reduced tumorigenicity found in gamma interferon secreting tumor cells may be due to nonspecific host immune responses and unrelated to changes in MHC surface expression by the tumor cells (Esumi *et al.*, 1991). Treatment of patients with multiple tumor types, including colorectal, renal cell, and ovarian cancer, with systemic recombinant gamma interferon has produced disappointing response rates in several clinical trials (O'Connell *et al.*, 1989; D'Acquisto *et al.*, 1988; Bruntsch *et al.*, 1990; Quesada *et al.*, 1987). In summary, the animal and human data generated using gamma interferon suggest decreased MHC expression is only one of several mechanisms by which tumors evade immune recognition and destruction.

An alternative mechanism to increase tumor immunogenicity is to transfer a gene encoding a strongly immunogenic protein directly into tumor cells. By forcing tumor cells to express a strong immunogen, potent CTLs and/or antibody responses may be generated. Foreign or allogeneic MHC molecules are strong stimulators of the immune system capable of generating both CTL and antibody responses. The transplant literature attests to the vigorous immune responses and CTL activity induced after transplantation of organs or bone marrow expressing mismatched or allogeneic MHC antigens.

III. PRECLINICAL STUDIES

Gary Nabel and his collaborators at the University of Michigan hypothesized that transfer of an allogeneic MHC gene into tumor cells may generate a tumor antigen-directed CTL response capable of rejecting tumors *in vivo*. In 1990, they tested this hypothesis in a murine tumor model by directly transferring the allogeneic murine MHC class I gene, H-2K^s, into murine tumors carrying an H-2K^d or H-2K^b MHC class I restriction (Plautz *et al.*, 1993). *In vivo* gene delivery of the allogeneic murine MHC directly into actively growing tumors resulted in generation of CD8[+] CTLs directed against gene-modified as well as unmodified, parental tumor cells. In mice pre-immunized with intraperitoneal injections of irradiated tumor cells genetically modified to express the allogeneic MHC class I gene, long-lasting CTL responses were generated. Tumor growth was also prevented after rechallenge with unmodified tumor cells in pre-immunized mice. To explain the observed immune mediated killing of unmodified tumor cells, Nabel's group hypothesized that the vigorous immune response generated against tumor cells expressing the foreign MHC proteins led to the release of immuno-stimulatory cytokines and stimulation of host immune effector cells. In this immunostimulatory microenvironment, unknown tumor-associated antigens were ingested and presented by antigen presenting cells to host effector cells primed to respond. In this way, the animal was sensitized to otherwise unrecognized tumor antigens producing the observed therapeutic effects.

In additional preclinical studies, a plasmid DNA construct encoding the human MHC class I gene, HLA-B7, was successfully transferred into arteries of pigs using cationic lipid as the transfer vehicle (Nabel *et al.*, 1992). A granulomatous mononuclear cell infiltrate was observed ten days after gene transfer, which persisted for at least 75 days. A specific CTL response against HLA-B7 was confirmed *in vitro* using peripheral blood obtained from transfected pigs. This study established that *in vivo* transfer of an allogeneic MHC gene encoded by plasmid DNA formulated with cationic lipid was capable of generating long-lasting and specific anti-HLA immune responses.

Most animal studies and clinical trials of active specific immuno-therapy with gene-modified cells have relied on *in vitro* delivery of the gene with selection of the targeted cells to ensure adequate transgene expression prior to *in vivo* delivery. While *in vitro* gene transfer and selection is ideal for optimizing expression of the transgene in targeted cells, it is not practical for many cancer patients. Procuring tumor cells, expanding cultures *in vitro*, transferring the gene of interest, and selecting for cells carrying the transgene are not only expensive and time-consuming, but often impossible for many tumor types that do not grow well in culture. By proving that *in vivo* gene transfer by direct injection of plasmid DNA complexed to cationic lipid produced immunologically active levels of the transgene, HLA-B7, these early porcine studies established the feasibility of this approach for human trials.

IV. CLINICAL TRIALS

In the first clinical trial of direct intratumoral gene transfer of an allogeneic MHC gene, five patients with metastatic malignant melanoma were treated with the allogeneic MHC class I gene, HLA-B7, complexed with cationic lipid (Nabel *et al.*, 1993). In this preliminary trial, dosing and schedule of gene injections were not standardized; however, HLA-B7 gene transfer was documented in the treated nodules of all patients. In one patient, injected cutaneous nodules (as well as several metastatic, non-injected tumor nodules) regressed completely. Tumor specific CTLs were detected in both of two patients tested for CTLs after therapy. Importantly, none of the patients developed significant toxicity or autoimmune reactivity. Because of these encouraging results, a multi-center clinical trial was initiated to determine safety, toxicities, gene transfer efficiency, and immunologic responses associated with intra-tumoral gene delivery of the allogeneic MHC gene, HLA-B7.

In the initial Phase I study, 49 patients were treated with intra-tumoral injection of the HLA-B7 gene formulated with the cationic lipids 1,2-dimyristyloxypropyl-3-dimethyl-hydroxyethyl ammonium bromide (DMRIE) and dioleoyl phosphatidyl-ethanolamine (DOPE). Seventeen patients with melanoma were treated at the Arizona Cancer Center (Stopeck *et al.*, 1997), seventeen patients with liver metastases from colorectal cancer were treated at the Mayo Clinic, Rochester, MN (Rubin *et al.*, 1997), and fifteen patients with metastatic renal cell carcinoma were treated at the University of Chicago, Chicago, IL (Vogelzang *et al.*, 1995). In general, patients had large tumor burdens and were heavily pre-treated with conventional chemotherapy. Patients were treated with one to three gene injections directly into a single tumor mass. Inclusion criteria included histologically confirmed tumor, bidimensional measurable tumor mass of at least 1 cm, Karnofsky performance status $\geq 70\%$, and adequate renal and hematologic function. All patients were HLA-B7 negative as determined by flow cytometric testing of their peripheral blood mononuclear cells. A normal lymphocyte proliferative response to the lymphocyte mitogen, PHA, was the only test of host immunologic function required for inclusion in the study.

All patients received the allogeneic MHC gene into a single tumor site by direct injection. Radiographic (CT scan or ultrasonic) guidance was required for the majority of patients as visceral metastases were predominantly treated. The trial was sponsored by Vical, Inc. of San Diego, CA. The study drug, Allovectin-7, was supplied by Vical in two separate vials. One vial contained the plasmid DNA (concentration 1.0 mg/ml) and the second vial contained the lipids DMRIE and DOPE. Dosing was determined by the amount (micrograms) of plasmid DNA injected into the tumor. The DNA to lipid mass ratio was formulated at 5:1 (charge ratio of –1:+0.1) for all administrations.

The plasmid encoded the genes for HLA-B7 as well as β_2-microglobulin driven by a single RSV (Rous Sarcoma Virus) promoter (Figure 1). In the initial clinical trial of Nabel *et al.* (1993), the HLA-B7 expression plasmid did not contain the β_2-microglobulin gene.

Figure 1. Diagram of plasmid with HLA-B7 and ß₂-microglobulin (ß₂m) genes driven by RSV promoter. Internal ribosomal entry site (CITE), kanamycin resistance gene (Kanʳ), bovine growth hormone poly A tail (Bgh polyA), and primer locations for PCR and RT-PCR (arrows) are shown *(reprinted, with permission, from Stopeck, et al., 1997).*

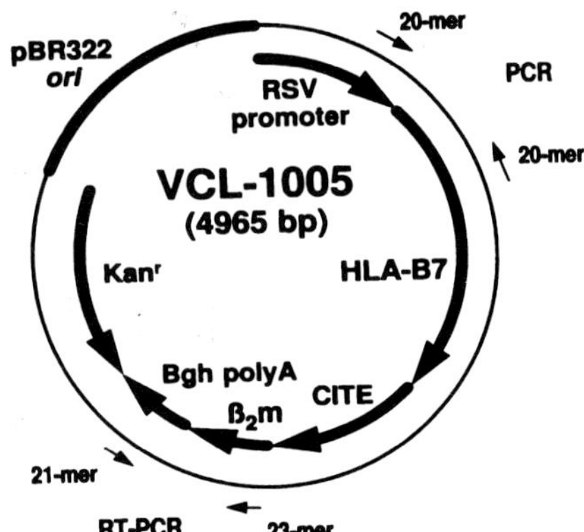

ß₂-microglobulin is necessary for surface expression of MHC class I. Deletions or mutations in the ß₂-microglobulin gene have been found in primary tumor biopsies and tumor cell lines (Momberg *et al.*, 1989; D'Urso *et al.*, 1991). Thus, by including the ß₂-microglobulin gene in the DNA plasmid a possible mechanism by which tumors escape immune recognition was eliminated.

Patients received either single injections of 10, 50, or 250 µg of plasmid DNA or multiple injections of the 10 µg DNA dose. Tumor biopsies were obtained pre-therapy and up to 8 weeks post-gene injection. Biopsies were used to confirm tumor diagnosis as well as HLA-B7 expression by tumor cells. Biopsies were also used to measure changes in tumor infiltrating T lymphocytes before and after therapy. Peripheral blood was obtained to measure immunoreactivity including specific CTL activity against HLA-B7 and nonspecific, natural killer (NK) and lymphokine activated killer (LAK) activity.

Responses in injected and non-injected tumor nodules were measured by physical examination or radiographically. An objective response to therapy was defined as a decrease of at least 25% in the product of the bidimensional measurements of the injected lesion. It is to be noted that this definition of response is different from the usual definition of remission, which necessitates tumor regression in all involved sites. Tumor responses were measured at approximately monthly intervals for three months after gene therapy and thereafter at three month intervals until progression.

The data for all three tumor types were similar regarding toxicities, gene transfer efficiencies, and immunologic responses. The only toxicities were the direct result of biopsies and injections and are expected complications of invasive procedures. Complete blood counts and serum biochemical parameters, including renal and liver function studies, were unaffected by gene transfer. None of the patients developed signs or symptoms suggestive of autoimmune diseases, and none developed antibodies directed against double stranded DNA or HLA-B7 antigen.

High gene transfer efficiencies were documented in all patients (Table 1). Persistence of the plasmid DNA by PCR or RT-PCR was documented up to 8 weeks (last biopsy obtained) after gene injection. As the number of transfected tumor cells was difficult to estimate by immunohistochemistry, tumor biopsies were also processed into single cell suspensions for flow cytometric analysis. The range and mean number of cells staining positive for HLA-B7 in tumor biopsies pre-gene injection was 0-7%, and 1.27±2.07%, respectively. Thus, tumors were considered positive for HLA-B7 expression if greater than 7% of the cells isolated from biopsies post gene injection expressed HLA-B7 by flow cytometry. In post-therapy biopsies considered positive for HLA-B7 by this definition, 8-30% (mean 15%) of the cells expressed HLA-B7 protein. Immunohistochemical staining confirmed that HLA-B7 was being expressed by tumor cells (Figures 2B, 3C). Increased numbers of tumor infiltrating T lymphocytes (TIL) were also observed in several tumor biopsies post gene therapy (Figures 2D, 3D). Generation of specific CTLs against target cells expressing HLA-B7 was analyzed *in vitro* from patient peripheral blood mononuclear cells before and after therapy. Fourteen of forty-five patients (31%) had an increased frequency of anti-HLA-B7 CTLs in their peripheral blood post-gene injection. Significant changes in NK or LAK cell frequencies before and after treatment were not observed at the time points analyzed.

The major differences observed among tumor types were in the clinical response data. Several patients treated with renal cell carcinoma or colon cancer had stabilization of their disease, but none of their injected tumors regressed by greater than 25%.

Table 1. Detection of HLA-B7 Expression in Tumor Biopsies

Tumor Type	Center	DNA	RNA	Protein	Positive for 1 or more
Melanoma	Arizona Cancer Center	9/14 (64%)	4/14 (29%)	7/14 (50%)	13/14 (93%)
Colorectal	Mayo Clinic	14/15 (93%)	5/15 (33%)	7/14 (50%)	15/16 (94%)
Renal Cell	University of Chicago	9/15 (60%)	4/15 (27%)	4/15 (27%)	12/15 (80%)
Melanoma	Univ. of British Columbia	5/7 (71%)	ND	ND	5/7 (71%)
	Totals	37/51 (73%)	13/44 (30%)	18/43 (42%)	45/52 (87%)

Values are expressed as the number of positive biopsies/number patients' post therapy biopsies analyzed. Positive value refers to detection of plasmid directed HLA-B7 expression by PCR for DNA, RT-PCR for RNA, and flow cytometric analysis for protein. The column labeled "Positive for 1 or more" indicates the post gene injection biopsy was positive for HLA-B7 expression by PCR, RT-PCR, or flow cytometry analysis. ND = not performed. *(Reprinted, with permission, from Stopeck, et al., 1997).*

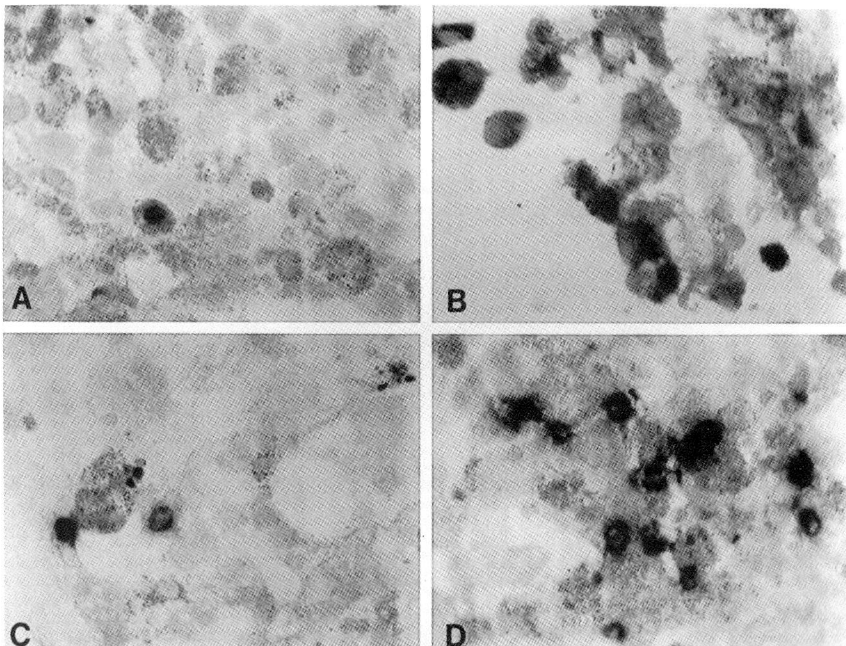

Figure 2. Immunohistochemical results from tumor biopsies removed from a patient with melanoma treated with two 10 μg DNA injections (day 1,15.)
A: day 43 biopsy, HLA-B7 negative control (irrelevant primary antibody); dark gray pigment is melanin in tumor cells. **B:** day 43 after initial gene injection, co-expression of HLA-B7 protein (black) and melanin (dark gray) in tumor cells. **C:** pre-gene injection biopsy stained for CD8[+] T lymphocytes (black.) **D:** increased CD8[+] T lymphocytes on day 43 biopsy *(reprinted, with permission, from Waddell, et al., 1997)*.

Of the 14 evaluable patients with metastatic melanoma, seven patients had objective responses insofar as the injected tumor nodule decreased by at least 25% (Table 2). Three of these responses were dramatic with complete or near complete resolution of the injected lesions (Figures 4,5). Responses were seen after injections at all dose levels and in nodules receiving single or multiple injections. Tumor responses typically were seen months after initial Allovectin-7 injection. The mean interval to tumor response was 3.6 months with a range of one to eleven months. Significant tumor responses were not observed in non-injected nodules in any of the patients treated.

Of the seven melanoma patients who had tumor regressions in response to Allovectin-7 injection, three are still alive. The duration of tumor response was difficult to measure. Three patients expired secondary to progressive disease within three months of documenting regression in the injected nodule. Three patients received retreatment after observing tumor regression following a single plasmid injection. One of these patients had a retrocaval lymph node as her only site of disease. She received two additional intratumoral HLA-B7 injections and is currently disease free 32 months from her initial therapy. Two other responders with lung nodules have recently received chemotherapy for progressive disease 23 and 30 months, respectively from their initial gene therapy treatments. The seventh responder, with ocular melanoma, had multiple large liver metastases from which he expired eight months after initiating therapy. The median survival of the 14 evaluable patients was 8.1 months.

In an effort to predict response to HLA-B7 gene injection, various parameters were compared between patients with injected nodules that regressed to therapy (responders) and those patients with non-responding tumors.

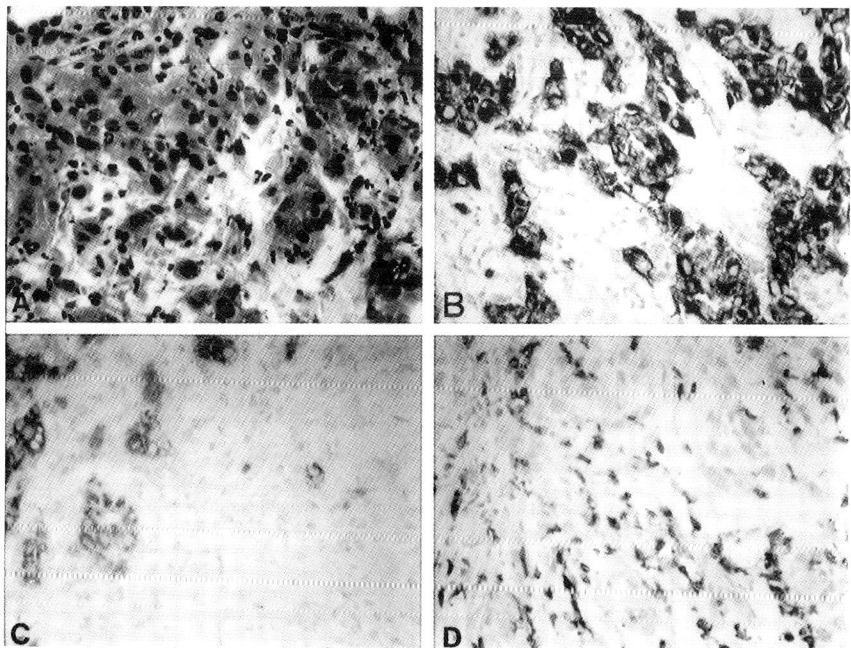

Figure 3. Immunohistochemical results from liver biopsies removed from a patient with metastatic colorectal cancer. **A:** Hematoxylin and eosin staining showing tumor replacing liver tissue. **B:** Keratin staining indicating tumor from epithelial origin. **C:** HLA-B7 staining (dark) revealing clusters of tumor cells expressing HLA-B7 protein. **D:** CD8$^+$ T cells (dark) infiltrating the tumor on post-gene therapy biopsy *(reprinted, with permission, from Stopeck, et al., 1997)*.

Table 2. Tumor Measurements Pre- and Post-Gene Injection in the 14 Evaluable Patients with Metastatic Melanoma Treated at the Arizona Cancer Center

Patient No./ Injected Site	DNA Dose (µg)	Tumor Size (mm^2) pre-Tx	Tumor Size (mm^2) post-Tx	Change in tumor size
1. retrocaval mass	10	598	77	-87%
2. lung nodule	10	500	110	-78%
3. mediastinal mass	10	5625	5625	0%
4. skin nodule	50	150	168	12%
5. lung nodule	50	2400	2862	19%
6. groin LN	50	900	3848	328%
7. lung nodule	250	1376	325	-76%
8. liver nodule	250	289	88	-70%
9. liver nodule	250	341	320	-6%
10. sc nodule	2 x 10	900	432	-52%
11. adrenal mass	2 x 10	720	513	-29%
12. sc nodule	2 x 10	225	1050	367%
13. sc nodule	3 x 10	400	600	150%
14. liver nodule	3 x 10	11875	8500	-28%

Abbreviations: sc, subcutaneous; LN, lymph node

Table 3. Response Correlations in Responding and Non-responding Patients with Metastatic Melanoma Treated at the Arizona Cancer Center

Patient Characteristics	Responders (7 patients)	Non-responders (7 patients)
Bx + for DNA by PCR	4/7 (57%)	5/7 (71%)
Bx + for mRNA by RT-PCR	2/7 (29%)	2/7 (29%)
Bx + for Protein by FACS	6/7 (86%)	4/7 (57%)
Visceral site	5/7 (71%)	3/7 (43%)
Cutaneous site	2/7 (29%)	4/7 (57%)
100% Performance Status	5/7 (71%)	5/7 (71%)
≥ 3 metastatic sites	5/7 (71%)	6/7 (86%)
Previous Immunotherapy	5/7 (71%)	7/7 (100%)
DTH response	5/7 (71%)	5/7 (71%)
anti-HLA B7 CTLs	1/7 (14%)	2/7 (29%)

Abbreviations: Bx, biopsy; +, HLA-B7 expression found in post-therapy tumor biopsy; DTH, delayed type hypersensitivity; CTLs, cytotoxic T lymphocytes. *(Reprinted, with permission, from Stopeck, et al., 1997)*

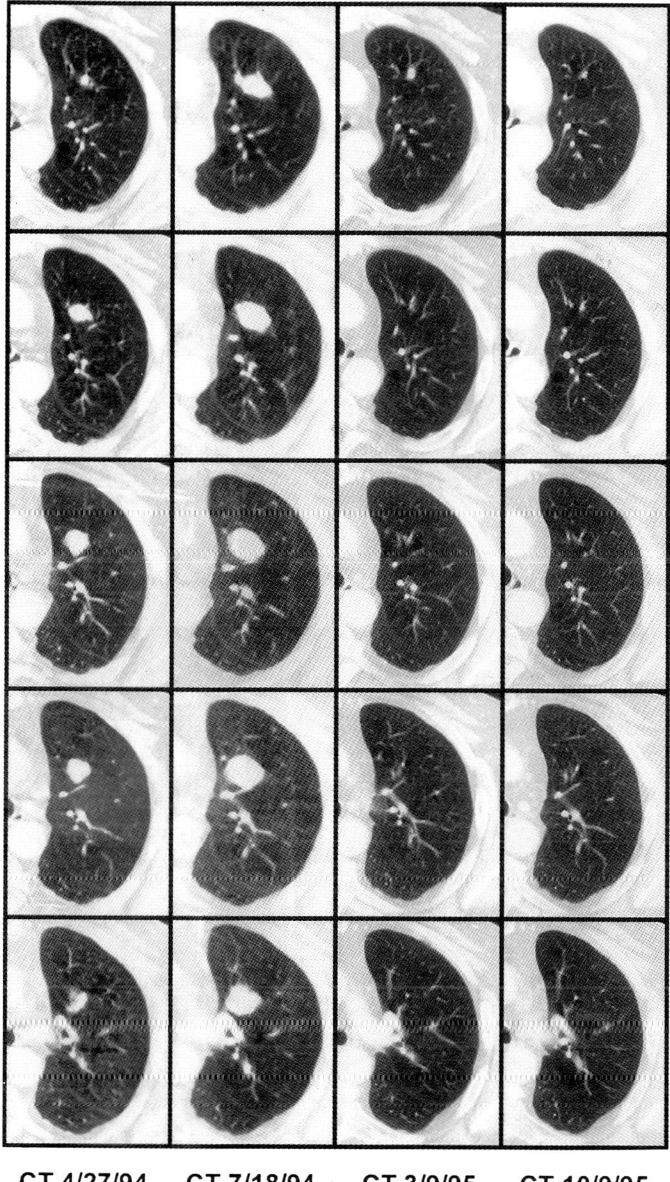

Figure 4. Composite chest CT scans from patient No. 2, Table 2. Patient was treated with 10 µg HLA-B7 plasmid DNA on 18 May, 1994. Tumor initially increased in size prior to disappearance of mass on later CT scans.

Documentation of successful gene transfer by DNA, RNA, or protein determination, generation of anti-HLA-B7 CTLs, increased numbers of tumor infiltrating lymphocytes (TIL) in post therapy biopsies, tumor burden, injection of visceral versus cutaneous tumor nodules, performance status, size of treated nodule, positive delayed type hypersensitivity reaction to Merieux skin tests, and previous therapy with an immunomodulatory agent were not found to predict response (Table 3). However, the low number of patients treated precluded statistical analysis or definitive conclusions.

The University of Michigan has continued to treat patients with metastatic melanoma with direct intratumoral injection of HLA-B7. In ten additional patients treated with multiple injections of plasmid DNA in cationic lipid at DNA doses ranging from 3-300 µg per injection, clinical responses have been seen in the injected nodules of two patients. All patients had HLA-B7 gene expression detected in post-therapy biopsies by PCR, RT-PCR, or immunohistochemical analysis. The number of cells expressing HLA-B7 in the injected tumors was estimated at 1-10% in the vicinity of the gene injection. One patient with an approximately 50% regression of the treated subcutaneous nodule also had regression of a non-injected lymph node. Isolation of TIL from this patient's tumor post-therapy revealed increased cytolytic activity and cytokine production in response to autologous tumor cells. TIL from this patient were subsequently expanded *ex vivo* and re-infused into the patient with IL-2. The patient had a complete response after her second course of TIL and IL-2 therapy. She is still in remission 21 months from her initial HLA-B7 intratumoral injection. This patient's clinical course is consistent with results obtained in murine tumor models where adoptive T cell therapy has been combined successfully with direct intratumoral gene transfer approaches (Wahl *et al.*, 1995).

Nabel and his collaborators studied host immune responses induced by intratumoral injection of HLA-B7 in TIL isolated from melanoma biopsies before and after gene injection (DeBruyne *et al.*, 1997). T cell receptor Vß usage in TIL isolated from melanoma nodules following HLA-B7 plasmid injection was more heterogeneous than the Vß repertoire of TIL isolated prior to therapy, determined by an increase in the total number of Vß families represented, as well as the number of Vß families required to comprise greater than 90% of the total repertoire. Increased heterogeneity also correlated with increased CD4[+] T cells in post-therapy tumor biopsies. Interestingly, the changes in Vß repertoire found in injected nodules were also seen in uninjected nodules in both patients analyzed. The specific antigen stimulating the T cells, as well as the MHC haplotype of the patient, determines T cell receptor Vß repertoires. Thus, these results provide further evidence that specific immune reactivity follows intratumoral injection of HLA-B7 plasmid DNA complexed to cationic lipids.

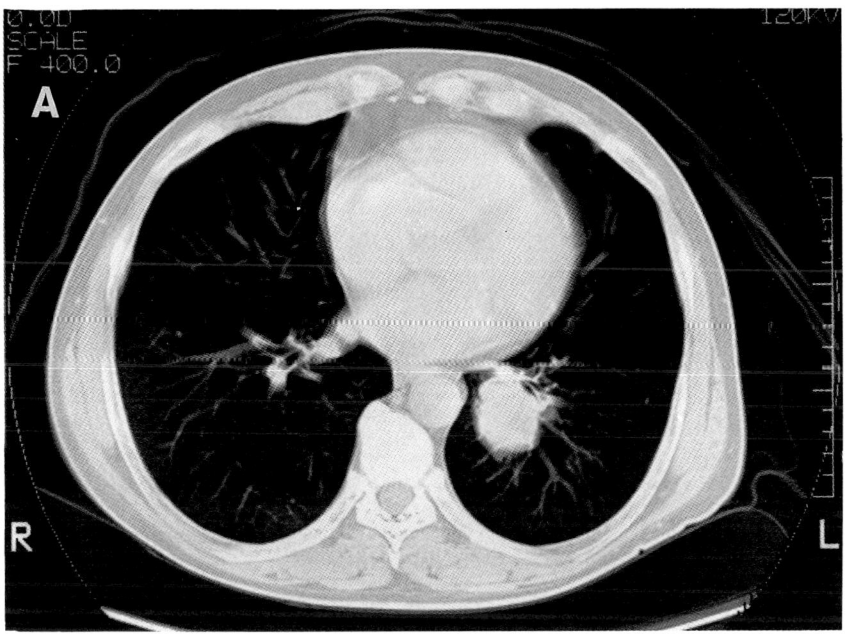

Figure 5. A: Baseline chest CT scan of patient No. 7 (Table 2) lying on back, demonstrating a large left lower lobe pulmonary mass. **B:** Chest CT scan of patient lying on stomach, showing placement of needle for HLA-B7 gene therapy injection into the left lower lobe mass. **C:** Followup chest CT scan of patient lying on back, approximately 16 weeks after initial gene injection, revealing capitation of the treated mass that decreased in size. There has also been interval increase in size in an adjacent untreated nodule.

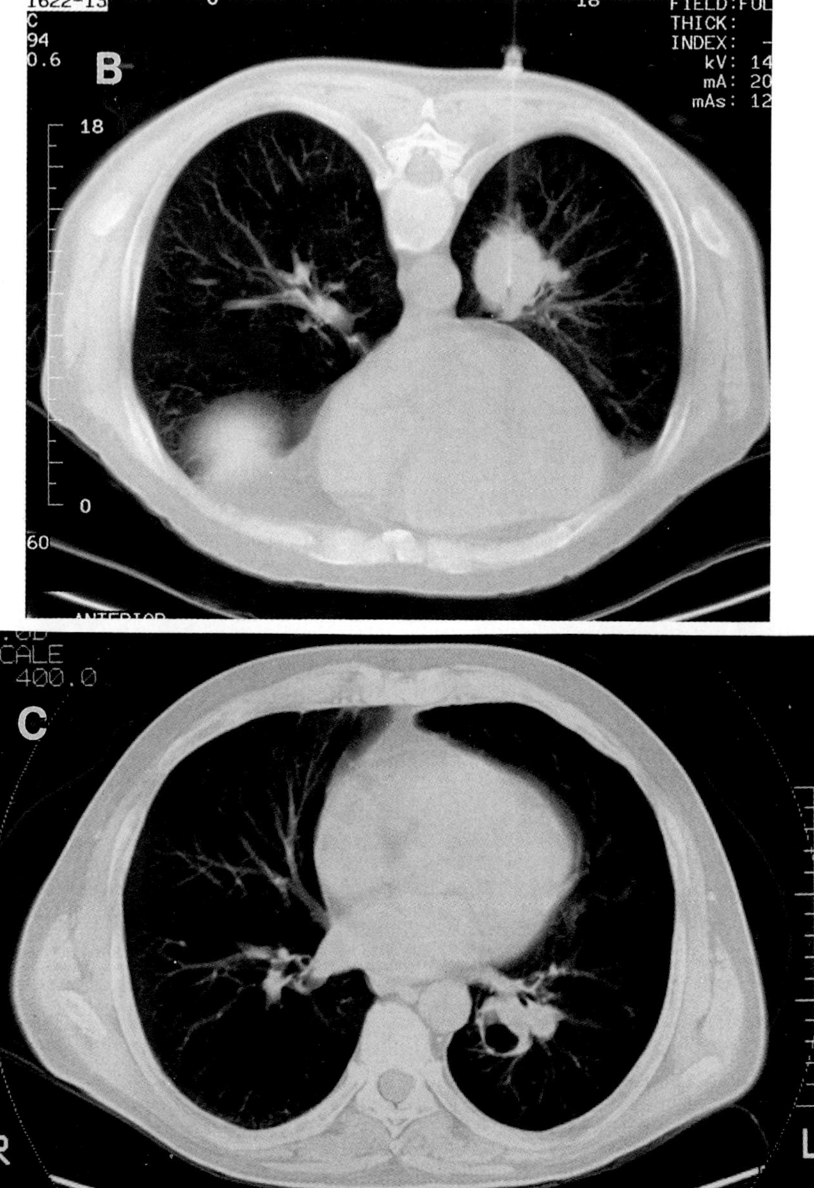

Figure 5. Continued.

In a preliminary trial performed at the British Columbia Cancer Agency, patients with melanoma were treated with 6 weekly intratumoral injections of Allovectin-7 regardless of their pre-study HLA-B7 typing (Silver et al., 1996). Three of the seven patients treated tested positive for HLA-B7 by flow cytometry of their peripheral blood mononuclear cells during their initial screening.

Three patients, including two of the patients who were initially found to be HLA-B7 positive, had clinical regressions in tumor nodules. In one of these patients, all three of his lesions decreased by greater than 50% seven weeks after gene injection into a single tumor nodule. The responses were short-lived; however, as tumor growth was noted in two nodules by week ten after injection. The other two patients with tumor shrinkage had mixed responses with some lesions progressing and others regressing simultaneously. Interestingly, the best responses were seen in non-injected nodules.

The results of this small study are notable in several respects. Patients had tumor responses observed in non-injected nodules after gene delivery into subcutaneous lesions. This agrees with results obtained at the University of Michigan where tumor responses were also observed in non-injected nodules after treating subcutaneous lesions (Nabel et al., 1993). The melanoma patients treated at the Arizona Cancer Center had predominantly visceral metastasis and responses in non-injected sites were not observed. Subcutaneous nodules may be better suited to vaccination strategies and generation of circulating CTLs secondary to their close proximity to antigen presenting dendritic cells. There is evidence to support this hypothesis as Nabel et al. (1996) found numerous HLA-DR positive dendritic cells around tumor cells in biopsies from subcutaneous melanoma nodules. Prior to HLA-B7 gene injection, dendritic cells rarely expressed the co-stimulatory molecules, CD80 or CD86 (B7.1 or B7.2) necessary for optimal T cell recognition and activation. After HLA-B7 gene injection, the number of dendritic cells expressing CD80 or CD86 increased markedly.

The British Columbia group also observed tumor responses in HLA-B7 positive patients, suggesting that syngeneic MHC gene injection could be successful. Murine and human tumors, including melanomas, often express reduced levels of surface MHC class I proteins (Plaksin et al., 1988; Hammerling et al., 1987; Ruiter et al., 1991). In fact, two of the three tumor biopsies from patients with pre-treatment HLA-B7 positive peripheral blood mononuclear cells had tumor biopsies negative for HLA-B7 expression by flow cytometry. In experimental models, transfection and expression of syngeneic MHC class I genes into murine tumors abrogate metastatic properties and tumorigenicity through the generation of CTLs (Plaksin et al., 1988; Wallich et al., 1985; Hui et al., 1984; Tanaka et al., 1985). By increasing the level of MHC class I expression in the tumors of HLA-B7 positive patients, tumor associated antigens may have been more effectively presented, resulting in immune recognition and the observed antitumor responses.

V. CONCLUSIONS

The trials discussed in this chapter are the first human trials using plasmid DNA formulated with cationic lipid to deliver genes directly into tumor cells *in vivo*. Biopsies from 90% of the injected tumors revealed HLA-B7 expression or plasmid persistence by PCR, RT-PCR, or flow cytometry (Table 4). Systemic immune recognition of the allogeneic protein as determined by increased frequency of HLA-B7 specific CTL was observed in 33% of the patients.

Furthermore, Nabel *et al.* (1996) and DeBruyne *et al.* (1997) documented changes in T cell receptor Vß usage, cytokine production after stimulation by autologous tumor cells, and increased cytolytic activity in TIL isolated after HLA-B7 plasmid injection. Thus, the "proof of principle" that gene delivery by direct intratumoral injection of plasmid DNA formulated with cationic lipid can produce immuno-modulatory levels of the transgene was established in humans. As these were phase I studies, toxicities, and not efficacy, were the true end points. These studies established the safety of this *in vivo* approach, as long-term or unexpected toxicities were not observed in any of the 66 patients treated.

Table 4. Response Data from Each Center After Intratumoral Injection of the HLA-B7 Gene

Tumor Type	Reference	Evaluable/ Entered	HLA-B7 Expression	Anti-HLA-B7 CTLs	Clinical Response
Melanoma	Nabel, 1993	5-May	5	1-Jan	1
Melanoma	Nabel, 1996	10-Oct	10	ND	2
Melanoma	Stopeck	14/17	13	14-Mar	7
Colorectal	Rubin	15/17	15	16-Jun	0
Renal Cell	Vogelzang	15/15	12	15-May	0
Melanoma	Silver	8-Jul	5	ND	3
	Totals	66/72 (92%)	60/67 (90%)	14/46 (33%)	13/66 (20%)

Evaluable/entered refers to patients entered (received at least one gene therapy injection) versus evaluable for clinical responses (on study for at least 4 weeks). HLA-B7 expression indicates biopsies contained HLA-B7 DNA, RNA, or protein. Anti-HLA-B7 CTLs imply that peripheral blood mononuclear cells obtained from patients post gene therapy had increased frequency of anti-HLA-B7 CTLs as measured in an *in vitro* assay as described (Stopeck *et al.*, 1997). Clinical responses were defined as a decrease of at least 25% in the product of the bidimensional measurement of any tumor lesion.
ND: not performed

In vivo gene delivery offers many advantages over *in vitro* gene transfer protocols. Using modern imaging and invasive radiological techniques, an *in vivo* approach is amenable to most cancer patients (Unger *et al.*, 1996). Extrapolating animal results and vaccination protocols used in treating infectious diseases, it is likely that multiple vaccinations will be necessary to evoke the required immune stimulation for active and specific antitumor immunity. Patients have received up to twelve injections of plasmid DNA formulated with cationic lipid without developing signs or symptoms of autoimmune disease or antibodies directed against double stranded DNA. Patients have been retreated with continued tumor responses, implying a lack of resistance to plasmid therapy or antibodies directed against the plasmid (Stopeck *et al.*, 1997; Nabel *et al.*, 1996). Recent insights into the role of co-stimulatory molecules such as the B7 ligand (June *et al.*, 1994; Clark *et al.*, 1994; Schwartz, 1992; Guinan *et al.*, 1994), Fas and Fas ligand interactions between tumor cells and host immune effector cells (Hahne *et al.*, 1996; Strand *et al.*, 1996; Nagata, 1996), and inhibitory proteins produced by tumor cells (Rozman *et al.*, 1991; Chen *et al.*, 1994) suggest multigene delivery may be required for successful cancer immunotherapy. The ease of plasmid preparation makes this approach ideally suited to simultaneously deliver combinations of genes in a single intratumoral injection. In addition, cationic lipid mediated gene transfer eliminates risks associated with viral vectors, including generation of replication competent virus, insertional mutagenesis, and vector immunogenicity which prevents repeat gene administrations (Blaese *et al.*, 1995).

The clinical response data confirm the inherent susceptibility of melanoma to immunotherapy approaches (Table 4) (Mukherji *et al.*, 1995; Hersey, 1994; Dalgleish, 1994). Combining the results of allogeneic gene injection in all melanoma patients treated, thirteen of thirty-six or 36% of the injected nodules regressed by at least 25%. In our data at the Arizona Cancer Center, four patients are still alive and one patient is in complete remission 32 months from her initial injection with Allovectin-7.

The absence of significant clinical responses in nodules of patients with renal cell and colon cancer is not surprising. Gene therapy cancer trials are still in their infancy. There is much to learn about generating specific active immunity capable of curing metastatic cancer. The pharmacokinetics of intratumoral gene injections including the optimal schedule, DNA dose, or duration of therapy are still undefined. Further research is needed to understand the role tumor burden, host immunocompetence, and tumor microenvironment play in affecting antitumor immunity. The role of other cytolytic immune effectors including NK cells, LAK, and macrophages may also be important or necessary to generate potent antitumor responses (Leong *et al.*, 1995). Superior clinical results may require increased gene expression through the use of alternative or improved methods of gene transfer. Intratumoral injection produced surface expression of the allogeneic MHC protein in approximately 15% of the tumor cells. Intraperitoneal, intravenous, or intra-arterial plasmid gene delivery may improve gene expression by increasing the volume of distribution of the plasmid. Undoubtedly, future developments in immunology and molecular biology will lead to improvements in gene delivery, targeting, expression, T cell responsiveness, and hopefully clinical responses.

These early phase I studies of direct intratumoral injection of an allogeneic MHC gene have established the safety and feasibility of plasmid DNA and cationic lipid gene delivery. Current phase II clinical trials involving gene transfer of HLA-B7 are underway in lymphoma, head and neck, prostate, renal, colon, and lung cancer. Hopefully, the results of these trials will further define the response rate, mechanism of action, and optimal dosing schedule for this innovative approach to cancer immunotherapy.

VI. REFERENCES

Blaese, M., Blankenstein, T., Brenner, M., Cohen-Haguenauer, O., Gansbacher, B., Russell, S., Sorrentino, B., and Velu, T. (1995) Vectors in cancer therapy: How will they deliver? *Cancer Gene Ther.* **2**:291-297.

Boyer, C.M., Dawson, D.V., Neal, S.E., Winchell, L.F., Leslie, D.S., Ring, D., and Bast, R.C. (1989) Differential induction by interferons of major histocompatibility complex-encoded and non-major histocompatibility complex-encoded antigens in human breast and ovarian carcinoma cell lines. *Cancer Res.* **49**:2928-2934.

Bruntsch, U., de Mulder, P.H., ten Bokkel Huinink, W.W., Clavel, M., Drozd, A., Kaye, S.B., Renard, J., and van Glabbeke, M. (1990) Phase II study of recombinant human interferon-gamma in metastatic renal cell carcinoma. *J. Biol. Res. Mod.* **9**:335-338.

Chen, L., Ashe, S., Brady, W., Hellstrom, I., Hellstrom, K., Ledbetter, J., McGowan, P., and Linsley, P. (1992) Costimulation of antitumor immunity by the B7 counterreceptor for the T lymphocyte molecules CD28 and CTLA-4. *Cell* **71**:1093-1102.

Chen, Q., Daniel, V., Maher, D.W., and Hersey, P. (1994) IL-10 is produced by melanoma cells and may have a role in immunosuppression mediated by melanoma. *Int. J. Cancer* **56**:1-6.

Clark, E.A. and Ledbetter, J.A. (1994) How B and T cells talk to each other. *Nature* **367**:425-428.

Colombo, M.P. and Forni, G. (1994) Cytokine gene transfer in tumor inhibition and tumor therapy: where are we now? *Immunol. Today* **15**:48-51.

D'Acquisto, R., Markman, M., Hakes, T., Rubin, S., Hoskins, W., and Lewis, J.L.Jr. (1988) A phase I trial of intraperitoneal recombinant gamma-interferon in advanced ovarian carcinoma. *J. Clin. Oncol.* **6**:689-695.

D'Urso, C.M., Wang, A., Cao, Y., Takate, R., Zeff, R.A., and Ferrone, S. (1991) Lack of HLA class I antigen expression by cultured melanoma cells FO-1 due to a defect in ß$_2$m gene expression. *J. Clin. Invest.* **87**:284-292.

Dalgleish, A.G. (1994) Cancer vaccines. *Eur. J. Cancer* **30A**:1029-1035.

DeBruyne, L.A., Chang, A.E., Cameron, M.J., Yang, Z., Gordon, D., Nabel, E.G., Nabel, G.J., and Bishop, D.K. (1996) Direct transfer of a foreign MHC gene into human melanoma alters T cell receptor Vß usage by tumor infiltrating lymphocytes. *Cancer Immunol. Immunother.* **43**:49-58.

Dranoff, G., Jaffee, E., Lazenby, A., Golumbek, P., Levitsky, H., Brose, K., Jackson, V., Hamada, H., Pardoll, D., and Mulligan, R. (1993) Vaccination with irradiated tumor cells engineered to secrete murine granulocyte-macrophage colony-stimulating factor stimulates potent, specific, and long-lasting anti-tumor immunity. *Proc. Natl. Acad. Sci. USA* **90**:3539-3543.

Esumi, N., Hunt, B., Itaya, T., and Frost, P. (1991) Reduced tumorigenicity of murine tumor cells secreting γ-interferon is due to nonspecific host responses and is unrelated to class I major histocompatibility complex expression. *Cancer Res.* **51**:1185-1189.

Fakhrai, H., Dorigo, O., Shawler, D.L., Lin, H., Mercola, D., Black, K.L., Royston, I., and Sobol, R.E. (1996) Eradication of established intracranial rat gliomas by transforming growth factor beta antisense gene therapy. *Proc. Natl. Acad. Sci. USA* **93**:2909-2914.

Fearon, E.R., Pardoll, D.M., Itaya, T., Golumbek, P., Levitsky, H.I., Simons, J.W., Karasuyama, H., Vogestein, B., and Frost, P. (1990) Interleukin-2 production by tumor cells bypasses T helper function in the generation of an antitumor response. *Cell* **60**:397-403.

Gansbacher, B., Bannerji, R., Daniels, B., Zier, K., Cronin, K., and Gilboa, E. (1990) Retroviral vector-mediated γ-interferon gene transfer into tumor cells generates potent and long lasting antitumor immunity. *Cancer Res.* **50**:7820-7825.

Germain, R.N. and Margulies, D.H. (1993) The biochemistry and cell biology of antigen processing and presentation. *Annu. Rev. Immunol.* **11**:403-450.

Gilboa, E., Lyerly, H.K., Vieweg, J., Saito, S. (1994) Immunotherapy of cancer using cytokine gene-modified tumor vaccines. *Semin. Cancer Biol.* **5**:409-417.

Golumbek, P.T., Lazenby, A.J., Levitsky, H.I., Jaffee, L.M., Karasuyama, H., Baker, M., and Pardoll, D.M. (1991) Treatment of established renal cancer by tumor cells engineered to secrete interleukin-4. *Science* **254**:713-716.

Goodenow, R.S., Vogel, J.M., and Linsk, R.L. (1985) Histocompatibility antigens on murine tumors. *Science* **230**:777-783.

Guinan, E.C., Gribben, J.G., Boussiotis, V.A., Freeman, G.J., and Nadler, L.M. (1994) Pivotal role of the B7:CD28 pathway in transplantation tolerance and tumor immunity. *Blood* **84**:3261-3282.

Hahne, M., Rimoldi, D., Schroter, M., Romero, P., Schreier, M., French, L.E., Schneider, P., Bornand, T., Fontana, A., Lienard, D., Cerottini, J., and Tschopp, J. (1996) Melanoma cell expression of Fas (Apo-1/CD95) ligand: implications for tumor immune escape. *Science* **274**:1363-1366.

Hammerling, G.J., Klar, D., Pulm, W., Momburg, F., and Moldenhauer, G. (1987) The influence of major histocompatibility complex class I antigens on tumor growth and metastasis. *Biochim. Biophys. Acta* **907**:245-259.

Hersey, P. (1994) Melanoma vaccines. Current status and future prospects. *Drugs* **47**:373-382.

Hui, K., Grosveld, F., and Festenstein, H. (1984) Rejection of transplantable AKR leukaemia cells following MHC DNA-mediated cell transformation. *Nature* **311**:750-752.

June, C.H., Bluestone, J.A., Nadler, L.M., and Thompson, C.B. (1994) The B7 and

CD28 receptor families. *Immunol. Today* **15**:321-331.

Kaklamanis, L., Leek, R., Koukourakis, M., Gatter, K.C., Harris, A.L., Vogelstein, B., and Frost, P. (1995) Loss of transporter in antigen processing 1 transport protein and major histocompatibility complex class I molecules in metastatic versus primary breast cancer. *Cancer Res.* **55**:5191-5194.

Kawakami, Y., Robbins, P.F., Wang, R.F., and Rosenberg, S.A. (1996) Identification of melanoma antigens recognized by T lymphocytes and their use in the immunotherapy of cancer. *PPO Updates* **10**, No. 12:1-20.

Lanzavecchia, A. (1993) Identifying strategies for immune intervention. *Science* **260**:937-943.

Leong, S.P.L., Zhou, Y-M., Granberry, M.E., Wang, T-F., Grogan, T.M., Spier, C., White, R., Mehta, A., and Lin, A.Y. (1995) Generation of cytotoxic effector cells against human melanoma. *Cancer Immunol. Immunother.* **40**:397-409.

Momberg, F. and Koch, S. (1989) Selective loss of $\Downarrow_2$-microglobulin mRNA in human colon carcinoma. *J. Exp. Med.* **169**:309-314.

Morton, D.L., Foshag, L.J., Hoon, D.S.B., Nizze, J.A., Waner, L.A., Chang, C., Davtyan, D.G., Gupta, R.K., and Elashoff, R. (1992) Prolongation of survival in metastatic melanoma after active specific immunotherapy with a new polyvalent melanoma vaccine. *Ann. Surg.* **216**:463-482.

Mukherji, B. and Chakraborty, N.D. (1995) Immunobiology and immunotherapy of melanoma . *Curr. Opin. Oncol.* **7**:175-184.

Nabel, E.G., Plautz, G.E., and Nabel, G.J. (1992) Transduction of a foreign histocompatibility gene into the arterial wall induces vasculitis. *Proc. Natl. Acad. Sci. USA* **89**:5157-5161.

Nabel, G.J., Nabel, E.G., Yang, Z.Y., Fox, B.A., Plautz, G.E., Gao, X., Huang, L., Shu, S., Gordon, D., and Chang, A.E. (1993) Direct gene transfer with DNA-liposome complexes in melanoma: expression, biologic activity, and lack of toxicity in humans. *Proc. Natl. Acad. Sci. USA* **90**:11307-11311.

Nabel, G.J., Gordon, D., Bishop, D.K., Nickoloff, B.J., Yang, Z., Aruga, A., Cameron, M.J., Nabel, E.G., and Chang, A.E. (1996) Immune response in human melanoma after transfer of an allogeneic class I major histocompatibility complex gene with DNA-liposome complexes. *Proc. Natl. Acad. Sci. USA* **93**:15388-15393.

Nagata, S. (1996) Fas ligand and immune evasion. *Nature Med.* **2**:1306-1307.

O'Connell, M.J., Ritts, R.A.Jr., Moertel, C.G., Schutt, A.J., and Sherwin, S.A. (1989) Recombinant interferon-gamma lacks activity against metastatic colorectal cancer but increases serum levels of CA 19-1. *Cancer* **63**:1998-2004.

Ogasawara, M. and Rosenberg, S.A. (1993) Enhanced expression of HLA molecules and stimulation of autologous human tumor infiltrating lymphocytes following transduction of melanoma cells with γ-interferon genes. *Cancer Res.* **53**:3561-3568.

Panelli, M.C., Wang, E., Shen, S., Schluter, S.F., Bernstein, R.M., Hersh, E.M., Stopeck, A., Gangavalli, R., Barber, J., Jolly, D., and Akporiaye, E.T. (1996) Interferon g gene transfer of an EMT6 tumor that is poorly responsive to IFNg stimulation: increase in tumor immunogenicity is accompanied by induction of a mouse class II transactivator and class II MHC. *Cancer Immunol. Immunother.* **42**:99-107.

Pardoll, D.M. (1993) Cancer vaccines. *Immunol. Today* **14**:310-314.

Plaksin, D., Gelber, C., Feldman, M., and Eisenbach, L. (1988) Reversal of the metastatic phenotype in Lewis lung carcinoma cells after transfection with syngeneic H-2K^b gene. *Proc. Natl. Acad. Sci. USA* **85**:4463-4467.

Plautz, G.E., Yang, Z.Y., Wu, B.Y., Gao, X., Huang, L., and Nabel, G.J. (1993) Immunotherapy of malignancy by *in vivo* gene transfer into tumors. *Proc. Natl. Acad. Sci. USA* **90**:4645-4649.

Porgador, A., Bannerji, R., Watanabe, Y., Feldman, M., Gilboa, E., and Eisenbach, L. (1993) Antimetastatic vaccination of tumor-bearing mice with two types of IFN-γ gene-inserted tumor cells. *J. Immunother.* **150**:1458-1470.

Quesada, J.R., Kurzrock, R., Sherwin, S.A., and Gutterman, J.U. (1987) Phase II studies of recombinant human interferon gamma in metastatic renal cell carcinoma. *J. Biol. Res. Mod.* **6**:20-27.

Restifo, N.P., Spiess, P.J., Karp, S.E., Mule, J.J., and Rosenberg, S.A. (1992) A nonimmunogenic sarcoma transduced with the cDNA for interferon γ elicits CD8$^+$ T cells against the wild-type tumor: correlation with antigen presentation capability. *J. Exp. Med.* **175**:1423-1431.

Rosenberg, S.A., Lotze, M.T., Muul, L.M., Chang, A., Avid, F.P., Leitman, S., Linehan, W.M., Robertson, G.N., Lee, R.E., Rubin, J.T., Seipp, C.A., Simpson, C.G., and White, D.E. (1987) A progress report on the treatment of 157 patients with advanced cancer using lymphokine-activated killer cells and interleukin-2 or high-dose interleukin-2 alone. *New Engl. J. Med.* **316**:889-905.

Rosenberg, S.A., Lotze, M.T., Yang, J.C., Aebersold, P.M., Linehan, W.M., Seipp, C.A., and White, D.E. (1989) Experience with the use of high-dose interleukin-2 in the treatment of 652 cancer patients. *Ann. Surg.* **210**:474-485.

Rozman, T., Elliott, L., and Brooks, W. (1991) Modulation of T-cell function by gliomas. *Immunol. Today* **12**:370-374.

Rubin, J., Galanis, E., Pitot, H.C., Richardson, R.L., Burch, P.A., Charboneau, J.W., Reading, C.C., Lewis, B.D., Stahl, S., Akporiaye, E.T., and Harris, D.T. (1997) Phase I study of immunotherapy of hepatic metastases of colorectal carcinoma by direct gene transfer of an allogeneic histocompatibility antigen, HLA-B7. *Gene Ther.* **4**:419-425.

Ruiter, D.J., Mattijssen, V., Broecker, E.B., and Ferrone, S. (1991) MHC antigens in human melanomas. *Semin. Cancer Biol.* **2**:35-45.

Saito, S., Bannerji, R., Gansbacher, B., Rosenthal, F., Romanenko, P., and Heston, W.D.W. (1994) Immunotherapy of bladder cancer with cytokine gene-modified tumor vaccines. *Cancer Res.* **54**:3516-3520.

Schwartz, R.H. (1992) Costimulation of T lymphocytes: the role of CD28, CTLA-4, and B7/BB1 in interleukin-2 production and immunotherapy. *Cell* **71**:1065-1068.

Silver, H.K.B., Klasa, R.J., Bally, M.B., LeRiche, J.C., Stahl, S., and Schreiber, A.B. (1996) Phase I gene therapy study of HLA-B7 transduction by direct injection in malignant melanoma. *Proc. Am. Assoc. Cancer Res.* **37**:342. (Abstract)

Steimle, V., Siegrist, C., Mottet, A., Lisowska-Grospierre, B., and Mach, B. (1994) Regulation of MHC Class II expression by interferon-γ mediated by the transactivator gene CIITA. *Science* **265**:106-109.

Stopeck, A.T., Hersh, E.M., Akporiaye, E.T., Harris, D.T., Grogan, T., Unger, E., Warncke, J., Schluter, S.F., and Stahl, S. (1997) Phase I study of direct gene transfer of an allogeneic histocompatibility antigen, HLA-B7, in patients with metastatic melanoma. *J. Clin. Oncol.* **15**:341-349.

Strand, S., Hofmann, W.J., Hug, H., Muller, M., Otto, G., Strand, D., Mariani, S.M.,

Stremmel, W., Krammer, P.H., and Galle, P.R. (1996) Lymphocyte apoptosis induced by CD95 (APO-1/Fas) ligand-expressing tumor cells-a mechanism of immune evasion. *Nature Med.* **2**:1361-1366.

Tanaka, K., Isselbacher, K.J., Khoury, G., and Jay, G. (1985) Reversal of oncogenesis by the expression of a major histocompatibility complex class I gene. *Science* **228**:26-30.

Townsend, S.E. and Allison, J.P. (1993) Tumor rejection after direct costimulation of CD8[+] T cells by B7-transfected melanoma cells. *Science* **259**:368-370.

Unger, E.C., Wright, W., Stopeck, A.T., and Hersh, E.M. (1996) Imaging's special role as gene therapy guide. *Diagnostic Imaging* (Nov.):157-161.

van Duinen, S.G., Ruiter, D.J., Broecker, E.B., van der Velde, E.A., Sorg, C., Welvaart, K., and Ferrone, S. (1988) Level of HLA antigens in locoregional metastases and clinical course of the disease in patients with melanoma. *Cancer Res.* **48**:1019-1025.

Vogelzang, N.J., Sudakoff, G., McKay, S., Lesting, T., and Stahl, S. (1995) A Phase I study of intra-lesional (IL) gene therapy in metastatic renal cell cancer (RCC). *Proc. Am .Soc. Clin. Oncol.* **14**:242. (Abstract)

Waddell, W. III, Wright, W., Unger, E., Stopeck, A., Akporiaye, E., Harris, D., Grogan, T., Schluter, S., Hersh, E., and Stahl, S. (1997) Human gene therapy for melanoma: CT-guided interstitial injection. *Am. J. Roentgenol.* **169**:63-67.

Wahl, W.L., Strome, S.E., Nabel, G.J., Plautz, G.E., Cameron, M.J., San, H., Fox, B.A., Shu, S., and Chang, A.E. (1995) Generation of therapeutic T-lymphocytes after *in vivo* tumor transfection with an allogeneic class I major histocompatibility complex gene. *J. Immunother.* **17**:1-11.

Wallich, R., Bulbuc, N., Hammerling, G.H., Katzav, S., Segal, S., and Feldman, M. (1985) Abrogation of metastatic properties of tumour cells by *de novo* expression of H-2K antigens following H-2 gene transfection. *Nature* **315**:301-305.

14

Defective Tumor Suppressor Gene Replacement and Oncogene Inactivation for the Treatment of Cancer

Jack Roth

University of Texas, Houston, Texas

I. INTRODUCTION

The strategy for inactivating or replacing cancer-causing genes is analogous to the classic concept of gene therapy for replacing defective or nonfunctioning genes. So far, this strategy has been applicable to two gene families implicated in carcinogenesis: dominant oncogenes and tumor suppressor genes. Previous studies have shown that regional administration of viral vectors expressing wild-type p53 protein and antisense K-*ras* RNA in orthotopic tumor models prevents the growth of tumors with the respective genetic lesions and mediates regression of large established tumors. These data provide the rationale for two novel clinical protocols recently approved by the NIH Recombinant DNA Advisory Committee and FDA: one of these will replace a defective *p53* gene with a normal *p53* encoded by a recombinant retrovirus injected intratumorally, and the other inactivate mutant K-*ras* mRNA via the expression of antisense K-*ras* RNA. If these methods are efficacious, their lack of toxicity may provide a sufficiently high therapeutic index to allow their use as adjuvants to surgery in patients with earlier stages of cancer or as agents to prevent second primary cancers in individuals with

279

genetically abnormal premalignant lesions. Although substantial research and development must be done, the possibility of specific gene targeting with a high therapeutic index makes gene activation or replacement a promising area for investigation.

The basic paradigm in gene therapeutic approaches to monogenic diseases is to treat and potentially cure disease by inserting and expressing a normal copy of the mutant or deleted gene in host cells. In this way, monogenic diseases such as adenosine deaminase deficiency or Gaucher's disease could be treated and potentially cured. Analogously, the identification of specific genes that contribute to the development of the cancer cell presents an opportunity to replace defective genes or inactivate genes that have gained transforming function, allowing the treatment and prevention of cancer. However, this is a considerably more complex problem than in monogenic disease, as the development of cancer is associated with multiple genetic abnormalities.

The gene families implicated in carcinogenesis include dominant oncogenes and tumor suppressor genes (Bishop, 1991; Weinberg, 1992) (Table 1). Proto-oncogenes (the normal homologues of dominant oncogenes) participate in critical cell functions, including signal transduction and transcription, and only a single mutant allele is required to induce malignant transformation. Primary modifications in the dominant oncogenes that confer gain of transforming function include point mutation, amplification, translocation, and rearrangement. In comparison, tumor suppressor genes usually require homozygous loss of function either by mutation, deletion, or both, although dominant p53 mutants have been reported. Some tumor suppressor gene products even

Table 1. Oncogenes and Tumor Suppressor Genes Altered in Lung Cancer

Small-Cell Lung Cancer	Non-Small-Cell Lung Cancer
Oncogenes	
c-*myc**, L-*myc*, N-*myc*	K-*ras**, N-*ras*, H-*ras*
c-*myb*, c-*raf*	c-*myc*, c-*raf*
c-*erbB*-1 (EGF-R)	c-*fur**, c-*fes*
c-*fms*, c-*rlf*	c-*erbB*-1 (EGF-R)
c-*erbB*-2 (*Her2, neu*)	
c-*sis*, bcl-1	
Tumor Suppressor Genes	
*p53**, Rb**, p16*	*p53**, p16*

*Most frequently altered genes in tumors or cell lines evaluated.
(Reprinted, with permission, from Greenblott et al., 1995)

appear to play a role in governing proliferation by their regulation of transcription. Alterations in the expression of dominant oncogenes and tumor suppressor genes may therefore influence certain characteristics of cells that contribute to the malignant phenotype.

Multiple genetic abnormalities are present in cancer cell lines and fresh tumor samples (Vogelstein et al., 1988, 1989; Yokota et al., 1987; Ibson et al., 1987). This is evident at the chromosomal level, where multiple chromosomal abnormalities have been identified. In addition, more and more oncogenes and tumor suppressor genes are being identified. Thus, some have felt that gene replacement cancer therapy would not be possible because of the difficulties associated with correcting so many genetic abnormalities in one cell. However, several observations by us and others suggest otherwise.

Several studies have shown that correction of a single genetic defect, such as eliminating expression of a dominant oncogene or adding a normal copy of a tumor suppressor gene to a cell with deleted or mutated copies, can reduce or eliminate such critical characteristics of the malignant phenotype as tumorigenicity or anchorage-independent growth (Takahashi, et al., 1992; Mukhopadhyay, et al., 1991; Bookstein et al., 1990). Studies done in our laboratory support this concept, and their results have been applied to animal models and clinical protocols, as described below. Essential to the mediation of any therapeutic effect however, is the efficient delivery of the therapeutic gene to the cancer cell. It was previously felt that most available vectors were very inefficient in transducing genes into cancer cells. However, studies in models of human tumors in vitro and in nu/nu mice show that uptake of a retroviral vector is efficient enough to mediate a therapeutic effect (Zhang, et al., 1993,1994; Georges, et al., 1993; Fujiwara, et al., 1994a). Retroviruses can indeed be integrated into a genome and express a transgene primarily in proliferating cells. Thus, this vector may be selective for cancer cells. Another type of vector, the adenovirus vectors, can be taken up by both dividing and non-dividing cells. Consequently, even a single infection in culture is sufficient to infect more than 90% of cancer cells (Zhang et al., 1994). Furthermore, both of these viral vector types can penetrate 3-dimensional tumor cell matrices (Fujiwara et al., 1993; Cusack et al., 1996).

Yet, despite the relatively high efficiencies of cancer cell transduction achieved by viral vectors, it is doubtful that all cancer cells in a tumor can be transduced. Fortuitously transduced cells can cause cell death in nontransduced cells by mediating a "bystander effect." This effect was first noted in brain tumor cells transduced with the herpes simplex thymidine kinase gene and then exposed to ganciclovir (Culver et al.,

1992), and was thought to be mediated by passage of toxic metabolites of the ganciclovir through gap junctions, by phagocytosis of apoptotic vesicles, or by immune responses (Freeman et al., 1993; Bi et al., 1993; Vile et al., 1994). Since then, the bystander effect has also been observed in lung cancer cells containing endogenous mutant p53 protein, which have been transduced with a retrovirus expressing the wild-type p53, although a different mechanism is probably involved (Cai et al., 1993). Together, these observations suggest that gene replacement in cancer has the potential for mediating therapeutic effects. Given the multiple genetic lesions in cancer cells, however, a critical question will be which genes to target.

II. DOMINANT ONCOGENES

One possible gene group to target is dominant oncogenes, and *ras* oncogenes in particular, since members of the *ras* oncogene family are frequently mutated in many common malignancies (Bos, 1989). These genes, which are homologous to the rat sarcoma virus, code for a protein called p21 that is located on the inner surface of the plasma membrane, has GTPase activity, and may participate in signal transduction. The oncogenes themselves are activated by point nucleotide mutations that alter the amino acid sequence of p21.

To test the idea of targeting one dominant oncogene for gene therapy, my laboratory used antisense technology to identify the effects of eliminating expression of a mutant K-*ras* oncogene in lung cancer cells (Mukhopadhyay *et al.*, 1991). In brief, a homozygous mutation at codon 61, i.e., substitution of a normal glutamine residue (CAA) by histidine (CAT), was detected in a clone of the NCI-H460a large cell undifferentiated human NSCLC cell line. An antisense K-*ras* RNA construct was then developed and transfected into H460a cells, which have, in addition to a K-*ras* mutation, five chromosomal deletions (chromosomes 1, 2, 9, 12, and 16). The construct selectively blocked the expression of mutant sense K-*ras* mRNA and reduced the growth rate of H460a tumors in *nu/nu* mice. Thus, although the cancer cells had multiple genetic abnormalities, the reversal of a single abnormality appeared to have profound effects on fundamental properties of the malignant phenotype, such as rapid proliferation and tumorigenicity.

Having established that gene therapy targeted to a single gene in cancer cells is efficacious, the question becomes how to deliver the therapy. Retroviruses have been extensively studied as delivery vehicles in gene transfer protocols and therefore, are a candidate vector for gene delivery *in vivo* (Danos and Mulligan, 1988). One approach was to create retroviral vectors that lack genes essential for replication. Such replication-defective vectors are capable of infecting cells and being integrated as proviruses which can then express recombinant genes in the cells.

Because gene constructs transduced by retroviruses are integrated preferentially in dividing cells, this technique gives proliferating cancer cells a selective advantage for expressing the gene construct. Moreover, retroviruses and cells modified by retroviral transduction have little acute toxicity, making multiple treatments with high-titer preparations feasible. Taking advantage of these traits, my laboratory developed a retroviral vector system for efficiently transducing a K-*ras* antisense construct into human cancer cells (Zhang *et al.*, 1993). The 1.8-kb K-*ras* gene fragment DNA, in antisense orientation to a ß-actin promoter, was inserted into the retroviral vector LNSX, and transduced into H460a cells. Expression of antisense K-*ras* RNA in the transduced H460a cells dramatically decreased colony formation in soft agarose and tumorigenicity in an orthotopic *nu/nu* mouse model (Georges *et al.*, 1993). We concluded that an antisense K-*ras* construct can be expressed effectively from a retroviral vector that efficiently transduces human cancer cells. Together, these studies in K-*ras* gene therapy showed that retroviruses expressing the appropriate therapeutic construct can mediate antitumor effects, that an antisense construct can mediate therapeutic tumor regression *in vivo*, and that therapeutic antitumor effects can occur following inhibition of oncogene expression.

III. TUMOR SUPPRESSOR GENES

Tumor suppressor genes are a second potential target for therapy, with the *p53* gene being a particularly good candidate. The *p53* gene, which encodes a 393 amino acid phosphoprotein that can form complexes with viral proteins such as large T antigen and E1B, is the most commonly mutated gene identified to date in human cancers. Missense mutations are common in the *p53* gene and in many cases will functionally impair the p53 protein (Raycroft *et al.*, 1990; Fields and Jang, 1990). In addition, the mechanism of *p53* transformation may vary depending on the type of *p53* mutation. The p53 protein appears to be multifunctional, with major domains that can transactivate other genes, bind other proteins, bind sequence-specific DNA, and oligomerize with other p53 proteins. Abnormalities in one or more of these functions could eliminate or reduce the tumor suppressor function of the p53 gene product. Failure of the mutant p53 protein to activate transcription of molecules essential for regulating the cell cycle and DNA repair, or the untimely expression of molecules transcriptionally enhanced by the mutant p53 protein, may make the cell more susceptible to genetic instability. Thus, the wild-type *p53* gene may suppress genes that contribute to uncontrolled cell growth and proliferation or activate genes that suppress uncontrolled cell growth, so that the absence or inactivation of the wild-type p53 protein may contribute to transformation. The *p53* gene also regulates cell-cycle progression. If DNA damage occurs, the cell will arrest at the G1 checkpoint until the damage is repaired. Failure to repair DNA damage may then trigger apoptosis (programmed cell death).

To study the effects of *p53* gene therapy, a retroviral vector was created for efficient transduction of the wild-type *p53* gene into human lung cancer cell lines H358a (deleted *p53*) and H322a (mutant *p53*) (Cai *et al.*, 1993). In brief, LNSX/*p53* constructs incorporating *p53* cDNA driven by a ß-actin promoter mediated the stable integration of *p53* into cancer cells. The consequent restoration of the wild-type *p53* gene suppressed growth in cell lines with a mutated *p53* but had no effect on transduced, H460a tumor cells, which have an endogenous wild-type *p53* gene. Furthermore, mixing experiments showed that transduced cells could reduce the growth rate of nontransduced cells, thus indicating a bystander effect that may have been mediated by factors shed into the supernatant of the transduced cell cultures.

Since mutations in the *p53* tumor suppressor gene are especially common in human lung cancers, and since the wild-type form of p53 protein is usually dominant over the mutant, restoration of wild-type p53 function in lung cancer cells may suppress their growth as tumors. Consequently, we also investigated the therapeutic efficacy of direct administration of a retroviral wild-type *p53* expression vector (LNp53B) in a mouse model of human orthotopic lung cancer (Fujiwara *et al.*, 1994a). H226Br cells, originally derived from a squamous lung cancer that metastasized to brain, carrying a point mutation (ATC to GTC) at exon 7, codon 254, of the *p53* gene, were used. Thirty days after inoculation with the H266Br cells, 63-80% of control mice showed macroscopic tumors of the right mainstem bronchus.

However, subsequent treatment with LNp53B suppressed H226Br tumor formation in 62-100% of the mice in a dose-dependent fashion. Thus, direct administration of a retroviral vector expressing wild-type p53 protein may inhibit local growth *in vivo* of human lung cancer cells with abnormal p53 expression. We concluded that development of gene-replacement treatment strategies based on the type of mutations found in target cancers is warranted and may lead to the development of new adjunctive therapies and gene-specific prevention strategies for lung cancer.

Unlike retroviral vectors, adenovirus vectors can transduce both dividing and non-dividing cells and may have a high affinity for infecting lung epithelium. Therefore, we also developed the adenovirus vector Ad5CMV-p53 for delivery of wild-type *p53* gene (Zhang *et al.*, 1993,1994). Using this vector, high level expression of exogenous p53 protein was achieved in H358 cells, which lack endogenous p53, at a multiplicity of infection (MOI) of 30 pfu/cell. In contrast, when H322 and H460 cells were infected at the same MOI, the level of expression of the exogenous p53 was three times higher than that of the endogenous mutated protein, and 14 times higher than that of the endogenous wild-type p53 protein in H460 cells.

In another study, the time course of exogenous p53 expression after a single infection at 10 pfu/cell was studied in H358 cells. The protein expression peaked at day 3 post-infection, sharply decreased after day 5, and lasted for at least 15 days. This is a critical point with respect to safety of the Ad5CMV-p53 vector, since it shows that 1) transient p53 expression is sufficient for mediating apoptosis but that 2) normal cells taking up the vector will express the exogenous p53 for only a short time. In short, Ad5CMV-p53 inhibited the proliferation of lung cancer cells with mutated or deleted p53 protein, but only minimally affected growth of cells with endogenous wild-type p53.

To study the efficacy of Ad5CMV-p53 in inhibiting tumorigenicity, we again used the mouse model of orthotopic human lung cancer and H226Br cells (Georges *et al.*, 1993). After 6 weeks, only 25% of Ad5CMV-p53-treated mice formed tumors, whereas in the vehicle or Ad5/RSV/GL2 control groups, 70-80% of treated mice formed tumors. Tumors in the Ad5CMV-p53 group were significantly smaller on average than those in the control groups. Together, these results indicated that Ad5CMV-p53 can prevent H226Br cells from forming tumors in the mouse model of orthotopic human lung cancer.

We also examined whether Ad-p53 and cisplatin given in a sequential combination could induce synergistic tumor regression *in vivo*. Following direct intratumoral injection of Ad-p53 on 3 consecutive days, H358a tumors subcutaneously transplanted into *nu/nu* mice showed a modest slowing of growth.

Ad-p53-injected tumors regressed, however, if cisplatin was first administered intraperitoneally on 3 consecutive days. Histologic examination revealed necrosis of tumor tissue in the area where Ad-p53 was injected into mice pretreated with cisplatin. *In situ* staining of tumor cells treated with Ad-p53 followed by cisplatin showed extensive areas of apoptosis. In contrast, *in situ* staining of tumors treated with either cisplatin or Ad-p53 alone showed no apoptosis.

IV. CLINICAL APPLICATIONS

The studies just described provide a rationale for two clinical protocols involving intratumoral injection of recombinant retroviruses and approved by the Recombinant DNA Advisory Committee (RAC) of the National Institutes of Health (NIH) and the Food and Drug Administration (FDA). One protocol aims to inhibit expression of mutant K-*ras* p21 with a vector expressing antisense K-*ras;* the other aims to replace a defective *p53* gene with a vector expressing normal p53 protein (Roth, 1996ab). Patients with unresectable lung cancer obstructing a bronchus will have their tumors directly injected with the appropriate retroviral supernatant, while patients with unresectable local tumors or isolated metastases may undergo radiologically guided injection of the vector.

The results of treatment of the first nine patients on the retroviral p53 protocol have been reported (Roth, *et al.*, 1996). All nine patients (median age, 68 years; range, 51-73) had a history of primary non-small cell lung carcinoma (NSCLC). Of these nine, four patients had recurrent endobronchial lesions (three squamous cell carcinomas and one adenocarcinoma); these were treated with bronchoscopic injections of the retroviral p53 expression vector ITRp53A. Another four patients had chest wall lesions (two large cell carcinomas, one squamous cell carcinoma, and one adenocarcinoma); these were treated with percutaneous injections under computed tomographic (n = 3) or fluoroscopic guidance. The ninth patient had a left adrenal metastasis from a large cell carcinoma; this was treated with percutaneous injection of the vector under CT guidance. All nine patients had recurrent or metastatic tumors that had progressed during prior treatment. Five of the nine patients had undergone surgical resection of either the primary lung cancer (n = 3) or a brain metastasis (n = 2). Five of the nine had received chemotherapy, and eight had received radiation therapy. Thus all patients in the study had failed existing treatments and had cancers that were progressively growing.

Polymerase chain reaction (PCR) analysis using primers specific for the retroviral transgene was used to detect the presence of vector sequences in tumor biopsies taken before and after treatment. While PCR revealed no retroviral sequences in any of the pretreatment biopsies, it did reveal sequences in DNA extracted from posttreatment biopsies of two patients and from a postmortem tumor of one patient (Table 2). Moreover, *in situ* hybridization using the *neo*[r] probe revealed nuclear localization of the vector by in six patients (Table 2).

Thus, all patients, with the exception of one who did not complete the treatment protocol (patient 9), showed some evidence of gene transfer. Six of the seven patients also showed increased terminal deoxynucleotidyl transferase biotin dUTP nick end-labeling (TUNEL) staining in posttreatment versus pretreatment biopsies, indicating apoptosis (programmed cell death) was occurring in cancer cells (Table 2).

Three of the seven patients evaluable for response showed evidence of tumor regression in the treated lesions. One patient had recurrence of a squamous carcinoma in the left mainstem bronchus at the bifurcation of the left upper and lower lobes. Bronchoscopy 1 month after retroviral injection showed marked regression of the tumor mass, which continued until the patient died 4 months after treatment secondary to tumor progression at untreated sites in the left lung and distant metastases (Figure 1).

Autopsy revealed no evidence of invasive cancer in the epithelium of the treated left mainstem bronchus. The second patient had an endobronchial recurrence that obstructed the right upper lobe, despite followup radiation therapy. Six separate biopsies one month after retroviral injection showed no evidence of viable tumor at the right upper lobe orifice. The third patient had tumor remaining after radiation therapy that obstructed the right upper lobe orifice. Six weeks after completion of radiation therapy, the patient's tumor was given a single cycle of five daily endobronchial injections of the p53 retroviral vector. One month later, the right upper lobe orifice was patent, with >50% regression of the treated endobronchial tumor (Figure 1).

In the remaining four evaluable patients, tumor growth stabilized in three for periods ranging from 8 to 9 weeks. The fourth patient had a chest wall lesion that continued to progress after the first cycle of treatment. Each of these four patients had other sites of disease not treated by gene replacement that continued to progress during treatment with the vector.

In each patient, progression at untreated sites was evident at each monthly evaluation, contrasting with the stabilization or regression noted in the treated lesions. The degree of inflammatory cell infiltrate from tumor biopsies showed no consistent changes between pretreatment and posttreatment biopsies, indicating that an immune response or inflammatory response was not responsible for the tumor regressions noted.

In our seven evaluable patients, no toxic effects were directly attributable to the vector, although three patients had complications related to the procedures involved in administering the vector. All seven patients had lymphocytes and sputum samples collected for up to 3 months after treatment. Three patients died during the study and were autopsied.

Table 2. Assessment of gene transfer by ITRp53A retroviral vector*

Patient No.	DNA-PCR	*In Situ* Hybridization Score[†]	%TUNEL Staining[‡]
1	+	3	17.0
2	-	3	34.0
3	+	ND[§]	ND
4	+	ND	27.6
5	-	2	10.2
6	-	2	10.2
7	-	2	0.8
8	-	1	23.0
9	ND	ND	ND

*Results shown are for indicated assays of posttreatment biopsy specimens. No pretreatment samples were positive for gene transfer in DNA-PCR and *in situ* hybridization assays. Examples are shown in Figures 1-3.

[†]All slides were coded and read by a single blinded observer without knowledge of the patient or collection date. The percentage of tumor cells with punctate nuclear staining was determined in 500 cells per slide (400 X magnification). The slides were evaluated on the following scoring system: no staining, 0; <5% of cells stained, 1; 5 - 20% of cells stained, 2; >20% of cells stained, 3. Slides with a score $\geq$1 were considered positive. The maximum score for each patient is given; significant increases in *in situ* hybridization were observed in 6 of 6 evaluable cases (p<.05, sign test).

[‡]Cells with morphologic features of necrosis were not included among the TUNEL-positive cells. The percentage of cells with nuclear staining was determined in 500 cells per slide (400X magnification). The background level of TDT staining in pretreatment samples was <4%. All slides were coded and read by a single blinded observer without knowledge of the patient or collection date. The maximum percentage of cells staining positively is given for each patient. The mean percentage $\pm$ SD of TUNEL stained cells was 1.0 $\pm$ 0.5 for all pretreatment biopsies and 10.6 $\pm$ 2.9 for posttreatment biopsies. Six of 7 biopsies showed definitive increases in TUNEL staining after treatment (p<.05, two-sided Wilcoxon signed rank test).

§ND, not done because specimen unsatisfactory or unavailable.

(Reprinted, with permission, from Roth et al., 1996).

None of the nontumor tissues analyzed by PCR, including lymphocytes, tracheal mucosa, brain, uninjected lung, gastrointestinal tract tissues, skeletal muscle, heart, spleen, liver, and testes, showed retroviral sequences. Since tumor regression in this study was noted only in endobronchial tumors and was limited to the local tumor, we plan to continue this study in endobronchial tumors, but with a cycle of three rather than five injections, to determine if the lower dose is also efficacious. Also, we plan future improvements in vector design and production techniques that may increase the efficacy of retroviral-mediated gene transduction and extend its clinical applications.

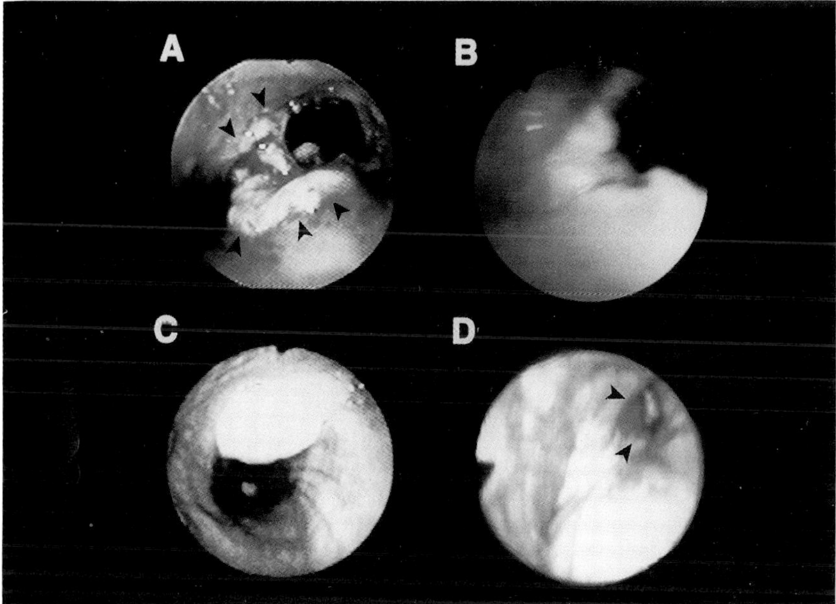

Figure 1. Pretreatment (**A** and **C**) and 30-day posttreatment (**B** and **D**) bronchoscopic images of patients 1 (**A** and **B**) and 5 (**C** and **D**). **A**: Lesion is situated in the left mainstem bronchus on the division of the upper and lower lobes; biopsies showed squamous cell carcinoma (arrows). **B**: Left mainstem bronchus 30 days following ITRp53A injection; five independent biopsies showed absence of viable tumor cells. **C**: Adenocarcinoma obstructing the right upper lobe orifice. **D**: Right upper lobe orifice (arrows) 30 days following treatment; biopsies in this region showed residual adenocarcinoma *(reprinted, with permission, from Roth et al., 1996)*.

Another protocol, which has also received NIH RAC and FDA approval, will test the vector Ad5CMV-p53 in an open-label upward-dose-ranging study. Patients for this protocol must have either an endobronchial tumor accessible by the bronchoscope, with some clinical evidence of bronchial obstruction, or advanced local-regional cancer that is unresectable. The study will be done in two phases.

In the first phase, which will assess toxicities related only to the vector, patients will receive one intratumor or intrapleural injection of Ad5CMV-p53 at an initial dose of 10^6 pfu. The second phase will evaluate Ad5CMV-p53 and cisplatin administered concurrently, based on evidence of the synergy between wild-type *p53* gene and cisplatin (Fujiwara, *et al.,* 1994b). Patients in this group will receive a single dose of cisplatin (80 mg/m^2) followed 3 days later by an intratumoral injection of Ad5CMV-p53. Three patients will be entered at each dose level except the maximum tolerated or maximum attainable dose; six patients will be entered at that level. The adenovirus dose will increase in 1 $\log_{10}$ increments for each group. A third RAC and FDA approved protocol, to be conducted in collaboration with Dr. Gary Clayman, will attempt to prevent local recurrences in patients with resectable recurrent head and neck cancers by treating the incision site with Ad5CMV-p53.

V. FUTURE STUDIES

It is clear from basic and clinical research that successful therapy and prevention interventions to reverse genetic lesions are possible. Genetic constructs could specifically inhibit expression of mutant proteins by dominant oncogenes and could replace the function of deleted or mutated tumor suppressor genes, provided they can be delivered with high efficiency to tumor cells *in vivo*. Viral vectors have this potential. The aerodigestive tract is especially suited to this approach because high concentrations of these relatively nontoxic vectors could be achieved with local instillation, thus avoiding the dilutional effects of intravenous injection. For example, restoration of normal p53 function in the tumor may increase radiosensitivity (Lee and Bernstein, 1993; Pardo, *et al.,* 1994).

Intervention to halt the progression of premalignant lesions to invasive cancer may also be possible. This is especially attractive since preventing the development of invasive cancers is clearly preferable to treating established cancer. Premalignant lesions such as bronchial dysplasia or Barrett's epithelium have tumor suppressor gene mutations that might be amenable to gene therapy (Casson, *et al.,* 1991; Bennett, *et al.,* 1993). However, there is also a potential role for viral vectors in the treatment of patients with more advanced cancer. Local recurrence or persistence of local disease is still a major problem in many cancers such as lung, head and neck, and pancreas. Intralesional injections or adjuvant use of gene constructs to prevent local recurrence after surgery could be considered. Limited metastatic disease could be injected with these agents percutaneously. Furthermore, if these agents prove efficacious, their lack of toxicity may provide a sufficiently high therapeutic index to allow their use as an adjuvant to surgery to treat patients with earlier stages of cancer or as agents for the prevention of second primary cancers in individuals with genetically abnormal premalignant lesions. The high titers achievable with adenovirus vectors also suggest that they could be used systemically, although immune responses to viral vectors have been reported. However, these can be markedly reduced with immunosuppressive agents (Yang, *et al.*, 1995; Fang, *et al.*, 1995). Vector targeting by expression of receptor ligands in the viral capsid is another possibility. Although much more additional research is required, the possibility of specific gene targeting with a high therapeutic index makes gene inactivation or replacement a promising area for investigation.

VI. REFERENCES

Bennett, W.P., Colby, T.V., Travis, W.D., Borkowski, A., Jones, R.T., Lane, D.P., Metcalf, R.A., Samet, J.M., Takeshima, Y., Gu, J.R., Vahakangas, K.H., Soini, Y., Paakko, P., Welsh, J.A., Trump, B.F., and Harris, C.C. (1993) p53 protein accumulates frequently in early bronchial neoplasia. *Cancer Res.* **53**:4817-4822.

Bi, W.L., Parysek, L.M., Warnick, R., Stambrook, P.J. (1993) *In vitro* evidence that metabolic cooperation is responsible for the bystander effect observed with HSV-tk retroviral gene therapy. *Hum. Gene Ther.* **4**:725-731.

Bishop, J.M. (1991) Molecular themes in oncogenesis. *Cell* 64: 235-248.

Bookstein, R., Shew, J.Y., Chen, P.L., Scully, P., and Lee, W.H. (1990) Suppression of tumorigenicity of human prostate carcinoma cells by replacing a mutated RB gene. *Science* **247**:712-715.

Bos, J.L. (1989) *Ras* oncogenes in human cancer: a review. *Cancer Res.* **49**:4682-4689.

Cai, D.W., Mukhopadhyay, T., Liu, Y.J., Fujiwara, T., and Roth, J.A. (1993) Stable expression of the wild-type p53 gene in human lung cancer cells after retrovirus-mediated gene transfer. *Hum. Gene Ther.* **4**:617-624.

Casson, A.G., Mukhopadhyay, T., Cleary, K.R., Ro, J.Y., Levin, B., and Roth, J.A. (1991) p53 gene mutations in Barrett's epithelium and esophageal cancer. *Cancer Res.* **51**:4495-4499.

Culver, K.W., Ram, Z., Wallbridge, S., Ishii, H., Oldfield, E.H., and Blaese, R.M. (1992) *In vivo* gene transfer with retroviral vector-producer cells for treatment of experimental brain tumors. *Science* **256**:1550-1552.

Cusack, J.C., Spitz, F.R., Nguyen, D., Zhang, W.W., Cristiano, R.J., and Roth, J.A. (1996) High levels of gene transduction in human lung tumors following intralesional injection of recombinant adenovirus. *Cancer Gene Ther.* **3**:245-249.

Danos, O., and Mulligan, R.C. (1988) Safe and efficient generation of recombinant retroviruses with amphotropic and ecotropic host ranges. *Proc. Natl. Acad. Sci. USA* **85**:6460-6464.

Fang, B., Eisensmith, R.C., Wang, H., Kay, M.A., Cross, R.E., Landen, C.N., Gordon, G., Bellinger, D.A., Read, M.S., Hu, P.C., Brinkhous, K.M., and Woo, S.L.C. (1995) Gene therapy for hemophilia B: host immunosuppression prolongs the therapeutic effect of adenovirus-mediated factor IX expression. *Hum. Gene Ther.* **6**:1039-1044.

Fields, S., Jang, S.K. (1990) Presence of a potent transcription activating sequence in the p53 protein. *Science* **249**:1046-1051.

Freeman, S.M., Abboud, C.N., Whartenby, K.A., Packman, C.H., Koeplin, D.S., Moolten, F.L., Abraham, G.N. (1993) The "bystander effect": tumor regression when a fraction of the tumor mass is genetically modified. *Cancer Res.* **53**:5274-5282.

Fujiwara, T., Grimm, E.A., Mukhopadhyay, T., Cai, D.W., Owen-Schaub, L.B., Roth, J.A. (1993) A retroviral wild-type p53 expression vector penetrates human lung cancer spheroids and inhibits growth by inducing apoptosis. *Cancer Res.* **53**:4129-4133.

Fujiwara, T., Cai, D.W., Georges, R.N., Mukhopadhyay, T., Grimm, E.A., Roth, J.A. (1994a) Therapeutic effect of a retroviral wild-type p53 expression vector in an orthotopic lung cancer model. *J. Natl. Cancer Inst.* **86**:1458-1462.

Fujiwara, T., Grimm, E.A., Mukhopadhyay, T., Zhang, W.W., Owen-Schaub, L.B., Roth, J.A. (1994b) Induction of chemosensitivity in human lung cancer cells *in vivo* by adenoviral-mediated transfer of the wild-type p53 gene. *Cancer Res.* **54**:2287-2291.

Georges, R.N., Mukhopadhyay, T., Zhang, Y.J., Yen, N., Roth, J.A. (1993) Prevention of orthotopic human lung cancer growth by intratracheal instillation of a retroviral antisense K-*ras* construct. *Cancer Res.* **53**:1743-1746.

Greenblott, M.S., Reddel, R.R., and Harris, C.C. (1995) Carcinogenesis, and cellular and molecular biology of lung cancer. In Roth, J.A., Ruckdeschel, J.C., and Weisenberger, T.H., eds., *Thoracic Oncology*, 2nd Ed. W.B. Saunders Company, New York, 5-25.

Ibson, J.M., Waters, J.J., Twentyman, P.R., Bleehen, N.M., Rabbitts, P.H. (1987) Oncogene amplification and chromosomal abnormalities in small cell lung cancer. *J. Cell Biochem.* **33**:267-288.

Lee, J.M., Bernstein, A. (1993) p53 mutations increase resistance to ionizing radiation. *Proc. Natl. Acad. Sci. USA* **90**:5742-5746.

Mukhopadhyay, T., Tainsky, M., Cavender, A.C., Roth, J.A. (1991) Specific inhibition of K-ras expression and tumorigenicity of lung cancer cells by antisense RNA. *Cancer Res.* **51**:1744-1748.

Pardo, F.S., Su, M., Borek, C., Preffer, F., Dombkowski, D., Gerweck, L., Schmidt, E.V. (1994) Transfection of rat embryo cells with mutant p53 increases the intrinsic radiation resistance. *Radiat. Res.* **140**:180-185.

Raycroft, L., Wu, H., Lozano, G. (1990) Transcriptional activation by wild-type but not transforming mutants of the p53 anti-oncogene. *Science* **249**:1049-1051.

Roth, J.A. (1996a) Modification of tumor suppressor gene expression in non-small cell lung cancer (NSCLC) with a retroviral vector expressing wild-type (normal) p53. *Hum. Gene Ther.* **7**:861-874.

Roth, J.A. (1996b) Modification of mutant K-*ras* gene expression in non-small cell lung cancer (NSCLC). *Hum. Gene Ther.* **7**:875-889.

Roth, J.A,, Nguyen, D., Lawrence, D.D., Kemp, B.L., Carrasco, C.H., Ferson, D.Z., Hong, W.K., Komaki, R., Lee, J.J., Nesbitt, J.C., Pisters, D.M.W., Putnam, J.B., Schea, R., Shin, D.M., Walsh, G.L., Dolormente, M.M., Han, C.-I., Martin, F.D., Yen, N., Stephens, L.C., McDonnell, T.J., Mukhopadhyay, T., and Cai, D. (1996) Retroviral-mediated wild-type p53 gene transfer to tumors of patients with lung cancer. *Nature Med.* **2** :985-991.

Takahashi, T., Carbone, D., Nau, M.M., Hida, T., Linnoila, I., Ueda, R., Minna, J.D. (1992) Wild-type but not mutant p53 suppresses the growth of human lung cancer cells bearing multiple genetic lesions. *Cancer Res.* **52**:2340-2343.

Vile, R.G., Nelson, J.A., Castleden, S., Chong, H., Hart, I.R. (1994) Systemic gene therapy of murine melanoma using tissue specific expression of the HSVtk gene involves an immune component. *Cancer Res.* **54**:6228-6234.

Vogelstein, B., Fearon, E.R., Hamilton, S.R., Kern, S.E., Preisinger, A.C., Leppert, M., Nakamura, Y., White, R., Smits, A.M.M., Bos, J.L. (1988) Genetic alterations during colorectal tumor development. *New Engl. J. Med.* **319**:525-532.

Vogelstein, B., Fearon, E.R., Kern, S.E., Hamilton, S.R., Preisinger, A.C., Nakamura, Y., White, R. (1989) Allelotype of colorectal carcinomas. *Science* **224**:207-211.

Wcinberg, R.A. (1992) Tumor suppressor genes. *Science* **254**:1138-1145.

Yang, Y., Trinchieri, G., Wilson, J.M. (1995) Recombinant IL-12 prevents formation of blocking IgA antibodies to recombinant adenovirus and allows repeated gene therapy to mouse lung. *Nature Med.* **1**:890-893.

Yokota, J., Wada, M., Shimosato, Y., Terada, M., Sugimura, T. (1987) Loss of heterozygosity on chromosomes 3, 13, and 17 in small-cell carcinoma and on chromosome 3 in adenocarcinoma of the lung. *Proc. Natl. Acad. Sci. USA* **84**:9252-9256.

Zhang, W.W., Fang, X., Branch, C.D., Mazur, W., French, B.A., Roth, J.A. (1993a) Generation and identification of recombinant adenovirus by lipsome-mediated transfection and PCR analysis. *Biotechniques* **15**:868-872.

Zhang, Y.J., Mukhopadhyay, T., Donchower, L.W., Georges, R.N., Roth, J.A. (1993b) Retroviral vector-mediated transduction of K-*ras* antisense RNA into human lung cancer cells inhibits expression of the malignant phenotype. *Hum. Gene Ther.* **4**:451 460.

Zhang, W., Fang, X., Mazur, W., French, B.A., Georges, R.N., Roth, J.A. (1994) High-efficiency gene transfer and high-level expression of wild-type p53 in human lung cancer cells mediated by recombinant adenovirus. *Cancer Gene Ther.* **1**:5-13.

15

The Molecular Basis of Bladder Cancer and Prospects for Gene Therapy Using Hammerhead Ribozymes

Eric J. Small and Mohammed Kashani-Sabet
University of California at San Francisco, San Francisco, California

David Y. Bouffard and Kevin J. Scanlon
Berlex Biosciences, Richmond, California

I. BLADDER CANCER: CLINICAL FEATURES

Bladder cancer is the most common malignancy of the urinary tract, and is the sixth most common cancer in men, and the tenth most common cancer in women. In 1996, it is anticipated that over 52,000 new cases of bladder cancer will be diagnosed in the United States, and that there will be nearly 12,000 deaths attributed to this disease (Wingo *et al.*, 1995). Transitional cell carcinoma (TCC) is the predominant type of urinary tract cancer in the United States, comprising more than 90% of all bladder cancers (Silverman *et al.*, 1992).

While superficial bladder cancer accounts for 60-70% of all cases of urothelial malignancies, approximately 30% of patients will present with muscle invasive disease at diagnosis (Wingo *et al.*, 1995). Gross lymph node involvement or distant metastases can generally be predicted on the basis of clinical stage of the tumor. Up to 50% of patients with muscle invasive disease will ultimately relapse with metastatic disease despite aggressive surgical and radiotherapeutic interventions (Lerner and Skinner, 1996).

Superficial bladder cancers are defined as neoplasms confined to the bladder mucosa, and include flat tumors remaining *in situ* (stage Tis: flat carcinoma *in situ* or CIS); outgrowing or papillary tumors arising from the epithelium (stage Ta: papillary carcinoma); and tumors with microinvasion of the lamina propria (stage T1). By contrast, invasive bladder cancer is defined as tumor invading into the muscle layer of the bladder or beyond (stages T2-T4).

Treatment of superficial bladder cancer attempts to eradicate or control the disease locally, and to prevent progression to invasive or metastatic disease. The primary modality is transurethral endoscopic resection of the bladder tumor (TURBT). Although it is usually possible to achieve removal of all identifiable tumor via TURBT, over 50% of patients will nevertheless still develop subsequent tumor recurrence within the bladder (Sidky *et al.*, 1986; Lamm *et al.*, 1993). Recurrences are often treated by repeat TURBT. However, 10-30% of superficial bladder tumors will also develop subsequent progression to higher stage muscle invasive bladder cancer (Lamm *et al.*, 1993; Sarosdy *et al.*, 1992). Muscle invasive tumors are associated with a high risk of occult metastases, and are, therefore, potentially life-threatening. Standard therapy for these tumors includes surgical removal of the bladder and adjacent tissues (radical cystectomy), sometimes combined with chemotherapy (Splinter and Scher, 1996).

For patients with superficial bladder cancer, intravesical therapy, in which chemotherapeutic or biological agents are instilled into the bladder, is used as an adjunct to TURBT, in order to treat and to prevent recurrences of superficial bladder cancer. The intravesical agent most commonly used in the United States is Bacillus Calmette-Guérin (BCG), which is believed to act as a nonspecific immunostimulant. BCG clearly reduces superficial bladder cancer recurrence. Its effect on tumor progression and survival is less clear. BCG therapy is usually well-tolerated, but can be associated with a number of toxicities, including significant cystitis, hematuria, fever, and BCG sepsis. Patients who fail BCG therapy, either because of continued tumor recurrence or because of toxicity, are often treated with intravesical chemotherapy, such as thio-tepa, mitomycin C, doxorubicin, or other intravesical biologic agents such as interferon α (Witjes *et al.*, 1996). More recently, an orally administered immunostimulant, bropirimine has become more popular. Nonetheless, despite these maneuvers, many patients continue to go on to muscle invasive disease, necessitating cystectomy.

II. BLADDER CANCER: MOLECULAR BIOLOGY

Cancer results from an accumulation of genetic injuries to biochemical pathways that regulate cell proliferation. Two key gene sets, the proto-oncogenes and the tumor suppressor genes, play vital balancing roles in guiding normal cell development and growth. Damage to tumor suppressor genes allows uncontrolled growth of cancer cells. The human retinoblastoma (*Rb*) gene is the prototype, as well as one of the best characterized, of the tumor suppressor genes (Goodrich and Lee, 1993). The human *p53* gene is also a tumor suppressor gene which has been extensively described (Levine *et al.*, 1991). Both *p53* and *Rb* appear to play important roles in the biology of transitional cell carcinoma of the uroepithelieum. The H-*ras* oncogene is an example of a proto-oncogene contributing to the malignant phenotype of transitional cell carcinoma following an activating mutation.

In the normal cell cycle, the *Rb* gene product, p110^{Rb}, in its hypophosphorylated state arrests cell proliferation in the G0 and G1 phases of the cell cycle by inhibiting the function of transcription factors such as E2F. Passage through the G1 checkpoint commits cells to DNA synthesis and cell division. Inappropriate passage through this checkpoint may cause cells to synthesize new DNA in the S phase from damaged templates before scheduled DNA repair can be completed in G1, or to miss signals to exit the cell cycle for differentiation or cell death. Therefore, p110^{Rb} functions as a critical gatekeeper in the normal cell cycle, preventing uncontrolled cellular proliferation and allowing terminal differentiation to occur.

Rb alterations have been associated with the initiation and progression of cancer. Although originally identified as the genetic defect responsible for the development of retinoblastoma, alterations in the *Rb* gene are also associated with many other malignant diseases, including bladder cancer, breast cancer, acute myelogenous leukemia, and lung cancer (Harbour *et al.*, 1988; Lee *et al.*, 1988; T'Ang *et al.*, 1988; Yokota *et al.*, 1988; Horowitz *et al.*, 1989,1990). The reintroduction and expression of the wild-type *Rb* gene in *Rb*-altered tumor cells suppresses tumor cell growth in *in vitro* and *in vivo* models. It is anticipated that *Rb* gene replacement therapy is a viable approach for the development of a widely applicable anticancer treatment.

In the normal cell cycle, p53 also plays an important role in cell cycle regulation. The p53 gene encodes a 53 Kd nuclear protein capable of arresting cells in G1 (Levine *et al.*, 1991). Alteration of the tumor suppressor gene p53, via mutation and/or allelic loss, appears to be a critical event in tumorigenesis in a variety of neoplasms, including colorectal carcinoma and bladder cancer. Because of the important role that p53 normally plays in cell cycle regulation, it is believed that loss of wild-type p53 function directly promotes tumorigenesis. Indeed, p53 alterations are common in bladder cancer and more advanced stages of bladder cancer are characterized by a higher likelihood of aberrant p53 expression (Esrig *et al.*, 1994).

The pathogenesis of TCC, as with other malignant diseases, appears to involve multiple molecular abnormalities affecting oncogenes and tumor suppressor genes. Of these, alterations of the retinoblastoma (*Rb*) gene appear to play an especially critical role in the development and progression of many bladder cancers. Patients with familial retinoblastoma and their relatives have an increased incidence of bladder cancer (Cordon-Cardo *et al.*, 1992a). Mutation or loss of expression of the *Rb* gene product can occur as an early genetic event in the development of superficial bladder cancer. *Rb* abnormalities have been reported in 10-50% of patients with superficial bladder cancer (Presti *et al.*, 1991; Xu *et al.*, 1993; Miyamoto *et al.*, 1995). In addition, *Rb* gene alterations appear to be associated with tumor progression, with 30-80% of invasive bladder cancers showing evidence of altered *Rb* expression or loss of heterozygosity at the *Rb* locus (Presti *et al.*, 1991; Xu *et al.*, 1993). Bladder cancers containing *Rb* alterations are associated with a particularly poor prognosis (Cordon-Cardo *et al.*, 1992b; Logothetis *et al.*, 1992; Lipponen and Luikkonen, 1995). Similarly, the role of p53 alterations in bladder cancer has been fairly well established. Mutation or loss of expression of *p53* gene also appears to occur as an early genetic event in the development of superficial bladder cancer. As with *Rb*, up to 60% of patients with bladder cancer will demonstrate altered p53 protein expression, and it is clear that the likelihood of mutant p53 being present is a function of the clinical stage (extent of invasion). As with *Rb*, increased expression of mutant *p53* gene product is a feature prognostic of worse outcome (Esrig *et al.*, 1994; Lipponen, 1993).

III. THE H-*RAS* GENE IN BLADDER CANCER

The *ras* oncogene family (consisting of the H-*ras*, K-*ras*, and N-*ras* genes) plays an important role in tumorigenesis; it is estimated to be involved in 10-15% of human cancers through single base mutations (Barbacid, 1987). Specifically, K-*ras* mutations are present in 90% of pancreatic, 50% of colon, and 33% of lung adenocarcinomas, respectively, whereas N-*ras* mutations have been demonstrated in 45% of malignant melanomas beyond Clark's level II. The H-*ras* gene appears to play a significant role in human bladder carcinoma. The *ras* oncogene was initially discovered as the transforming gene of the T24/EJ bladder carcinoma cell line (Reddy *et al.*, 1982; Shih and Weinberg, 1982). Several studies have examined the potential role of activated H-*ras* genes in bladder carcinoma. Overall, H-*ras* mutations have been demonstrated in 6-76% of primary bladder tumors (Burchill *et al.*, 1991; Czerniak *et al.*, 1992; Knowles and Williamson, 1993; Levesque *et al.*, 1993; Burchill *et al.*, 1994). It is useful to analyze these reports in greater detail in order to appreciate that some of the discrepancies noted are in part due to sample size, population, assay sensitivity, and methodological differences. In one study of 67 bladder carcinomas, 45% were shown to contain H-*ras* mutations at codon 12, one of the hot spots at which H-*ras* mutations occur (Czerniak *et al.*, 1992).

Specifically, there were no H-*ras* mutations in grade I tumors, with a sharp increase to 44% in grade II tumors. The mutations were particularly frequent in grade III, invasive tumors (65%) and in aneuploid tumors (58%), which clinically exhibit a more aggressive behavior. All mutations detected converted the glycine-encoding GGC at codon 12 to a GUC sequence, encoding valine. This particular study used the polymerase chain reaction (PCR) assay in order to amplify a 63 base-pair region flanking the codon 12 of the H-*ras* gene. These results suggested the high frequency of H-*ras* mutations in bladder carcinomas and the potential association of these mutations with tumor progression.

Other studies probing this same question have shown variable results. On the positive side, a study by Burchill *et al.* (1991) investigated 50 bladder tumors and showed a 76% prevalence of H-*ras* codon 12 mutations, with 54% occurring as a GUC. This report utilized PCR analysis and probing with H-*ras* cDNA in order to detect the H-*ras* mutations. However, in a followup study, when direct sequencing was used to analyze these 50 tumors, only 9 were shown to have mutations, with 8 occurring as a GUC (Burchill *et al.*, 1994). On the negative side, a study of 152 bladder tumors using the single-stranded conformation polymorphism analysis (SSCP) modification of the PCR assay to detect single base mutations, followed by direct sequencing, revealed 9 mutations, a prevalence of 6% (Knowles and Williamson, 1993). Finally, in a study of 111 tumors, 33 were found to have a mutation, with 26 occurring as a GUC (Levesque *et al.*, 1993).

This report also utilized the PCR-SSCP technique. Moreover, the amplified samples were then digested with Msp I, which cleaves the normal (but not mutated) H-*ras* sequence. Following digestion, the DNA samples were sequenced. One of the technical problems associated with studies analyzing archived samples is the isolation of pure tumor cells, as opposed to potentially significant contamination with normal tissue. The Msp I digestion is an attempt to circumvent this important problem.

In summary, even though the exact incidence of H-*ras* gene mutations in bladder carcinoma remains controversial, it is clear that it plays an important role in urothelial tumorigenesis. This conclusion is supported by additional evidence from functional studies which revealed that over-expression of the mutated H-*ras* gene confers an invasive phenotype on bladder cancer cells (Theodorescu *et al.*, 1990). In this study, stable transfection of the mutated H-*ras* gene into the non-invasive bladder cell line RT4 resulted in an invasive phenotype resembling that of EJ cells in an orthotopic model. These studies supported the previously suggested correlation between H-*ras* gene mutations and invasion (Czerniak *et al.*, 1992). Finally, the converse has also been examined; our group has shown that inhibition of mutated H-*ras* gene expression reverses the invasive phenotype of malignant bladder cancer cells (Kashani-Sabet *et al.*, 1992, described below).

IV. HAMMERHEAD RIBOZYMES AND INHIBITION OF GENE EXPRESSION

The presumed role of H-ras expression in bladder tumorigenesis has been the basis for the experimental suppression of H-*ras* gene expression. At the molecular level, this suppression has entailed the use of antisense molecules or hammerhead ribozymes. Ribozymes are catalytic RNAs that bind to an RNA sequence selectively and cleave at predetermined sequences. Ribozymes were initially discovered in the self-splicing of *Tetrahymena* ribosomal RNA (Grabowski *et al.*, 1981) and in *Escherichia coli* RNase P (Guerrier-Takada *et al.*, 1983). The characterization of a potentially trans-acting RNA was achieved by Forster and Symons (1987) in the virusoid from lucerne transient streak virus. This ribozyme was termed hammerhead owing to its secondary structure. The consensus sequences for target RNA cleavage of the ribozyme were initially defined by Haseloff and Gerlach (1988). Thus, hammerhead ribozymes were shown to cleave 3' to XUN sequences, where X is any base and N is A, C, or U. An example of a target sequence is the GUC sequence of the mutated codon 12 of the H-*ras* oncogene mRNA (Figure 1). The ribozyme molecule consists of a catalytic core (including the hammerhead domain) and three helices. Two of these helices (stems I and III) act as flanking sequences which confer specificity by binding to the target message in antisense fashion. Ribozymes therefore have the potential advantage of acting as antisense RNAs as well as of cleaving the target RNA directly. The biochemistry of hammerhead ribozymes has been extensively reviewed elsewhere (Bratty *et al.*, 1993; Kashani-Sabet and Scanlon, 1995).

Extensive studies have been carried out to demonstrate the efficacy of hammerhead ribozymes as inhibitors of human gene expression. Hammerhead ribozymes have been proposed as agents to inhibit HIV replication, reverse drug resistance, and suppress the malignant phenotype in human tumors. With respect to human cancer, ribozymes have been targeted to various oncogenes (such as H-*ras*, c-*fos*, c-*myc*), mutated tumor suppressor genes (such as *p53*), and drug-resistance genes (such as the multidrug resistance-1 or MDR-1 gene; reviewed by Kashani-Sabet and Scanlon, 1995). The first utility of ribozymes in cancer entailed a hammerhead ribozyme designed against c-*fos* as an agent to reverse resistance to the chemotherapeutic agent cisplatin in human ovarian carcinoma cells (Scanlon *et al.*, 1991). More recently, the anti-*fos* ribozyme has been demonstrated to reverse the MDR phenotype *in vitro*, with greater efficacy than an anti-MDR-1 ribozyme (Scanlon *et al.*, 1994).

V. HAMMERHEAD RIBOZYMES AND GENE THERAPY OF BLADDER CANCER

The frequent mutation in codon 12 of the H-*ras* gene prevalent in bladder carcinomas converts a GGC sequence (encoding glycine) into a

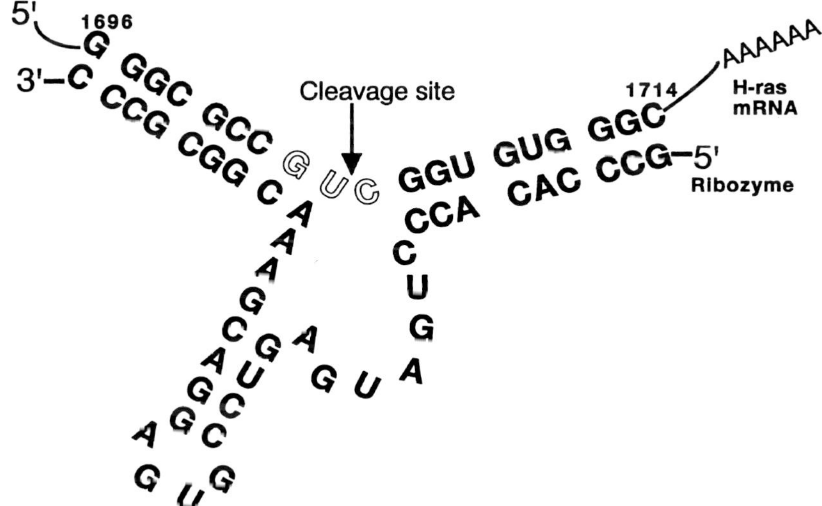

Figure 1. Structure of a hammerhead ribozyme designed to target the mutant GUC (Val) codon 12 of H-*ras* oncogene. The wild type codon 12 is GGC (Gly).

GUC sequence (encoding valine), thereby creating a ribozyme cleavage sequence. As mentioned previously, from 80-90% of H-*ras* mutations in bladder cancer consist of codon 12 GUC sequences. A ribozyme designed to cleave H-*ras* RNA at codon 12 would therefore be effective in cleaving mutated and not normal H-*ras* RNA, essentially acting as a tumor-specific agent. This has been confirmed by data showing the ribozyme to be incapable of cleaving normal H-*ras* RNA and of altering NIH3T3 cell growth *in vitro* (Funato *et al.*, 1994). Studies in which the anti-H-*ras* ribozyme was expressed in EJ bladder carcinoma cells have represented the cornerstone of attempts to develop this strategy in gene therapy of bladder carcinoma.

Initial attempts to examine the cellular effects of the anti-*ras* ribozyme in EJ human bladder carcinoma cells relied on stable trans-fection of the ribozyme in EJ cells (Kashani-Sabet *et al.*, 1992). The anti-*ras* ribozyme was cloned into the mammalian expression vector pHß Apr-1-neo, which possesses the ß-actin promoter to drive the expression of cloned sequences (Gunning *et al.*, 1987). Following transfection into EJ cells, ribozyme expression resulted in proportionally decreased H-*ras* RNA and protein levels; clones with higher levels of ribozyme gene expression exhibited more reduced H-*ras* gene expression. *In vitro*, ribozyme-mediated reduction of H-*ras* gene expression led to altered morphology, sluggish cell growth (as reflected by decreased colony formation in soft agar and increased generation time) and decreased DNA synthesis (as determined by [³H]thymidine incorporation into DNA). *In vivo*, control EJ cells transfected with the plasmid sequences only were tumorigenic in 5/6 mice, with a median survival of 47 days (Kashani-Sabet *et al.*, 1992). When the highest expressing ribozyme clone was orthotopically implanted into the bladders of nude mice, only 1 out of 5 mice developed tumors. Moreover, the tumors formed by the ribozyme-containing clones failed to demonstrate invasion into vascular spaces, similar to the non-invasive cell line RT4. Cytologically, these cells were characterized by fewer hyperchromatic nuclei and lack of anaplastic or invasive properties. Survival of the mice was significantly improved when the EJ cells expressed the anti-*ras* ribozyme. These studies established a role for ribozyme-mediated suppression of H-*ras* gene expression as a viable strategy for inhibiting the malignant phenotype of human bladder carcinoma cells *in vivo*.

Second generation studies have further characterized the antineo-plastic properties of anti-*ras* ribozymes. The anti-*ras* ribozyme has been extensively studied in the NIH3T3 murine fibroblast system, and has been shown to inhibit and protect from EJ-*ras*-mediated transformation (Funato *et al.*, 1994). Specifically, NIH3T3 cells transfected with the activated H-

ras gene exhibit increased clonogenicity in soft agar, along with shortened generation time and increased DNA synthesis. These signs of transformation are reversed by expression of the anti-*ras* ribozyme in transformed cells. Interestingly, parental NIH3T3 cells were initially transfected with the anti-*ras* ribozyme and then supertransfected with the activated H-*ras gene*. Expression of the ribozyme alone in NIH3T3 cells was shown to have no deleterious effects on normal cell growth, suggesting its discrimination of normal versus mutated H-*ras* transcripts. And the ribozyme was able to protect from subsequent mutated H-*ras* mediated transformation *in vitro*.

The potential utility of activated H-*ras* gene inhibition by hammerhead ribozymes may not be limited to bladder carcinomas. The anti-*ras* ribozyme has also been shown to abrogate the neoplastic phenotype of NIH3T3 cells transfected with DNA of a human melanoma cell line containing a mutated H-*ras* gene *in vivo* (Kashani-Sabet *et al.*, 1994). This suppression was achieved using both subcutaneous and intravenous injection of transformed cells into nude mice. The anti-*ras* ribozyme has also been shown to alter the transformed phenotype of human melanoma cells *in vitro* (Ohta *et al.*, 1994).

Finally, other studies have established a role for catalytic activity of the hammerhead ribozyme when introduced into tumor cells. Cellular extracts of ribozyme-containing cells have been shown to cleave H-*ras* transcripts *in vitro* (Kashani-Sabet *et al.*, 1992). Moreover, the products of ribozyme cleavage have been detected by PCR (in the case of H-*ras*; Kashani-Sabet *et al.*, 1992) and by Northern analysis in the case of MDR-1 (Scanlon *et al.*, 1991). The ribozyme has been demonstrated to exert greater suppressive effects on H-*ras* gene expression and tumorigenicity than a control ribozyme with a mutation in the catalytic core, which limits the construct to antisense activity (Funato *et al.*, 1994). This confirms the theoretical advantages of ribozymes over antisense methods, given the catalytic activity of the hammerhead moiety.

VI. ADENOVIRAL-MEDIATED DELIVERY OF HAMMERHEAD RIBOZYMES

Recent studies have optimized the delivery of ribozymes to tumor cells for gene therapy using recombinant adenoviruses. Recently, adenoviruses have become an attractive alternative for *in vivo* gene therapy trials. Adenoviruses are advantageous as a delivery vector because of 1) long term clinical experience with a proven safety profile, 2) generation of high viral titers (10^{10} or 10^{11} pfu/ml) resulting in efficient delivery of target genes, and 3) presence of the virus in an episomal form (i.e., non-integrated), thereby easing concerns regarding safety and toxicity from random integration and long-term gene expression. Live, wild-type (non-recombinant) adenoviruses have been used as vaccines for the prophylaxis of adenoviral upper respiratory infection. These vaccines, which were at one time given routinely to high-risk populations such as military recruits, are well-tolerated and are considered non-oncogenic. In addition, oral recombinant versions of adenovirus vaccines have entered clinical trials. Thus far, they too appear to be without significant toxicity. Recently, clinical trials using recombinant, replication-defective adenoviral vectors for gene therapy have been initiated. In these trials, recombinant adenoviral vectors carrying the gene for the cystic fibrosis transmembrane conductance regulator (CFTR) have been delivered via intra-airway administration to patients with cystic fibrosis (Crystal *et al.*, 1994). Recombinant adenoviral vectors (rAd) are also being developed for gene therapy of cancer. For example, a recombinant adenovirus carrying the human wild-type *p53* tumor suppressor gene has been constructed, and is being developed for the treatment of hepatocellular cancer and metastatic colorectal cancer (Wills *et al.*, 1994).

In our studies, a recombinant adenovirus was constructed to encode the anti-*ras* ribozyme (Feng *et al.*, 1995). The resultant adenoviral vector contained the CMV promoter, the anti-*ras* ribozyme, and deletions of adenoviral E1 sequences (Figure 2). The adenoviral vector was shown to be highly efficient, since it successfully transfected >95% of EJ cells. The recombinant adenovirus was packaged in the EIA trans-complementing cell line 293, and EJ cells were infected with the ribozyme-containing virus at a multiplicity of infection (MOI) of 500 pfu/cell. At this dose, no non-specific vector-associated cytotoxicity was observed. Infected EJ cells exhibited significant inhibition in growth kinetics; no viable cells could be identified 5 days post-infection. A control adenoviral vector lacking the ribozyme did not exert any suppression of cell growth parameters, implying that the significant growth suppression was a specific effect of the ribozyme. Moreover, EJ cells infected with the ribozyme-containing adenoviral vector were not tumorigenic upon heterotopic transplantation into nude mice; no tumor nodules were noted after 35 days post-implantation (Feng *et al.*, 1995). These results indicate that the adenoviral vector efficiently delivers the anti-*ras* ribozyme to human bladder carcinoma cells, resulting in profound anti-neoplastic effects.

Finally, other studies have examined the efficiency of adenoviral-mediated gene transfer to human bladder epithelial cells (Bass *et al.*, 1995), which would be of obvious importance in attempts to deliver potentially therapeutic genes in a local/regional fashion.

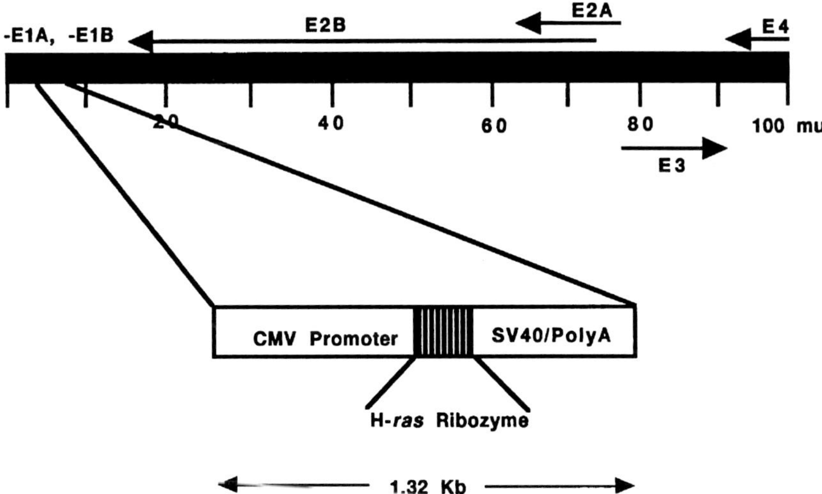

Figure 2. Recombinant adenovirus type 5 vector construct (36 kbp) encoding the anti-H-*ras* ribozyme.

These studies have shown that several bladder carcinoma cell lines were efficiently transduced *in vitro* by an adenoviral vector expressing the luciferase reporter gene. Expression was noted at 24 hours following viral transduction and was detectable at seven days. Moreover, intravesical administration of the luciferase-encoding vector to both mice and rats resulted in luciferase gene expression in bladder epithelial cells, peaking at 24 hours post-transduction.

In our most recent preclinical studies, we have begun to probe the utility of a recombinant adenoviral vector encoding an anti-*ras* ribozyme for the therapy of experimental bladder carcinoma. The results of these studies should further support the notion that delivering the anti-*ras* ribozyme with a recombinant adenoviral vector is a safe, viable, and efficient gene transfer strategy for treatment of advanced bladder tumors.

Because of its unique anatomic, pathophysiologic, and clinical aspects, bladder cancer is an ideal target for loco-regional gene therapy. Drug (or gene) delivery to superficial bladder cancer cells is direct and allows extremely high local drug concentrations, while systemic exposure is usually limited. Hence, intravesical gene therapy should impact on bladder cancer cells and nonmalignant urothelial cells that are transduced, and is anticipated to result in reversion of the malignant phenotype and reduction of tumor recurrence. In superficial bladder cancer, especially, intravesical administration of an adenovirus-based gene vector may result in efficient access to malignant cells. A phase I gene therapy trial utilizing a recombinant adenoviral vector has been planned at the University of California at San Francisco in order to determine the safety and potential biological efficacy of administration of a single dose of such an agent to patients with high risk superficial bladder carcinoma.

VI. REFERENCES

Barbacid, M. (1987) *Ras* genes. *Ann. Rev. Biochem.* **56**:779-827.

Bass, C., Cabrera, G., Elgavish, A., Robert, B., Siegal, G.P., Anderson, S. C., Maneval, D. C., and Curiel, D.T. (1995) Recombinant adenovirus-mediated gene transfer to genitourinary epithelium *in vitro* and *in vivo*. *Cancer Gene Ther.* **2**:97-104.

Bratty, J., Chartrand, P., Ferbeyre, G., and Cedergren, R. (1993) The hammerhead RNA domain, a model ribozyme. *Biochem. Biophys. Acta.* **1216**:345-359.

Burchill, S. A., Neal, D.E., and Lunec, J. (1994) Frequency of H-ras mutations in human bladder cancer detected by direct sequencing. *Br. J. Urol.* **73**:516-521.

Burchill, S. A., Lunec, J., Mellon, K., and Neal, D. E. (1991) Analysis of Ha-ras mutations in primary human bladder tumours. *Br. J. Cancer* **63** (Suppl. 13):62.

Cordon-Cardo, C., Dalbagni, G., and Richon, V. M. (1992a) Significance of the retinoblastoma gene in human cancer. In DeVita, V. T., Hellman, S., and Rosenberg, S. A., eds., *Principles and Practice of Oncology Update*, Vol. 6. J. B. Lippincott Co., Philadelphia, PA, 1-10.

Cordon-Cardo, C., Warlinger, D., Petrylak, D., Dalbagni, G., Fair, W. R., Fuks, Z., and Reuter, V. E. (1992b) Altered expression of the retinoblastoma gene product: Prognostic indicator in bladder cancer. *J. Natl. Cancer Inst.* **84**:1251-1256.

Crystal, R. G., McElvaney, N. G., Rosenfeld, M. A., Chu, C. S., Mastrangeli, A., Hay, J. G., Brody, S.L., Jaffe, H. A., Eissa, N. T., and Danel, C. (1994) Administration of an adenovirus containing the human CFTR cDNA to the respiratory tract of individuals with cystic fibrosis. *Nature Gen.* **8**:42-51.

Czerniak, B., Cohen, G.L., Etkind, P., Deitch, D., Simmons, H., Herz, F., and Koss, L. G. (1992) Concurrent mutations of coding and regulatory sequences of the Ha-*ras* gene in urinary bladder carcinomas. *Human Pathol.* **23**:1199-1204.

Esrig, D., Elmajian, D., Groshen, S., Freeman, J.A., Stein, J.P., Chen, S.C., Nichols, P.W., Skinner, D.G., Jones, P.A., and Cote, R.J. (1994) Accumulation of nuclear p53 and tumor progression in bladder cancer. *New Engl. J. Med.* **331**:1259-1264.

Feng, M., Cabrera, G., Deshane, J., Scanlon, K. J., and Curiel, D. T. (1995) Neoplastic reversion accomplished by high efficiency adenoviral mediated delivery of anti-*ras* ribozymes. *Cancer Res.* **55**:2024-2028.

Forster, A. C., and Symons, R. H. (1987) Self-cleavage of plus and minus RNAs of a virusoid and structural model for the active sites. *Cell* **49**:211-220.

Funato, T., Shitara, T., Tone, T., Jiao, L., Kashani-Sabet, M., and Scanlon, K. J.

(1994) Suppression of H-ras-mediated transformation in NIH 3T3 cells by a *ras* ribozyme. *Biochem Pharm.* **48**:1471-1475.

Goodrich, D. W., and Lee, W. H. (1993) Molecular characterization of the retinoblastoma susceptibility gene. *Biochim. Bioshys. Acta.* **1155**:43-61.

Grabowski, P. J., Zaug, A. J., and Cech, T. R. (1981) The intervening sequence of the ribosomal RNA precursor is converted to a circular RNA in isolated nuclei of Tetrahymena. *Cell* **23**:467-476.

Guerrier-Takada, C., Gardiner, K., March, J., Pace, N., and Altman, S. (1983) The RNA moiety of ribonuclease P is the catalytic subunit of the enzyme. *Cell* **35**:849-857.

Gunning, P., Leavitt, J., Muscat, G., Ng, S.-y., and Kedes, L. (1987) A human §-actin expression vector system directs high-level accumulation of antisense transcripts. *Proc. Natl. Acad. Sci. USA* **84**:4831-4835.

Harbour, J. W., Lai, S. L., Whang-Peng, J., Gazdar, A. F., Minna, J. D., and Kaye, F. J. (1988) Abnormalities in structure and expression of the human retinoblastoma gene in SCLC. *Science* **241**:353-357.

Haseloff, J., and Gerlach, W. L. (1988) Simple RNA enzymes with new and highly specific endoribonuclease activities. *Nature* **334**:585-591.

Horowitz, J. M., Yandell, D. W., Park, S. H., Canning, S., Whyte, P., Buchkovich, K., Harlow, E., Weinberg, R. A., and Dryja, T. P. (1989) Point mutational inactivation of the retinoblastoma antioncogene. *Science* **243**:937-940.

Horowitz, J. M., Park, S. H., Bogenmann, E., Cheng, J. C., Yandell, D. W., Kaye, F. J, Minna, J. D., Dryja, T. P., and Weinberg, R. A. (1990) Frequent inactivation of the retinoblastoma anti-oncogene is restricted to a subset of human tumor cells. *Proc. Natl. Acad. Sci. USA* **87**:2775-2779.

Kashani-Sabet, M., and Scanlon, K. J. (1995) Application of ribozymes to cancer gene therapy. *Cancer Gene Ther.* **2**:213-223.

Kashani-Sabet, M., Funato, T., Florenes, V.A., Fodstad, O., and Scanlon, K. J. (1994) Suppression of the neoplastic phenotype *in vivo* by an anti-*ras* ribozyme. *Cancer Res.* **54**:900-902.

Kashani-Sabet, M., Funato, T., Tone, T., Jiao, L., Wang, W., Yoshida, E., Kashfian, B. I., Shitara, T., Wu, A. M., Moreno, J. G., Traweek, S. T., Ahlering, T. E., and Scanlon, K. J. (1992) Reversal of the malignant phenotype by an anti-*ras* ribozyme. *Antisense Res. Dev.* **2**:3-15.

Knowles, M. A., and Williamson, M. (1993) Mutation of H-*ras* is infrequent in bladder cancer: Confirmation by single-strand conformation polymorphism analysis, designed restriction fragment length polymorphisms, and direct sequencing. *Cancer Res.* **53**:133-139.

Lamm, D. L., DeHaven, J. I., Riggs, D. R., Delgra, C., and Burrell, R. (1993) Keyhole limpet hemocyanin immunotherapy of murine bladder cancer. *Urol. Res.* **21**:33-37.

Lee, E. Y., To, H., Shew, J. Y., Bookstein, R., Scully, P., and Lee, W. H. (1988) Inactivation of the retinoblastoma susceptibility gene in human breast cancers. *Science* **241**:218-221.

Lerner, S. P., and Skinner, D. G. (1996) Radical cystectomy for bladder cancer. In Vogelzang, N. J., Scardino, P. T., Shipley, W. U., and Coffey, D. S., eds.,

Comprehensive Textbook of Genitourinary Oncology. Williams & Wilkins, Baltimore, MD, 442-463.

Levesque, P., Ramchurren, N., Saini, K., Joyce, A., Libertino, J., and Summerhayes, I. C. (1993) Screening of human bladder tunors and urine sediments for the presence of H-*ras* mutations. *Int. J. Cancer* **55**:785-790.

Levine, A. J., Momand, J., and Finlay, C. A. (1991) The p53 tumor suppressor gene. *Nature* **351**:453-456.

Lipponen, P. K. (1993) Over-expression of p53 nuclear oncoprotein in transitional-cell bladder cancer and its prognostic value. *Int. J. Cancer* **53**:365-370.

Lipponen, P. K., and Liukkonen, T. J. (1995) Reduced expression of retinoblastoma (Rb) gene protein is related to cell proliferation and prognosis in transitional-cell bladder cancer. J. *Cancer Res. Clin. Oncol.* **121**:44-50.

Logothetis, C. J., Xu, H. J., Ro, J. Y., Hu, S. X., Sahin, A., Ordonez, N., and Benedict, W. F. (1992) Altered expression of retinoblastoma protein and known prognostic variables in locally advanced bladder cancer. *J. Natl. Cancer Inst.* **84**:1256-1261.

Miyamoto, H., Shuin, T., Torigoe, S., Iwasaki, Y., and Kubota, Y. (1995) Retinoblastoma gene mutations in primary human bladder cancer. *Br. J. Cancer* **71**:831-835.

Ohta, Y., Tone, T., Shitara, T., Funato, T., Jiao, L., Kashfian, B. I., Yoshida, E., Horng, M., Tsai, P., Lauterbach, K., Kashani-Sabet, M., Florenes, V. A., Fodstad, O. Y., and Scanlon, K. J. (1994) H-*ras* ribozyme mediated alteration of the human melanoma phenotype. *Ann. N.Y. Acad. Sci.* **716**:242-253.

Presti, J. C., Reuter, V. E., Galan, T., Fair, W. R., and Cordon-Cardo, C. (1991) Molecular genetic alterations in superficial and locally advanced human bladder cancer. *Cancer Res.* **51**:5405-5409.

Reddy, E. P., Reynold, R. K., Santos, E., and Barbacid, M. (1982) A point mutation is responsible for the acquisition of transforming properties of the T24 human bladder carcinoma oncogene. *Nature* **360**:149-152.

Sarosdy, M.F., Lamm, D. L., Williams, R. D., Moon, T. D., Flanigan, R. C., Crawford, E. D., Wilks, N. E., Earhart, R. H., and Merrit, J. A. (1992) Phase I trial of oral bropirimine in superficial bladder cancer. *J. Urol.* **147**:31-33.

Scanlon, K. J., Ishida, H., and Kashani-Sabet, M. (1994) Ribozyme-mediated reversal of the multidrug-resistant phenotype. *Proc. Natl. Acad. Sci. USA* **91**:11123-11127.

Scanlon, K. J., Jiao, L., Funato, T., Wang, W., Tone, T., Rossi, J. J., and Kashani-Sabet, M. (1991) Ribozyme-mediated cleavage of c-*fos* mRNA reduces gene expression of DNA synthesis enzymes and metallothionein. *Proc. Natl. Acad. Sci. USA* **88**:10591-10595.

Shih, C., and Weinberg, R. A. (1982) Isolation of a transforming sequence from a human bladder carcinoma cell line. *Cell* **29**:161-169.

Sidky, Y. A., Borden, E. C., Wierenga, W., and Bryan, G. T. (1986) Inhibitory effects of interferon-inducing pyrimidinones on the growth of transplantable mouse bladder tumors. *Cancer Res.* **46**:3798-3802.

Silverman, D. T., Hartge, P., Morrison, A. S., and Devesa, S. S. (1992) Epidemiology of bladder cancer. *Hematol. Oncol. Clin. North Am.* **6**:1-30.

Splinter, T. A. W., and Scher, H. I. (1996) Adjuvant and neoadjuvant chemotherapy for invasive (T3-T4) bladder cancer. In Vogelzang, N. J., Scardino, P. T., Shipley, W. U., and Coffey, D. S., eds., *Comprehensive Textbook of Genitourinary Oncology.* Williams & Wilkins, Baltimore, MD, 464-471.

Tõang, A., Varley, J. M., Chakraborty, S., Murphree, A. L., and Fung, Y. K. (1988) Structural rearrangement of the retinoblastoma gene in human breast carcinoma. *Science* **242**:263-266.

Theodorescu, D., Cornil, I., Fernandez, B. J., and Kerbel, R. S. (1990) Overexpression of normal and mutated forms of H-*ras* induces orthotopic bladder invasion in a human transitional cell carcinoma. *Proc. Natl. Acad. Sci. USA* **87**:9047-9051.

Wills, K. N., Maneval, D. C., Menzel, P., Harris, M. P., Sutjipto, S., Vaillancourt, M. T., Huang, W. M., Johnson, D. E., Anderson, S. C., and Wen, S. F. (1994) Development and characterization of recombinant adenoviruses encoding p53 for gene therapy of cancer. *Human Gene Ther.* **5**:1079-1088.

Wingo, P. A., Tong, T., and Bolden, S. (1995) Cancer statistics, 1995. *CA Cancer J. Clin.* **45**:8-30.

Witjes, J. A., Oosterhof, G. O. N., and Debruyne, F. M. J. (1996) Management of superficial bladder cancer Ta/T1/TIS: Intravesical chemotherapy. In Vogelzang, N. J., Scardino, P. T., Shipley, W. U., and Coffey, D. S., eds., *Comprehensive Textbook of Genitourinary Oncology.* Williams & Wilkins, Baltimore, MD, 416-427.

Xu, H.-J., Cairns, P., Hu, S. X., Knowles, M. A., and Benedict, W. F. (1993) Loss of *Rb* protein expression in primary bladder cancer correlates with loss of heterozygosity at the *Rb* locus and tumor progression. *Int. J. Cancer.* **53**:781-784.

Yokota, J., Akiyama, T., Fung, Y. K., Benedict, W. F., Namba, Y., Hanaoka, M., Wada, M., Terasaki, T., Shimosato, Y., Sugimura, T., and Terada, M. (1988) Altered expression of the retinoblastoma (*Rb*) gene in small-cell carcinoma of the lung. *Oncogene* **3**:471-475.

16

In Situ Gene Insertion for Immunotherapy Using Vaccinia Virus Vectors

Edmund C. Lattime, Laurence C. Eisenlohr, and Michael J. Mastrangelo
Thomas Jefferson University, Philadelphia, Pennsylvania

I. INTRODUCTION

The concept that tumor growth could be susceptible to control by the host's immune system dates to the early 1900s. However, it was not until 1957 that Prehn and Main convincingly demonstrated that active specific immunization could induce a functionally effective antitumor immune response. These investigators immunized syngeneic mice with a chemically induced sarcoma using a technique whereby the transplanted tumor was allowed to grow but was excised before it killed the host. These now surgically cured mice were able to reject a subsequent transplant of the same tumor yet accept a skin graft from a mouse of the same strain as that in which the tumor had arisen. This experiment provided a scientific foundation for the pursuit of active specific immunization as a cancer treatment (i.e., antigens existed on tumor cells which could be targeted to effect tumor rejection). Early trials of active specific tumor immunotherapy, both in animals and in man, utilized whole tumor cells as a source of immunogen. Clinical trials focused mainly on melanoma and to a lesser extent renal cell, colon and lung cancers. It is encouraging that active specific immunization with tumor cell-based vaccines, when optimally formulated, induced objective tumor regression in 10-20% of patients (reviewed by Mastrangelo *et al.*, 1995a). However, the limited availability of relevant tumor cells, and the cumbersome processing required, curtailed the application of such vaccines. In the search for a more user friendly vaccine, most investigators are exploring chemically defined tumor associated antigens. We continue to focus on autologous tumor cells because these cells display not one but all of the potential antigens, both known and as yet unidentified, expressed by the patient's tumor. Our goals are to facilitate the exploitation of autologous tumor cells as an antigen source and to improve their immunogenicity by manipulation *in situ in vivo*.

II. ENHANCEMENT OF IMMUNOGENICITY

Cytokines provide costimulatory signals important for T lymphocyte activation. Enhancement of the cytokine milieu at the site of immunization can promote the acquisition of cellular immunity to microbial antigens (Russo *et al.*, 1989). Leong *et al.* (1996) treated twenty melanoma patients with cyclophosphamide, followed by irradiated autologous tumor cells admixed with Bacillus Calmette-Guerin (BCG). The cytokine granulocyte macrophage colony stimulating factor (GM-CSF) was injected concurrently at the vaccination site, as well as daily for four days thereafter. Four patients (20%) had substantial tumor regression. However, this result seems not unlike those results reported by others using similar regimens without GM-CSF. Reasoning that higher and perhaps more effective cytokine levels could be sustained at the immunization site by local production rather than local injection, tumor cells or fibroblasts genetically engineered to produce the molecule(s) of interest were included in vaccine preparations. Tepper *et al.* (1989) were the first to transduce a cytokine gene into a tumor cell as an anti-tumor strategy. Dranoff *et al.* (1993) compared the ability of ten murine cytokines, the interleukins IL-2, IL-4, IL-5, and IL-6, IL-1 receptor antagonist, GM-CSF interferon (IFN) γ; tumor necrosis factor (TNF) α, intracellular adhesion molecule (ICAM) 1, and CD2, to enhance the immunogenicity of B16 melanoma cells in an immunoprophylaxis model. They found that murine GM-CSF most effectively stimulated potent, long lasting and specific antitumor immunity, requiring CD8[+] T lymphocytes. All tumor challenges were rejected. Less frequently, regression of established murine tumors has been achieved by immunization with cytokine secreting tumor cells (Golumbek *et al.*, 1991; Tahara *et al.*, 1995). Critical to success in these established murine tumor treatment models was a very limited tumor burden.

Rosenberg (1992) and Rosenberg *et al.* (1992) were the first to use tumor cells engineered to secrete cytokines to immunize patients. Autologous melanoma cells were established in culture, transduced with either the IL-2 or TNFα genes and then used to immunize the donors. Several weeks later the lymph nodes draining the immunization site were removed. The cytotoxic lymphocytes were expanded in culture and then reintroduced into the patient. The therapeutic efficacy of these lymphocytes remains to be defined. Jaffee *et al.* (1996) conducted a phase I trial in seventeen patients with renal cancer to compare autologous irradiated, tumor cells with autologous irradiated GM-CSF secreting tumor cells for safety and for the induction of tumor immunity. The GM-CSF secreting vaccine produced delayed type hypersensitivity (DTH) responses which were four-fold greater than those achieved with the control vaccine. One partial tumor response was observed. As tumor cells are often difficult to transduce, Lotze *et al.* (1994) resorted to readily available fibroblasts for genetic modification and reintroduced these IL-4 producing cells admixed with the unmodified tumor cells. Outcome results have not yet been reported. The use of *ex vivo* gene transfer techniques in the clinic is severely limited by the requirement that autologous tumor or fibroblasts be surgically accessible, excised, grown *in vitro*, transfected with the gene(s) of interest, etc. We are attempting to effect gene transfer *in situ in vivo*. Our efforts focus on melanoma and urinary bladder cancer, the two tumors that have been successfully treated by intralesional immunotherapy in man.

III. INTRALESIONAL IMMUNOTHERAPY

III.A. BCG

Regression of neoplastic disease had been reported in patients with acute bacterial infections. Thus, W. B. Coley (1914) initiated treatment of human cancers with the toxic products of bacterial metabolism. Tumor regressions were reported for a large percentage of patients with histologically confirmed malignant disease. Therapy appeared most effective when toxin was administered directly into tumors. Despite these encouraging early results, this area of therapeutic research remained dormant until the animal studies of Zbar *et al.* (1971) stimulated interest in the use of intralesional BCG in the treatment of cancer. BCG is an attenuated form of the bovine tuberculosis bacillus *Mycobacterium bovis* that is used to immunize against the human pathogen. Working with an inbred strain of guinea pigs and a transplantable hepatocarcinoma, these investigators demonstrated that intralesional BCG can induce regression of established dermal tumor transplants, as well as stimulate tumor-specific transplantation immunity. They found that tumor regression requires an intact immune system and direct contact between BCG and tumor cells. Subsequent studies in this system indicated that intralesional BCG treatment of dermal tumor transplants can control nodal and visceral metastases in selected situations (Hanna and Peters, 1975). We and others have successfully employed intralesional BCG in the treatment of primary lesions, as well as dermal, subcutaneous and lymph node metastases from cutaneous melanoma. These data have been reviewed by Tems-Reines and Mastrangelo (1980) and are summarized in Table 1. Overall, 60+% of injected dermal/subcutaneous metastases regressed. In about 20% of cases, uninjected metastases regional to treatment sites regress. Durable complete remissions were achieved in 20-25% of patients.

In almost all cases, patients with visceral metastases have responded poorly to BCG therapy. However, regression of distant visceral metastases (liver, lung) subsequent to intralesional BCG therapy of dermal metastases has been reported in two cases (Morton, 1974; Mastrangelo *et al.*, 1975).

Table 1. Immunotherapy of Melanoma with Intralesional BCG

BCG Source	Regression of Nodules[a]		Complete Responders	Investigators
	Injected	Uninjected		
Glaxo, Tice	25/29[b]	6/36	6/36	Morton *et al.* (1970)
?	1/4	1/4	1/4	Krementz *et al.* (1971)
Tice	7/9	2/9	2/9	Nathanson (1972)
Glaxo	0/1	0/1	0/1	Levy *et al.* (1972)
Tice, Glaxo	15/25	2/25	2/25	Pinsky *et al.* (1973)
Tice, Glaxo	11/14	10/13c	6/14	Mastrangelo et al (1976)
Tice	3/7	1/7?	?	Smith *et al.* (1973)
?	2/8?	?	?	Minton (1973)
Tice	1/2	0/2	?	Baker and Taub (1973)
Connaught	2/3	1/3	?	Klein *et al.* (1973)
Tice	4/6	2/6?	3/6	Lieberman *et al.* (1975)
Pasteur	7/11?	0/11	?	Israel *et al.* (1973)
Glaxo	1/2	0/2	0/2	Grant *et al.* (1974)
Glaxo, Tice	5/5	?	?	Cohen *et al.* (1975)
	85/127(66%)	26/120(21%)	26/98(27%)	

[a]Number of patients showing regression of nodules per total number patients treated. [b]Regression on a per patient basis available on only 29 patients.
[c]One patient had only a single nodule.

The most clearly documented of these was the regression of a solitary pulmonary metastasis concurrent with regression of injected and uninjected dermal melanoma metastases (Mastrangelo *et al.*, 1975). The precise effect of intralesional BCG therapy of dermal melanoma metastases on nondermal lesions remains to be defined.

In the discussion appended to a paper on immunotherapy of acute myelogenous leukemia (Powles, 1973), Thomas described a case of melanoma metastatic to the bowel, which regressed following 4 months of oral BCG therapy. In view of the location of the metastases and the route of BCG administration, there is the likelihood of close approximation of BCG organisms with the visceral lesion, resulting in regression. Silverstein *et al.* (1974) have noted regression of an injected bladder melanoma metastasis. The emergence of new percutaneous, imaging-guided methods for introducing BCG into visceral tumor deposits has rekindled interest in exploring the role of intralesional BCG in the treatment of a wide variety of metastases.

Morales *et al.* (1976) were the first to employ BCG in the prophylaxis and treatment of superficial bladder cancer. Patients who, after resection or fulguration, had at least two recurrences in the preceding twelve months were eligible for study. Within ten days after cystoscopy and documentation of recurrence, BCG (Institut Armand Frappier) treatment was started. BCG (120 mg in 50 ml of saline) was instilled into the bladder through a catheter and retained for at least two hours. At the same time, 5 mg of BCG were administered intradermally, using a heat gun. This treatment was repeated weekly for six consecutive weeks. Followup was by cystoscopy every three months. As can be seen from Table 2, there was a sharp decrease in episodes of tumor recurrence following BCG treatment, and 14 of 16 patients remained disease free. Lamm *et al.* (1980) and Pinsky *et al.* (1985) used the same strain, dose, route, and schedule of BCG to treat patients with recurrent superficial bladder cancer. Experimental design differed in that a randomized conventionally treated control group was included. This permitted comparison of pre- and post-study episodes of recurrence in individual patients, as was done by Morales, and also afforded an opportunity to compare post-study recurrence rates in BCG and conventionally treated patients.

The results of these two studies, which are detailed in Table 2, indicated that BCG increases the mean time to recurrence, and decreases the number of recurrences, confirming the observations by Morales *et al.* (1976). Thus, BCG is of benefit, when combined with conventional therapy, in the management of recurrent superficial bladder cancer. Indeed, intravesical BCG was approved by the Food and Drug Administration for the treatment of superficial bladder cancer.

Table 2. Intravesical BCG in the Treatment of Recurrent Superficial Bladder Cancer

Morales *et al.* (1976)		
Treatment	Pre-BCG	Post-BCG
Patients (no.)	16	16
Median follow up (months)	11	11
Disease free (no.)	0	14
Episodes recurrence	47	2
Lamm *et al.* (1980)		
Treatment	BCG	control
Patients (no.)	23	24
Disease free (no.)	18	13
Episodes recurrence	8	19
Mean time to recurrence (months)	19.4	12
		P = .039
Pinsky *et al.* (1985)		
Treatment	BCG	Control
Patients (no.)	22	22
Tumors per month		
prestudy	3.6	2.97
poststudy	0.75	2.37
	P = .001	P = .59

III.B. Vaccinia

Wild-type vaccinia virus has been employed intralesionally in the treatment of primary and metastatic malignant melanoma. The results of published reports are presented in Table 3 and can be summarized as follows. Dermal and subcutaneous metastases were most frequently injected, with seventeen complete and three partial responses among twenty-eight patients with dermal lesions and three complete and five partial regressions in fourteen patients with subcutaneous melanoma deposits. Uninjected dermal and subcutaneous lesions rarely regressed. Visceral metastases were not injected. However, a mediastinal mass did regress partially and a histologically unconfirmed liver metastasis regressed completely. Injection of 0.1 to 1.0 immunizing dose (ID) of virus into a tumor was effective in producing tumor regression and in these individuals, who were pre-vaccinated, resulted in only modest infection/inflammation at the injection site and modest systemic complaints. Indeed, the only serious local and systemic side effects were seen in one patient who received 400 IDs distributed among multiple lesions at one treatment. Repeated intralesional vaccinia injections (over periods ranging from three weeks to two years) were variably successful in "taking" in the four patients so treated (Burdick, 1960; Burdick and Hawk, 1964; Hunter-Craig *et al.*, 1970; Roenigk *et al.*, 1974). As immunity became progressively stronger, the local and systemic toxicities diminished. The reports cited above date from 1960-1974 and are not as detailed as desired.

In an effort to confirm the toxicity and lesional response rates reported in the above cited trials, and to gain personal experience with intralesional vaccinia virus therapy, a feasibility study was conducted (Mastrangelo *et al.*, 1995b). The wild type-virus (Dryvax; Wyeth Laboratories, Inc.) was injected into the lesions melanoma metastases in five patients. The clinical characteristics of the patients studied are presented in Table 4. Our conclusions were as follows: 1) Despite historical and physical evidence of prior vaccination, all patients experienced a major reaction with pustule formation at the initial cutaneous inoculation site. 2) In the immunocompetent patient, very large amounts of vaccinia can be administered safely (10^7 pock forming units [pfu] per injection; 12×10^7 pfu total). 3) It is possible to locally infect tumor cells at virus injection sites (with viral protein production), albeit transiently, even in the face of anti-vaccinia immunity. A biopsy of patient #1's melanoma metastasis taken on day 70 (six hours post-vaccinia injection; anti-vaccinia antibody titer of 1/3200), on immunohistochemical staining with antibody TW2.3 showed 2+ cytoplasmic staining for EL3 (an early gene product in vaccinia replication).

Administration of vaccinia virus also has been employed as an adjunct to the surgical treatment of melanoma. In twenty-three patients, Everall *et al.* (1975) injected the primary melanoma with vaccinia virus two weeks prior to definitive surgical therapy. The disease-free intervals for these patients were compared (actuarial method) to those achieved by a control group of twenty-five patients who received only standard surgical therapy. At four years, 85% of the treated group was disease free as compared with only 60% of the control group. As expected, an inflammatory response developed at the injection site, which resolved substantially by day 14. Transient regional lymphadenopathy developed in 20% of patients. Systemic symptoms consisted of malaise, fever, chills and rarely nausea or vomiting. In 66% of patients these symptoms subsided in less than 48 hours. There was no serious morbidity.

In summary, intralesional treatment of primary or metastatic melanoma with vaccinia virus has been quite safe and reasonably effective locally and regionally. The regression of distant metastases in some cases suggest the generation of a systemic anti-tumor immune response. We hypothesize that if this immune response can be intensified, the magnitude of local and distant tumor regression can be markedly enhanced.

Table 3. Intralesional Therapy of Melanoma with Vaccinia

Investigators	Burdick	Belisario & Milton	Burdick & Hawk	Milton & Lane Brown	Hunter-Craig *et al.*	Roenigk *et al.*	Everall *et al.*
Year	1960	1961	1964	1966	1970	1974	1975
#patients	1	7	1	4	19	18	23
Pre-vaccinated	?	?	1	4	19	18	23
Re-vaccinated	0	0	0	0	6/19	0	0
Vaccine/Source	?	?	?	Bp/cSL	2×10^8 pfu/ml Lister/SL	calf lymph	$\geq 10^8$ pfu/ml Lister/SL
Dose/Lesion	?	1/ID	.1-.5 vial	?	0.1-1/ID	1 tube	1/ID
Total dose/Rx	?	3/ID	5vials	3-400/ID	<5/ID	?	1/ID
Repeat Rx	q 3-6wk.	N/A	q1-12 wk.×2yr.	N/A	2 pts	4 pts	N/A

Site				Response			
Primary	N/A	N/A	N/A	1 PR/1	N/A	N/A	23/23
Dermal	CR	7/7	CR	3 CR/3	6 CR, 3 PR/10	6 CR/13	N/A
SC	PR	?	N/A	3 CR/3	4 PR/9	?	N/A
LN	N/A	?	N/A	1 PR/1	N/A	0/1	N/A
Visceral	N/A	N/A	N/A	N/A	N/A	0	N/A
Uninjected	PR mediastinal mass	1 CR, lymph nodes;2?	?CR liver	N/A	0/1	0	N/A

Toxicity							
Injection Site	1-2+	1+	1-2+	4+	1-2+	1-2+	1-2+
Systemic	2-3+	NR	1-2+	3+	1-2+	1-2+	1-2+
Fever	2+	NR	1+	NR	1+	rare	1-2+
Malaise	1-2+	NR	1-2+	NR	1+	NR	1-2+
Nausea	1+	NR	1+	NR	NR	NR	1+
Vomiting	0	NR	1+	NR	NR	NR	1+
Lymphadeno-pathy	NR	NR	NR	NR	NR	NR	5/14
Vitiligo	NR	NR	1	NR	NR	1	NR

Abbreviations: N/A, not applicable; NR, not reported; ID, immunizing dose; SL, sheep lymph; CR, complete response; PR, partial response; Rx, treatment.

IV. VECTOR SELECTION

Both BCG and vaccinia have demonstrated antitumor effects when used intralesionally and both have been explored as vectors for gene transfer (Gicquel, 1995; Moss, 1991). These microorganisms share many important characteristics. They are inexpensive, live vaccines that induce

Table 4. Clinical Characteristics of Patients Receiving Intralesional Wild-Type Vaccinia Therapy

Number	Age/Sex	Mctastases	Prior treatment
1	66/F	One, exophytic, 3.2 × 3.5 × 0.6 cm; one subcutaneous, 1 cm; both right leg	IL BCG, IL IFN-γ R24, BCDT
2	81/M	>20, intradermal and polypoid, <2 cm, left scalp and cheek	IL IFN-α IL BCG, BCDT
3	56/M	One, subcutaneous, 0.7 × 0.9 × 0.6 cm, right arm	IL BCG
4	84/M	Eight, intradermal 2-6 mm, right leg	None
5	70/M	One, subcutaneous, 1 cm; 1, inguinal lymph node 2.5 cm;	IL IFN-α, IL BCG, Bec-2 immunization

Abbreviations: IL, intralesional; BCG, Bacillus Calmette-Guerin; IFN, interferon; R24, murine anti-GD3 monoclonal antibody; BCDT, BCNU + cisplatin + DTIC + tamoxifen; Bec-2, murine monoclonal antibody directed against the idiotype of R24 (a surrogate GD3)

long lasting immunity. Hundreds of millions of individuals have been vaccinated with each microorganism and the incidence of important side effects is low. BCG has been used as an adjuvant in various immunization protocols. We have chosen vaccinia virus as our vector for gene insertion.

Vaccinia virus is highly infectious and will efficiently lyse infected tumor cells, thus directly reducing tumor burden. The wild-type virus alone will trigger a cytokine cascade and recruit helper T lymphocytes to the injection site (Mitchison, 1970). The virus has been successfully employed for gene insertion in patients at risk for AIDS (Cooney *et al.*, 1991) and in cancer patients (Hamilton *et al.*, 1994). The large size of the virus allows it to serve as a carrier for multiple genes, if needed. Passenger genes are delivered into, and function in, the cytoplasm of the host cell. The lytic property of the virus, as well as the host anti-vaccinia immune response, will prevent unrestricted (and potentially dangerous) cytokine production. However, as discussed above, we have demonstrated that local production of viral gene products can be sustained by repeated injection of vaccinia.

V. CYTOKINE SELECTION

The successful induction of a cellular immune response depends upon the interaction of a complex of cytokines, both known and yet to be discovered. For this trial, a vaccinia/GM-CSF recombinant will be evaluated. GM-CSF is a 18-30 kDa monomeric glycoprotein first characterized in the mouse by Burgess *et al.* (1977). The biology of GM-CSF has been reviewed by Rapoport *et al.* (1992). The human gene encoding this protein is located on chromosome 5 (Huebner *et al.*, 1985).

GM-CSF is a multilineage stimulating factor (Sieff *et al.*, 1985) for granulocytes, macrophages, megakaryocytes, erythroid colonies and B and T lymphocytes. Costello (1993) reviewed the therapeutic use of GM-CSF. Sargramostim® is a recombinant yeast-derived form of GM-CSF that has been approved for use in myeloid reconstitution after autologous bone marrow transplantation. The recommended dose (Physician's Desk Reference, 1993) is 250 $\mu g/m^2$/day x 21 days. The maximum tolerated dose has not been determined, as toxicity is modest.

A review by Steinman (1991) suggested that GM-CSF may play an important role in the maturation and/or function of specialized antigen presenting cells. Thus, the localized expression of GM-CSF by tumor cells might specifically enhance tumor-antigen presentation by host antigen presenting cells and thus promote the acquisition of functionally effective anti-tumor immunity. This may be the mechanism for the results observed by Dranoff *et al.* (1993) and Leong *et al.* (1996).

VI. RECOMBINANT VACCINIA IN PRECLINICAL MODELS

VI.A. Rationale

As alluded to above, support for the role of localized immune enhancement in the induction of anti-tumor immunity can be found in studies using tumors transfected to produce a variety of cytokines. A number of laboratories have taken the approach that specific anti-tumor immunity might be enhanced by the use of gene transfection aimed at inducing cytokine/lymphokine production at the local tumor site. While these studies supported the principle of enhanced immunogenicity with cytokine expression, the approach required the *in vitro* manipulation of the host tumor and reinjection of transfected tumor which, if not treated to assure lack of growth and continued cytokine production might be ineffective on the one hand and lead to metastatic growth on the other. To overcome these limitations, we have examined the feasibility of direct *in situ* cytokine gene transfection using vaccinia virus vectors.

As noted above, we and others had demonstrated that intralesional administration of vaccinia virus in patients with melanoma resulted in regression of injected lesions. As part of these studies, we had demonstrated vaccinia gene function in injected lesions (Mastrangelo *et al.,* 1995b). Nevertheless, the suitability of vaccinia virus vectors in transfecting tumors *in situ* had yet to be demonstrated. Therefore, prior to launching a major clinical effort, it was necessary to undertake a number of *in vitro* and preclinical studies to demonstrate the feasibility of introducing cytokine genes into tumors *in situ* utilizing vaccinia virus recombinants. More specifically, the following questions were addressed: 1) would cytokine-encoding vaccinia virus recombinants infect/transfect tumor cells *in vitro* with resultant cytokine production; 2) would vaccinia-virus recombinants transfect tumor cells *in vivo* following intralesional or intravesical (bladder) administration; and 3) would prior immunity to vaccinia which would be expected in immunized patients or following multiple injections prevent tumor transfection following intralesional/intravesical administration of the viral recombinants?

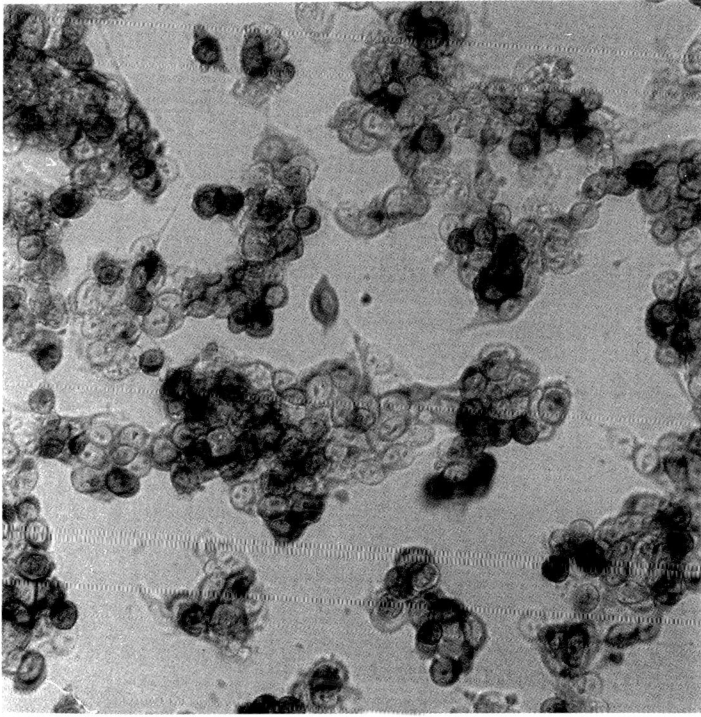

Figure 1. Recombinant vaccinia virus infects/transfects tumor cells *in vitro*. The murine bladder carcinoma MBT2 was infected *in vitro* with a vaccinia recombinant encoding for the reporter gene HA and the cells stained for expression of the HA antigen. Darkly stained cells represent infected/transfected cells.

VI.B. *In vitro* tumor transfection and cytokine gene expression using vaccinia virus vectors

As a first step in the above analysis, we obtained vaccinia recombinants produced in the Bennink and Yewdell laboratories (Yewdell *et al.*, 1981) which encoded the reporter hemagglutinin (HA) and nuclear protein (NP) genes from human influenza, as well as monoclonal antibodies specific for the encoded proteins. Using a panel of tumor cells, including murine and human melanomas and bladder carcinomas, we showed that following *in vitro* infection, all tumors tested expressed the encoded antigens, thus demonstrating high efficiency transfection of tumor cells using vaccinia recombinants (Figure 1).

Regarding cytokine-producing recombinants, studies from Ramshaw *et al.* (1992) initially demonstrated that cytokine-expressing vaccinia recombinants enhanced anti-viral immunity in mice. To extend this analysis to tumors, we generated vaccinia (VAC) recombinants encoding murine IL4, IFNγ, and GM-CSF. For IFNγ, a 620 bp DNA fragment encoding the entire sequence was obtained from Dr. Eli Gilboa with permission from Dr. A. Morris (Morris and Ward, 1987). For IL4, the cDNA encoding a 140 amino acid protein prior to cleavage of the signal sequence was obtained from Dr. T. Honjo of the Kyoto University Faculty of Medicine (Sideras, *et al.*, 1988).

For GM-CSF, we obtained the cDNA encoding the molecule in Bluescript from Immunex, Seattle WA. Cytokine cDNA was directionally ligated into the pSC11/NS plasmid under the control of the vaccinia early/late $P_{7.5}$ promoter and the fragment containing the viral promoter and cytokine gene integrated into the vaccinia virus genome (Western Reserve strain) at the thymidine kinase (TK) locus by homologous recombination. Recombinant viruses were selected based on TK deficient phenotype, screened by ß-galactosidase expression and amplified (following 4 rounds of plaque purification) (Chakrabarti *et al.*, 1985; Mackett *et al.*, 1984). Upon infection with the vaccinia recombinants, significant levels of the encoded cytokines were measured in the supernate, thus confirming the Ramshaw findings and extending them to tumors (Ramshaw *et al.*, 1992). Figure 2 demonstrates IL4 production by tumor cells infected with the VAC-IL4 produced in our laboratory.

VI.C. *In vivo* infection and tumor transfection using recombinant vaccinia vectors

Having demonstrated that vaccinia recombinants infected tumor cells *in vitro* and were capable of inducing cytokine production, we next examined their ability to infect/transfect tumors *in vivo*. To demonstrate

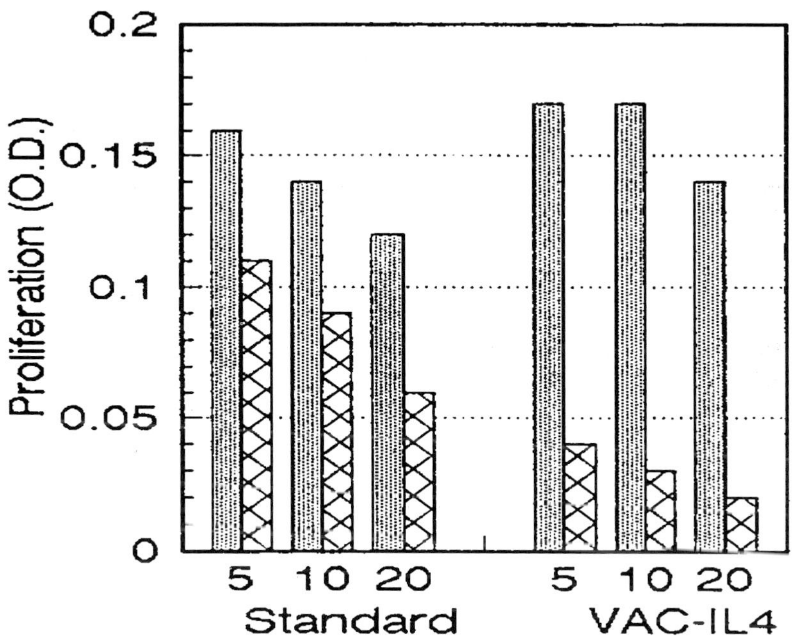

Figure 2. Infection of tumor cells with VAC-IL4 results in secretion of biologically active IL4 into the supernatant. Supernatants from standard human osteosarcoma cells and from VAC-IL4 infected human osteosarcoma cells were diluted 5X, 10X, and 20X, then assayed for IL4 activity using IL4-dependent CT4 cells in an MTT assay (shaded bars). IL4 dependence was demonstrated by neutralization using anti-IL4 antibody (crosshatched bars).

that vaccinia was able to recruit lymphocytes to the tumor site, in this case bladder, and generate a systemic immune response, graded numbers of wild type VAC from 10^1 to 10^3 pfu were instilled into bladders of C57BL/6 mice. Following 2 weeks incubation, spleen cells were removed from the mice, restimulated *in vitro* with VAC for 7 days and the resultant cells tested for their ability to lyse the VAC infected MB49 tumor target. We found that as few as 10 pfu instilled intravesically resulted in significant VAC immunity, demonstrating its high degree of infectivity (Lee *et al.,* 1994).

To assess tumor transfection *in vivo*, and the effects of established anti-vaccinia immunity, C57BL/6 mice were first immunized to vaccinia via an intraperitoneal injection of wild type virus at a dose shown to induce an anti-vaccinia cytotoxic T lymphocyte response. Two weeks later, bladders of vaccinia-immune mice were instilled with the syngeneic bladder tumor MB49. Following one week of tumor growth, the bladders were instilled with the above described NP and HA reporter constructs. At a series of times, the bladders were removed, and frozen sections stained for the reporter gene product. Bladders from these vaccinia-immune mice expressed a high level of reporter expression (Figure 3) demonstrating that pre-existing anti-vaccinia immunity did not prevent tumor infection/transfection by the recombinant vaccinia vector (Lee *et al.*, 1994). These results verified our demonstration of vaccinia gene function in patients whose melanoma lesions were injected repeatedly with wild type vaccinia as described above (Mastrangelo *et al.*, 1995b).

To expand these reporter gene studies to the cytokine system, we have recently assessed recombinant vaccinia encoded cytokine gene function (GM-CSF) following intralesional injection of subcutaneous melanoma lesions in the B16 model system. Using RT-PCR and primers designed to identify virally encoded GM-CSF as distinct from GM-CSF produced by secondary lymphoid cell recruitment and activation, we have assessed the kinetics of virally encoded GM-CSF mRNA and the effects of the development of viral immunity on the kinetics of recombinant gene expression.

To characterize the kinetics of *in vivo* expression, syngeneic C57BL/6 mice with palpable subcutaneous B16.F10 melanoma tumors were injected intralesionally with PBS, wild type vaccinia, or vaccinia encoding GM-CSF. The tissue was recovered at various times following injection. Total RNA was isolated, and RT-PCR was performed using the primers for ß-actin (control), and our vaccinia-specific primers. The results demonstrated that virally encoded GM-CSF mRNA was measurable as early as 7 hrs following injection. In mice not previously immunized with vaccinia, GM-CSF expression was maintained for as long as 5 days. When mice were preimmunized to vaccinia as above, or following multiple injections of the recombinant vaccinia vector resulting in enhanced anti-vaccinia immunity, recombinant GM-CSF gene expression was always seen in the injected tumors at 24 hrs., but the duration of expression was reduced.

V. SUMMARY

Our preclinical studies strongly support the utility of recombinant vaccinia virus as a vector for the *in situ* transfection of tumor. We have shown both *in vitro* and *in vivo* that vaccinia recombinants infect/transfect murine melanoma (B16) and bladder carcinoma (MB49) as well as a panel of human melanoma and bladder tumor cell lines *in vitro*.

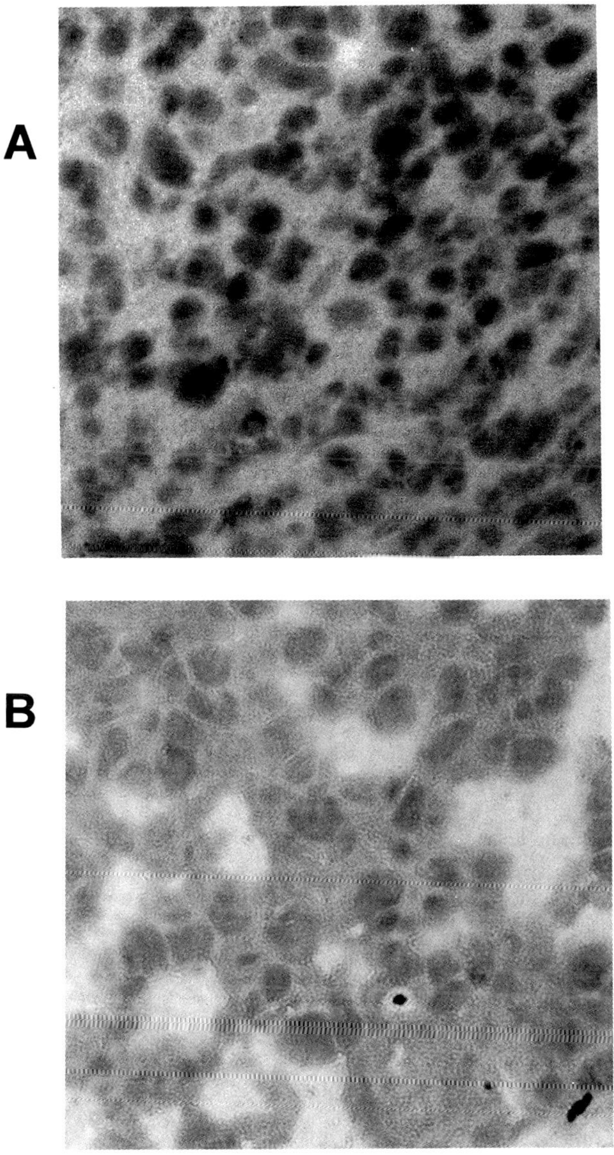

Figure 3. Recombinant vaccinia virus infects/transfects murine tumors *in vivo*. The murine bladder tumor MB49 cell line growing intravesically expresses the reporter construct HA after intravesical administration of the virus encoding HA. **A:** staining with antibody control; **B:** staining with anti-HA *(reprinted, with permission, from Lee et al., 1994).*

We have produced and tested a series of cytokine-encoding recombinant vaccinia viruses and demonstrated high levels of cytokine production *in vitro* and cytokine mRNA expression *in vivo*. Finally, we have shown that while preimmunization with vaccinia virus (either as a result of vaccination or repeated intratumor injections) results in anti-vaccinia immunity, as expected, such immunity does not prevent *in situ* transfection, although it does reduce the duration of cytokine expression.

VII. REFERENCES

Baker, M.A., and Taub, R.N. (1973) BCG in malignant melanoma. *Lancet* **1**:1117-1118.

Belisario, J.C., and Milton, G.W. (1961) Experimental local therapy of cutaneous metastases of malignant melanoblastoma with cowpox vaccine or colcemid. *Austral. J. Dermatol.* **6**:113-118.

Burdick, K.H. (1960) Malignant melanoma treated with vaccinia injections. *Arch. Dermatol.* **82**:438-439.

Burdick, K.H., and Hawk, W.A. (1964) Vitiligo in a case of vaccinia virus-treated melanoma. *Cancer* **17**:708-712.

Burgess AW, Camakaris, J., and Metcalf, D. (1977) Purification and properties of colony stimulating factor from mouse lung conditioned medium. *J. Biol. Chem.* **252**:1998-2003.

Chakrabarti, S., Brechling, K., and Moss, B. (1985) Vaccinia virus expression vector: coexpression of ß-galactosidase provides visual screening of recombinant virus plaques. *Mol. Cell Biol.* **5**:3403-3409.

Cohen, M.H., Felix, E., Jessup, J., and Rosenberg, S. (1975) Treatment of metastatic melanoma by intralesional injection of BCG, organic chemicals and *C. parvum*. In Crispen, R.G., ed., *Neoplasm Immunity Mechanism*. R.G. Franklin Institute Press, Philadelphia, 121-134.

Coley, W.B. (1914) *The Treatment of Malignant Inoperable Tumors with the Mixed Toxins of Erysipelas and Bacillus prodigiosus*. M. Weissenbruch, Brussels.

Cooney, E.L., Collier, A.C. Greenberg, P.D., Coombs, R.W., Zarling, J., Arditti, D.E., Hoffman, M.C., Hu, S.L., and Corey, L. (1991) Safety of and immunological response to a recombinant vaccinia virus vaccine expressing HIV envelope glycoprotein. *Lancet* **337**:567-572.

Costello, R.T. (1993) Therapeutic use of granulocyte-macrophage colony-stimulating factor. A review of recent experience. *Acta Oncol.* **32**:403-408.

Dranoff, G., Jaffee, E., Lazenby, A., Golumbek, P., Vevitsky, H., Brose, K., Jackson, V., Hamada, H., Pardoll, D., and Mulligan, R.C. (1993) Vaccination with irradiated tumor cells engineered to secrete murine granulocyte-macrophage colony-stimulating factor stimulates potent, specific and long-lasting anti-tumor immunity. *Proc. Natl. Acad. Sci. USA* **90**:3539-3543.

Everall, J.D., O'Doherty, C.J., Ward J., and Dowd, P.M. (1975) Treatment of primary melanoma by intralesional vaccinia before excision. *Lancet* **2**:583-586.

Gicquel, B. (1995) BCG as a vector for the construction of multivalent vaccines *Biologicals* **23**:113-118.

Golumbek, P.T., Lazenby, A.J., Levitsky, H.I., Jaffee, L.M., Karasuyama, H., Baker, H., and Pardoll, D.W. (1991) Treatment of established renal cell cancer by tumor cells engineered to secrete interleukin-4. *Science* **254**:713-716.

Grant, R.M., Mackie, R., Cochran, A.J., Murray, E.L., Hoyle, D., and Ross, C. (1974) Results of administering BCG to patients with melanoma. *Lancet* **2**:1096-1098.

Hamilton, J.M., Chen, A.P., Nguyen, B., Grem, J., Abrams, S., Chung, Y., Kantor, J., Phares, J.C., Bastian, A., Brooks, C., Morrison, G., Allegra, C.J., and Schlom, J. (1994) Phase I study of recombinant vaccinia virus that expresses human carcinoembryonic antigen in adult patients with adenocarcinomas. *Proc. Am. Soc. Clin. Oncol.* **13**:295. (Abstr. #961)

Hanna, M.G., and Peters, L.C. (1975) Efficacy of intralesional BCG therapy in guinea pigs with disseminated tumors. *Cancer* **36**:1298-1304.

Huebner, K., Isobe, M., Croce, C.M., Nathan, D.G., Wang, E.A., Wong, G.G., and Clark, S.C. (1985) The human gene encoding GM-CSF is at 5q21-q32, the chromosome region deleted in the 5q-anomaly. *Science* **230**:1282-1285.

Hunter-Craig, I., Newton, H.A., Westbury, G., and Lacey, B.W. (1970) Use of vaccinia virus in the treatment of metastatic melanoma. *Br. Med. J.* **2**:512-515.

Israel, L., Depierre, A., and Edelstein, R. (1973) Effect of intranodular BCG in 22 melanoma patients. *Proceedings IV International Symposium on the Locoregional Treatment of Tumors,* Turin, Italy, UICC, 9/19-21.

Jaffee, E.M., Marshall, F., Weber, C., Pardoll, D.M., Levitsky, H., Nelson, W., Carducci, M., Mulligan, R., and Simons, J. (1996) Bioactivity of a human GM-CSF tumor vaccine for the treatment of metastatic renal cell carcinoma. *Proc. Am. Soc. Clin. Oncol.* **15**:237 (Abstr).

Klein, E., Holterman, O.A., Papermaster, B., Milgrom, II., Rosner, D., Klein, L., Walter, M.J., and Zbar, B. (1973) Immunologic approaches to various types of cancer with the use of BCG and purified protein derivatives. *Natl. Cancer Inst. Monogr.* **39**:229-239.

Krementz, E.T., Samuels, M.S., and Wallace, J.H. (1971) Clinical experience in the immunotherapy of cancer *Surg. Gynecol. Obstet.* **133**:209-217.

Lamm, D.L., Thor, D.E., Harris, S.C., Reyna, J.A., Stogdill, V.D., and Radwin, H.M. (1980) Bacillus Calmette-Guerin immunotherapy of superficial bladder cancer. *J. Urol.* **124**:38-42.

Lee, S.S., Eisenlohr, L.C., McCue, P.A., Mastrangelo, M.J., and Lattime, E.C. (1994) Intravesical gene therapy: *In vivo* gene transfer using vaccinia vectors. *Cancer Res.* **54**:3325-3328.

Leong, S.P.L., Enders-Zohr, P., Zhou, Y.M., Allen, R.E., Jr., Sagebiel, R.W., Glassberg, A.B., and Hayes, F.A. (1996) Active specific immunotherapy with GM-CSF as an adjuvant to autologous melanoma vaccine in metastatic melanoma. *Proc. Am. Soc. Clin. Oncol.* **15**:437. (Abstr)

Levy, N.L., Mahaley, M.S., Jr., and Day, E.D. (1972) Serum-mediated blocking of cell-mediated antitumor immunity in a melanoma patient: Association with BCG immunotherapy and clinical deterioration. *Intl. J. Cancer* **10**:244-248.

Lieberman, R., Wybran, J., and Epstein, W. (1975) The immunologic and histopathologic changes in BCG-mediated tumor regression in patients with malignant melanoma. *Cancer* **35**:756-777.

Lotze, M.T., Rubin, J.T., Carty, S., Edington, H., Ferson, P., Landreneau, R., Pippin, B., Posner, M., Rosenfelder, D., and Watson, C. (1994) Gene therapy of cancer: A pilot study of IL-4 gene modified fibroblasts with autologous tumor to elicit an immune response. *Hum. Gene. Ther.* **5**:41-55.

Mackett, M., Smith, G.L., and Moss, B. (1984) General method for production and selection of infectious vaccinia virus recombinants expressing foreign genes. *J. Virol.* **49**:857-864.

Mastrangelo, M.J., Bellet, R.E., Berkelhammer, J., and Clark, W.H., Jr. (1975) Regression of pulmonary metastatic disease associated with intralesional BCG therapy of dermal melanoma metastases. *Cancer* **36**:1305-1308.

Mastrangelo, M.J., Sulit, H.L., Prehn, L.M., Bornstein, R.S., Yarbro, J.W., and Prehn, R.T. (1976) Intralesional BCG in the treatment of metastatic malignant melanoma. *Cancer* **37**:684-692.

Mastrangelo, M.J., Maguire, H.C., Jr., Lattime, E.C., and Berd, D. (1995a) Whole cell vaccines. In DeVita, V.T., Hellman, S., Rosenberg, S.A., eds., *Biological Therapy of Cancer, 2nd Edition*. Lippincott, Philadelphia, 648-658.

Mastrangelo, M.J., Maguire, H.C., Jr., McCue, P., Lee, S.S., Alexander, A., Nazarian, L., Eisenlohr, L.C., Nathan, F., Berd, D., and Lattime E.C. (1995b) A pilot study demonstrating the feasibility of using intratumoral vaccinia injections as a vector for gene transfer. *Vaccine Res.* **4**:55-69.

Milton, G.W., and Lane Brown, M.M. (1966) The limited role of attenuated smallpox virus in the management of advanced malignant melanoma. *Austral. NZ. J. Surg.* **35**:286-290.

Minton, J.R. (1973) Mumps virus and BCG vaccine in metastatic melanoma, *Arch. Surg.*, **106**: 503-506.

Mitchison, N.A. (1970) Immunologic approach to cancer. *Transplant. Proc.* **2**:92-103.

Morales, A., Eidinger, D., and Bruce A.W. (1976) Intracavitary Bacillus Calmette-Guerin in the treatment of superficial bladder tumors. *J. Urol.* **116**:180-183.

Morris, A.G., and Ward, G. (1987) Production of recombinant interferon by expression in heterologous mammalian cells. In M.J. Clemens, A.G. Morris and A.J.H. Gearing, eds., *Lymphokines and Interferons: A Practical Approach*. IRL Press, Washington, DC, 61-72.

Morton, D.L., Eilber, F.R., Joseph, W.L., Wood, W.C., Trahan, E., and Ketcham, A.S. (1970) Immunological factors in human sarcomas and melanomas. *Ann. Surg.* **172**:740-747.

Morton, D.L. (1974) Cancer Immunotherapy: An overview. *Semin. Oncol.* **1**:297-310.

Moss, B. (1991) Vaccinia virus: A new tool for research and development. *Science* **252**:1660-1667.

Nathanson, L. (1972) Regression of intradermal melanoma after intralesional injection of *Mycobacterium bovis* strain BCG, *Cancer Chemother. Rep.*, **56**:659-665.

Physician's Desk Reference (1993) Medical Economics, Montvale, NJ, 1153-1156.

Pinsky, C.M., Hirshaut, Y., and Oettgen, H.F. (1973) Treatment of malignant melanoma by intratumoral injection of BCG. *Natl. Cancer Inst. Monogr.* **39**:225-228.

Pinsky, C.M., Camacho, F.J., Kerr, D., Geller, N.L., Klein, F.A., Herr, H.A., Whitmore, W.F., Jr., and Oettgen, H.F. (1985) Intravesical administration of bacillus Calmette-Guerin in patients with recurrent superficial carcinoma of the urinary bladder: Report of a randomized prospective trial. *Cancer Treat. Rep.* **69**:47-53.

Powles, R. (1973) Immunotherapy of acute myelogenous leukemia. *Natl. Cancer Inst. Monogr.* **39**:243-247.

Prehn, R.T., and Main, J.M. (1957) Immunity to methylcholanthrene induced sarcomas. *J. Natl. Cancer Inst.* **18**:769-778.

Ramshaw, I., Ruby, J., Ramsay, A., Ada, G., and Karupiah, G. (1992) Expression of cytokines by recombinant vaccinia viruses: A model for studying cytokines in virus infections *in vivo*. *Immunol. Rev.* **127**:157-182.

Rapoport, A.P., Abboud, C.N., and DiPersio, J.F. (1992) Granulocyte-macrophage colony-stimulating factor (GM-CSF): Receptor biology, signal transduction and neutrophil activation. *Blood Rev.* **6**:43-57.

Roenigk, H.H., Jr., Deodhar, S., St. Jacques, R., and Burdick, K. (1974) Immunotherapy of malignant melanoma with vaccinia virus. *Arch. Dermatol.* **109**:668-673.

Rosenberg, S.A. (1992) The immunotherapy and gene therapy of cancer. *J. Clin. Oncol.* **10**:180-199.

Rosenberg, S.A., Anderson, W.F., Blaese, R.M., Ettinghausen, S.E., Hwu, P., Karp, S.E., Kasid, A., Mule, J.J., Parkinson, D.R., Salo, J.C., Schwartzentruber, D.J., Topalian, S.L., Weber, J.S., Yannelli, J.R., Yang, J.C., and Linehan, W.M. (1992) Immunization of cancer patients using autologous cancer cells modified by the insertion of the gene for interleukin-2. *Hum. Gene Ther.* **3**:75-90.

Russo, D.M., Sundy, J.S., Young, J.F., Maguire H.C., Jr., and Weidanz, W.P. (1989) Cell mediated immune responses to vaccine peptides derived from the circumsporozoite protein of *Plasmodium falciparum*. *J. Immunol.* **143**:655-659.

Sideras, P., Noma, T., and Honjo, T. (1988) Structure and function of Interleukin 4 and 5. *Immunological Rev.* **102**:189-212.

Sieff, C.A., Emerson, S.G., Donahue, R.E., Golde, D.W., Kaufman, S.E., and Gasson, J.C. (1985) Human recombinant granulocyte-macrophage colony-stimulating factor: A multi-lineage hematopoietin. *Science* **230**:1171-1173.

Silverstein, M.J., deKernion, J., and Morton, D.L. (1974) Malignant melanoma metastatic to the bladder. Regression following intratumor injection of BCG vaccine. *J. Am. Med. Assoc.* **229**:688.

Smith, G.V., Morse, P.A., Jr., Deraps, G.D., Raju, S., and Hardy, J.D. (1973) Immunotherapy of patients with cancer. *Surgery* **74**:59-68.

Steinman, R.M. (1991) The Dendritic cell system and its role in immunogenicity. *Annu. Rev. Immunol.* **9**:271-296.

Tahara, H., Zitvogel, L., Storkus, W.J., Zeh, H.J., III, McKinney, T.G., Schreiber, R.D., Gubler, U., Robbins, P.D., and Lotze, M.T. (1995) Effective eradication of established murine tumors with IL-12 gene therapy using a polycistronic retroviral vector. *J. Immunol.* **154**:6466-6474.

Tems-Reines, S.J., and Mastrangelo, M.J. (1980) Malignant melanoma. *Clin. Immunother.* **11**:7-29.

Tepper, R.I., Pattengale, P.K., and Leder, P. (1989) Murine interleukin-4 displays potent anti-tumor activity *in vivo. Cell* **57**:503-512.

Yewdell, J.W., Frank, E., and Gerhard, W. (1981) Expression of influenza A virus internal antigens on the surface of infected P815 cells. *J. Immunol.* **126**:1814-1819.

Zbar, B., Bernstein, I.D., and Rapp, H.J. (1971) Suppression of tumor growth at the site of infection with living Bacillus Calmette-Guerin. *J. Natl. Cancer Inst.* **46**:831-839.

17

Antisense Oligonucleotide-Based Therapy for HIV-1 Infection from Laboratory to Clinical Trials

Sudhir Agrawal

Hydridon, Inc., Cambridge, Massachusetts

I. INTRODUCTION

Human immunodeficiency virus (HIV-1), the etiologic agent of acquired immunodeficiency syndrome (AIDS), presents a challenging target for therapeutic intervention. Various events in the replication cycle of HIV-1 have been targeted, therefore, for drug design and discovery. A number of drugs and drug candidates have been designed and studies as inhibitors of virus-specific enzymes, such as reverse transcriptase (DeClerq, 1992) and protease (Debouck, 1992).

An attractive alternative to conventional drugs is an emerging new paradigm in drug design and discovery involving the use of oligonucleotides (Zamecnik *et al.*, 1986). These molecules, called "antisense oligonucleotides", are designed to be complementary to a unique segment of the viral genome, or RNA derived from it. The antisense oligonucleotides selectively interfere with processes dependent on that segment of the viral genome by a phenomenon called "hybridization arrest" (Zamecnik, 1996); the oligonucleotide binds to its target nucleic acid by Watson-Crick base pairing or by Hoogsteen base pairing. In Watson-Crick base pairing, the heterocyclic bases of the antisense oligonucleotide form hydrogen bonds with complementary bases (A-T, G-C) of the target single-stranded DNA or RNA, whereas in Hoogsteen base pairing the heterocyclic bases of the oligonucleotide form hydrogen bonds with a target double-stranded DNA.

The concept underlying antisense oligonucleotide-based therapy of HIV-1 infection is that an oligonucleotide complementary to a unique segment of a viral genome, or to the RNA derived from the viral genome, can inhibit the production of key structural or functional proteins necessary for maintenance of the viral life cycle. A combination of the following properties makes antisense oligonucleotides notably effective therapeutic agents: 1) stability to nucleases in cells and body fluids; 2) efficient cellular uptake; 3) specific hybridization to complementary nucleic acids; and 4) strong affinity for the target sequence. In addition, favorable pharmacokinetic and safety profiles of an oligonucleotide are very important for its use as a therapeutic agent.

Several reports have described the use of antisense oligonucleotides as inhibitors of HIV-1 replication (Agrawal *et al.*, 1988, 1989, 1992a, 1992b, 1993; Sarin *et al.*, 1988; Matsukura *et al.*, 1989; Shibahara *et al.*, 1989; Letsinger *et al.*, 1989; Agrawal, 1991; Stein *et al.*, 1991; Agrawal and Tang, 1992, Lisziewicz *et al.*, 1992, 1993, 1994; Li *et al.*, 1993; Balotta *et al.*, 1993; Degas *et al.*, 1994; Agrawal and Lisziewicz, 1994; Weichold *et al.*, 1995). In this chapter, the oligodeoxynucleotide phosphorothioate GEM[®]91[1] is discussed in detail.

GEM[®]91 is a 25-mer oligodeoxynucleoside phosphorothioate designed to bind to the initiation site of *gag* mRNA of HIV-1 (Figure 1) (Agrawal and Tang, 1992; Agrawal *et al.*, 1993; Lisziewicz *et al.*, 1994; Agrawal

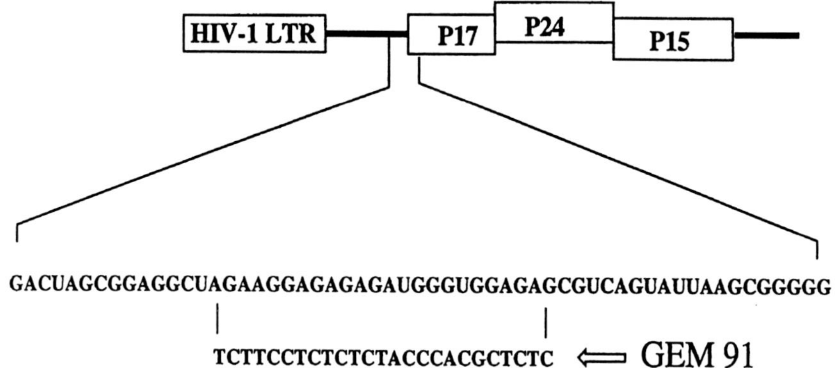

Figure 1. Complementary target site and sequence of GEM[®]91. The internucleotide linkages of the sequence are phosphorothioates, in which sulfur is substituted for a non-bridged oxygen.

---

[1] GEM[®]91 is a registered trademark of Hybridon, Inc.

and Lisziewicz, 1994). The nucleotide sequence around the initiation site of the HIV-1 *gag* gene is highly conserved, and is essential for viral packaging (Lever *et al.*, 1989; Harrison *et al.*, 1992; Baudin *et al.*, 1993); deletion of segments at this site causes insufficient packaging (Aldovini and Young, 1990) of the virus, thereby inhibiting production of infectious viral particles.

II. TARGETS FOR GEM®91 IN HIV-1 REPLICATION CYCLE

The targets for GEM®91 are, in general, single-stranded RNA such as mRNA and genomic viral RNA. After HIV-1 enters the cell, its genome is reverse transcribed in the cytoplasm to a double-stranded DNA intermediate. This step is followed by the formation of a pre-integration complex, which carries the viral DNA into the nucleus.

GEM®91 can act at several steps in this process to prevent replication of the virus. The viral genomic RNA represents an early target for GEM®91. By hybridizing to the incoming RNA, GEM®91 may block the reverse transcription and may degrade viral RNA by activation of RNase H at the hybridization site. RNase is an endonuclease that recognizes the RNA:DNA duplex and cleaves the RNA at the duplex site, thereby inhibiting translation of the mRNA (Hostomsky and Mathews, 1993). Once the proviral DNA is integrated in the host cell genome, it stays in the double-stranded DNA form. Following the transcription step, GEM®91 may bind to the appropriate site on pre-mRNA to inhibit translation of pre-mRNA in the cytoplasm. Additionally, in the cytoplasm, GEM®91 may 1) inhibit translation by binding to the AUG site (initiation site) of the target mRNA and blocking ribosomal association or 2) activate RNase H, to cleave the mRNA/oligonucleotide duplex, thereby preventing the synthesis of the viral proteins. In the absence of the *gag* gene product, which is essential for packaging of the virus, the production of viral particles is inhibited.

III. ANTIVIRAL ACTIVITY OF GEM®91 *IN VITRO*

The anti-HIV activity of GEM®91 was measured using various cell types, including peripheral blood mononuclear cells (PBMC) and macrophages, which are susceptible to infection and allow for high levels of HIV-1 replication. Virus replication was measured by cell viability, levels of the viral p24 protein, quantitative syncytia, and a reverse transcriptase assay. In general, the IC_{50} (effective concentration to cause 50% inhibition of HIV-1 replication) of GEM®91 in the acute infection assay was in the range of 0.1 - 0.3 µM (Agrawal and Tang, 1992). The main target cells of

HIV-1 infection *in vivo* are CD4$^+$ T lymphocytes and macrophages. To demonstrate anti-HIV activity, activated PBMCs from normal donors were infected with HIV-1 and cultured in the presence of GEM®91. GEM®91 inhibited HIV-1 replication in a dose-dependent manner, as evaluated by the measurement of viral p24 levels. In addition to inhibiting HIV-1 replication, GEM®91 also protected CD4$^+$ T cells from the cytopathic effects induced by HIV-1. This protection was dependent on the concentration of GEM®91 and was inversely related to the level of virus production. Similar results were obtained when GEM®91 was studied for its antiviral activity using two primary HIV-1 isolates, HIV-1 MN and HIV-1/571 (Table 1) (Lisziewicz *et al.*, 1994).

GEM®91 also demonstrated antiviral activity in primary cells isolated from an HIV-infected patient, which may be the case most relevant to the clinical situation. Effective inhibition of HIV-1 replication coincided with an increase in the percentage of CD4$^+$ T-cells in cell culture, suggesting that GEM®91 at 0.25 µM was effective in inhibiting viral replication in this

Table 1. HIV-1 Replication in Primary PBMCs Treated with Oligonucleotide Phosphorothioates

Isolate	Sequence	Conc.	p24, ng per 10^6 cells		
		µM	Day 7	Day 11	Day 14
HIV-1/IIIB	GEM®91	1.0	6.4	1.0	3.7
		0.5	11.8	7.3	23.5
		0.25	115.9	19.4	12.8
	Random	1.0	149.3	34.6	29.2
		0.5	111.9	16.9	14.4
		0.25	100.0	20.9	19.2
	None	-	153.0	11.4	12.7
HIV-1/MN	GEM®91	1.0	0.06	0.1	0.03
		0.5	6.0	22.6	18.3
		0.25	26.0	69.6	17.3
	Random	1.0	36.5	50.2	22.1
		0.5	169.7	44.6	10.7
		0.25	148.5	47.8	8.0
	None	-	53.7	37.0	12.6
HIV-1/571	GEM®91	1.0	0	0.08	0.01
		0.5	1.0	7.0	7.2
		0.25	26.1	18.8	9.9
	Random	1.0	2.7	5.0	18.5
		0.5	23.7	25.7	13.8
		0.25	27.4	23.4	10.4
	None	-	48.5	19.2	8.6

(Reprinted, with permission, from Lisziewicz et al., 1994).

patient (Agrawal and Lisziewicz, 1994). These results show that GEM®91 is effective at a low dose against an already established *in vivo* infection.

GEM®91 has been tested on various field HIV-1 isolates and showed antiviral activity against strains with syncytia-inducing or non-syncytia-inducing phenotypes and against strains with documented moderate to severe resistance to 3'-azidothymidine (AZT) and dideoxyinosine (ddI). GEM®91 inhibited all HIV-1 isolates obtained so far, although the inhibitory concentration was variable (Byrn *et al.*, 1995). In a study designed to generate escape mutants of HIV-1, GEM®91 did not allow the emergence of resistant HIV-1 mutants *in vitro* (Papp *et al.*. 1995). Clinical evaluation should determine if GEM®91-resistant HIV-1 will emerge *in vivo*.

Although GEM®91 may act primarily by inhibiting the processing and translation of HIV-1 RNA through hybridization arrest, "random" oligodeoxynucleoside phosphorothioates also inhibited HIV-1 replication in short-term assays, most likely by inhibition of reverse transcriptase activity (Stein *et al.*, 1989). Association of such "random" sequences with the V3 loop of gp120 has also been reported to inhibit $CD4^+$-gp120 binding (Stein *et al.*, 1991,1993).

This non-sequence specific inhibition produced by oligonucleotides was a transient effect, which only delayed the HIV-1 infection and did not stop virus replication in infected cells (Lisziewicz *et al.*, 1992). Multiple mechanisms of action have been proposed for inhibition of HIV-1 by antisense phosphorothioate oligonucleotides, including interference with viral adsorption and reverse transcription in sequence-specific and non-sequence-specific manners, as well as antisense activity (Zalpheati *et al.*, 1994).

IV. ADMINISTRATION OF GEM®91 IN RATS

IV.A. Single-Dose Pharmacokinetics in Rats

Pharmacokinetic studies of GEM®91 were carried out in rats following intravenous (IV) bolus administration of a single dose of 30 mg/kg [^{35}S]GEM®91 (Zhang *et al.*, 1995). Plasma disappearance curves for GEM®91 (based on radioactivity levels) could be described by the sum of two exponentials, with half-lives of 0.95 and 47 h (Figure 2). Following administration, plasma concentration (C_{max}) was 418 μg/ml (52 μM), mean residence time (MRT) was 42 h, and clearance was 21 ml/kg/h. Urinary excretion represented the major pathway of elimination of GEM®91 with 26.6% of the administered dose excreted over 24 hours and 58.1% over 240 hours after GEM®91 administration. Renal retention of GEM®91 was minor, with only 1.4% of the administered dose found in the kidney at 24 h, and 8.5% over 240 h.

A wide tissue distribution of GEM®91 was also observed (Figure 2). During the initial 30 min, the highest levels of radioactivity were found in the kidney, liver, spleen, lungs, and heart; radioactivity was retained over longer time periods in the kidney, liver, spleen, bone marrow, and intestine.

Polyacrylamide gel electrophoretic analysis (PAGE) of the extracted radioactivity at various time points from plasma, kidney, liver and other tissue showed the presence of both intact GEM®91 and shorter, partially degraded oligonucleotides. Radioactivity in urine was found to be predominantly associated with the partially degraded GEM®91 fragments, and rate of degradation was time-dependent. Figure 2C shows that GEM®91 was distributed to various tissues, and that the majority of the administered dose was retained in tissues for an extended period of time.

The pharmacokinetics of GEM®91 have also been studied following subcutaneous (s.c.) administration of a 30 mg/kg dose of [^{35}S]GEM®91 (Agrawal *et al.*, 1995). Following s.c. administration, the plasma disappearance curve of GEM®91 (based on radioactivity levels) could be described by the sum of two exponentials with half lives of 5.4 and 50 hours (Figure 3). The C_{max} achieved following subcutaneous (s.c.) administration was about five-fold lower than achieved following intravenous (i.v.) administration. The tissue disposition and elimination were quite similar to those obtained following i.v. administration.

IV.B. Multiple-Dose Pharmacokinetics in Rats

We have also determined the plasma concentration of GEM®91 after daily i.v. administration to rats for 8 days (Agrawal *et al.*, 1995). Doses of 1 mg/kg/day (1 µCi) were administered via the tail vein to 5 male Sprague-Dawley rats (weight range 311 to 392 g). During the 8-day administration period, samples of blood (approximately 0.1 ml) were taken, via the orbital sinus, immediately prior to the next administration of oligonucleotide.

Slightly larger aliquots (0.5 ml) of blood were taken on days 9, 10, and 11. Plasma was then separated from white and red blood cells by centrifugation. Saline was added to the red blood cells, which were then hemolyzed by freezing and thawing.

Figure 4 shows the plasma concentration of oligonucleotide 24 hours after administration of the dose, just prior to administration of the next dose. Plasma concentrations of oligonucleotide remained relatively flat (0.2 µg equivalents/ml) during the 8-day administration period.

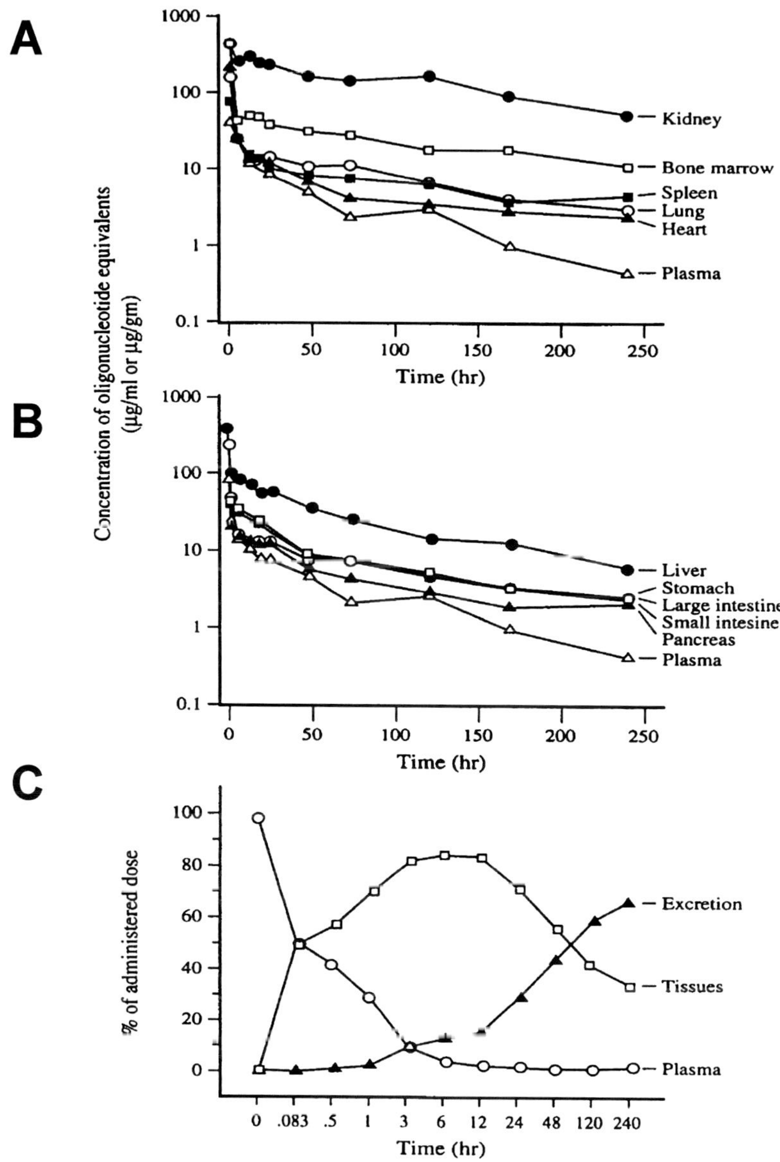

Figure 2. Plasma clearance, tissue disposition, and excretion of GEM®91 following intravenous administration in rats. **A,B**: disposition of GEM®91 in selected tissues; **C**: combined concentration in all tissues *(reprinted, with permission, from Zhang et al., 1995).*

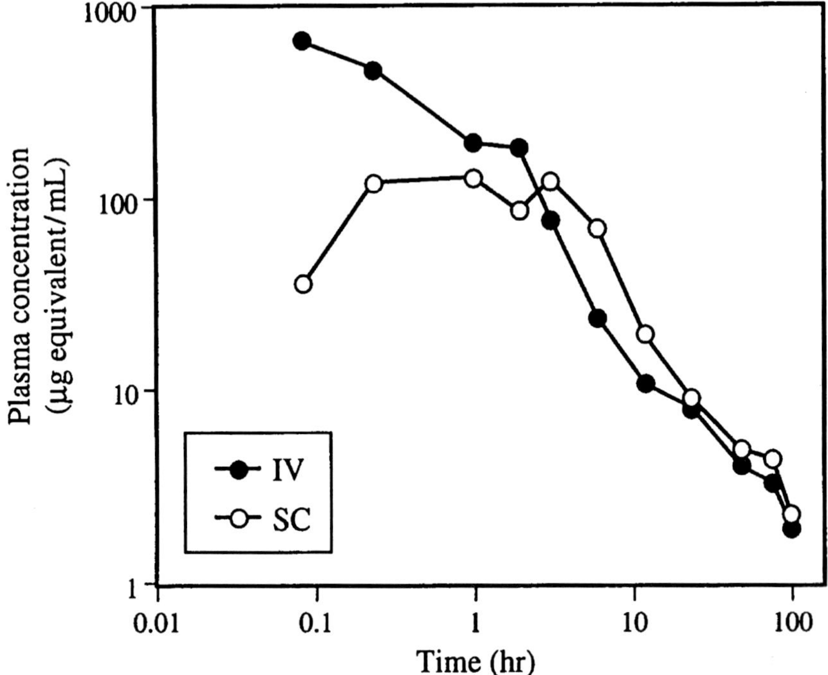

Figure 3. Comparative plasma clearance of GEM®91 following intravenous and subcutaneous administration of 30 mg/kg dose. Plasma elimination half-lives following subcutaneous administration were 5.4 hr ($t_{1/2}\ \alpha$) and 50.3 hr ($t_{1/2}\ \beta$), and maximum plasma concentration obtained was 134.9 μg equivalent/ml *(reprinted, with permission, from Agrawal et al., 1995).*

Plasma concentrations of oligonucleotide remained relatively flat (0.2 μg equivalents/ml) during the 8-day administration period. After the last dose (day 8), a decrease in the plasma concentration was observed on the second and third days following (days 10 and 11).

To evaluate oligonucleotide accumulation in red blood cells, radio-activity in red blood cell hemolysates was also measured. The concentration of oligonucleotide in the red blood cell pellet was approximately five-fold lower than that in plasma. Urine and feces were also collected from each animal for 11 days. The excretion pattern of [^{35}S]oligonucleotide is shown in Figure 5. The major route of excretion was via the urine. The percentage of the dose excreted in urine increased from 42% on day 2 to 62% on day 4 and remained around 60% during days 4 through 9. The percentage decreased on days 10 and 11. A similar trend of excretion, but at much lower levels, was seen in the feces.

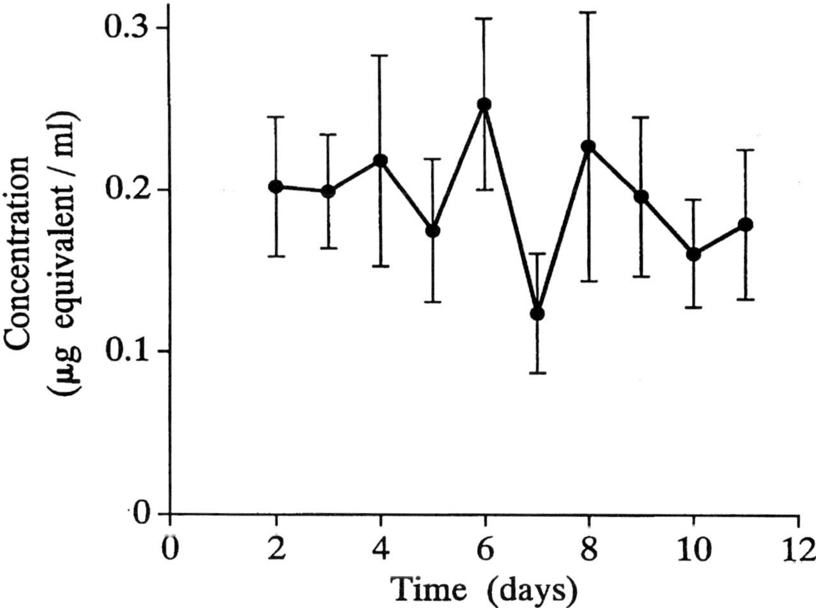

Figure 4. Concentration of GEM®91 in plasma following daily intravenous administration (1 mg/kg) in rats. The concentrations are in µg equivalents, on the basis of radioactivity levels, and were taken 24 hours after administration. The bars represent the standard deviation of the mean values from 5 animals *(reprinted, with permission, from Agrawal et al., 1995.)*

IV.C. Safety Studies in Rats

To study the safety profile of GEM®91 following repeated administration in rats, 6 male and 6 female rats received the oligonucleotide at three dose levels. The dose levels were 3, 10, and 30 mg/kg, administered intravenously via the tail vein, daily for seven days. After 7 days of administration, rats were sacrificed, and clinical pathology and histopathology of selected organs were performed.

The rats in the control group gained about 20.7±2.6 g in weight, whereas rats that received GEM®91 gained 17.2±4.2 g in the 3 mg/kg dose group and 16.0±0.81 g in the 30 mg/kg dose group after 7 days. There was no indication of a dose-dependent effect on body weight gain. Clinical pathology of the rats receiving 3, 10, or 30 mg/kg GEM®91 showed dose-dependent changes. Male rats that received the 30 mg/kg dose showed about 40% decrease in platelet count and about 35% increase in lymphocytes compared to rats in the control group.

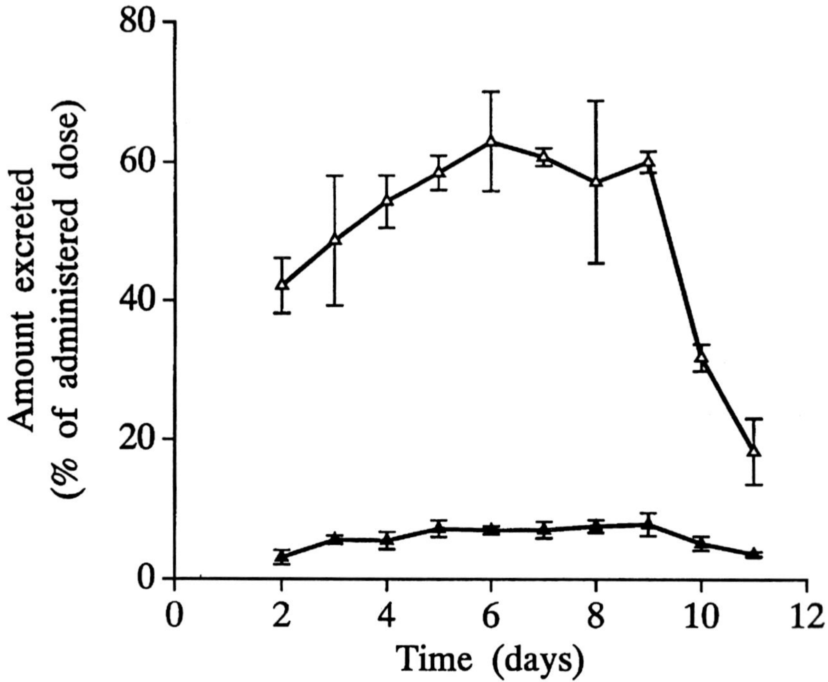

Figure 5. Urinary (Δ) and fecal (▲) elimination of GEM®91 after intravenous administration in rats, daily for 8 days. The bars represent the standard deviation of the mean values for 5 animals. *(reprinted, with permission, from Agrawal et al., 1995)*.

Clinical chemistry analysis showed dose-dependent increases in aspartate and alanine aminotransferase (AST and ALT) levels in rats receiving GEM®91. Male rats that received the oligonucleotide at 30 mg/kg exhibited about 35% increase in ALT levels and about 50% increase in AST levels compared to rats in the control group.

Rats receiving GEM®91 also exhibited dose-dependent histopathological changes in various organs, as compared to rats in the control group. In the liver, hepatic cell infiltrates, Kupffer cell vacuolation, and individual cell necrosis were observed. The cytoplasmic vacuolation of Kupffer cells was a morphologic change indicative of Kupffer cell activation and proliferation. In kidneys, changes observed included tubular epithelial cell degeneration of the cortex accompanied by mononuclear cell infiltration. In spleens, hematopoietic cell proliferation and/or white pulp lymphoid hyperplasia were observed.

V. ADMINISTRATION OF GEM®91 IN MONKEYS

V.A. Pharmacokinetics in Monkeys

A pharmacokinetic study of GEM®91 was carried out in monkeys (Agrawal *et al.*, 1995). Following i.v. administration of 1 mg/kg of [³⁵S]-GEM®91, the plasma disappearance curves for oligonucleotide-derived radioactivity could be described by the sum of two exponentials with half-lives of 0.8 and 0.7 h and 56 and 46 h in two individual animals (Figure 6).

Urinary excretion represented the major pathway of elimination of GEM®91, with 28% of the dose excreted in the 96 h after administration. The C_{max} was dose-dependent. The plasma C_{max} value for the 1 mg/kg dose was 38 µg/ml (4.7 µM), and for the 5 mg/kg dose it was 150 µg/ml (18.7 µM) (Agrawal *et al.*, 1995).

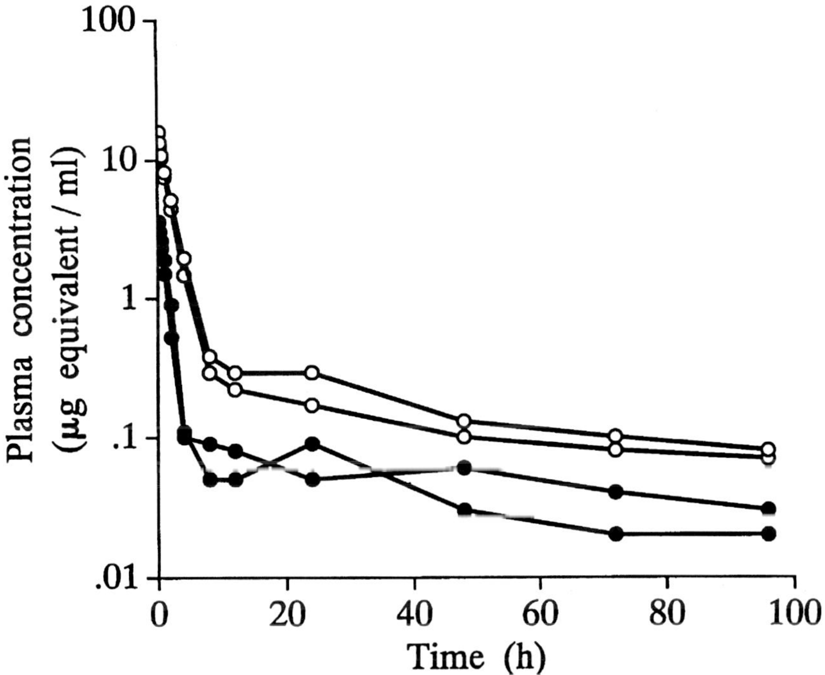

Figure 6. Plasma clearance of GEM®91 following intravenous administration of 1 mg/kg (●) and 5 mg/kg (O) dose of the oligonucleotide in two monkeys each. *(reprinted, with permission, from Agrawal et al., 1995).*

V.B. Safety Study in Monkeys

The side effects observed following bolus i.v. administration of GEM®91 in monkeys were significantly different from those observed in rats. Bolus i.v. administration of the oligonucleotide in monkeys produced a transient decrease in peripheral total white blood cell and neutrophil counts, hematocrit, and a brief increase followed by a prolonged decrease in arterial blood pressure. These changes were dependent on the dose of GEM®91 administered and its rate of infusion (Galbraith *et al.*, 1994).

Administration of GEM®91 at 5 mg/kg and 20 mg/kg produced these changes, but administration of the same dose over a 120 minute period produced no significant changes. Detailed studies carried out to identify the nature of these side effects in monkeys showed changes in complement concentration and prolongation of partial thromboplastin time (aPTT), in addition to some changes in hematological values.

As shown in Figure 7, serum complement CH50 concentrations decreased at a dose ≥10 mg/kg (infused over 10 min), beginning at 5 min from the start of infusion, and the C5a split products of complement increased, beginning within 2 min of infusion, at a dose of ≥5 mg/kg (infused over 10 min). The higher the dose, the earlier C5a appeared. At doses ≥2 mg/kg no changes were observed. Similarly, administration of 5, 10, or 20 mg/kg GEM®91 to monkeys by 10 min infusion induced prolongation of aPTT in dose-dependent manner (Figure 8).

Both complement activation and prolongation of aPTT were transient, and returned to normal values immediately or a few hours after the end of infusion. Both of these changes were dose-dependent, and rate-of-infusion-dependent, suggesting that they were due to plasma concentration of GEM®91 achieved.

The complement activation and prolongation of aPTT observed in monkeys are common to all phosphorothioate oligodeoxynucleotides tested so far and are not sequence-specific. To avoid these side effects, clinical studies in humans are being performed in which phosphorothioate oligodeoxynucleotides are administered by 2-hour i.v. infusion.

VI. CLINICAL STUDIES OF GEM®91

The protocol for human use of [³⁵S]GEM®91 was approved by the University of Alabama at Birmingham Institutional Review Board and Radiation Committee and the U.S. Food and Drug Administration. Table 2 illustrates the characteristics of six HIV-infected individuals entered into a phase I clinical trial. Six male homosexual or bisexual patients had positive laboratory tests for hepatitis B infection.

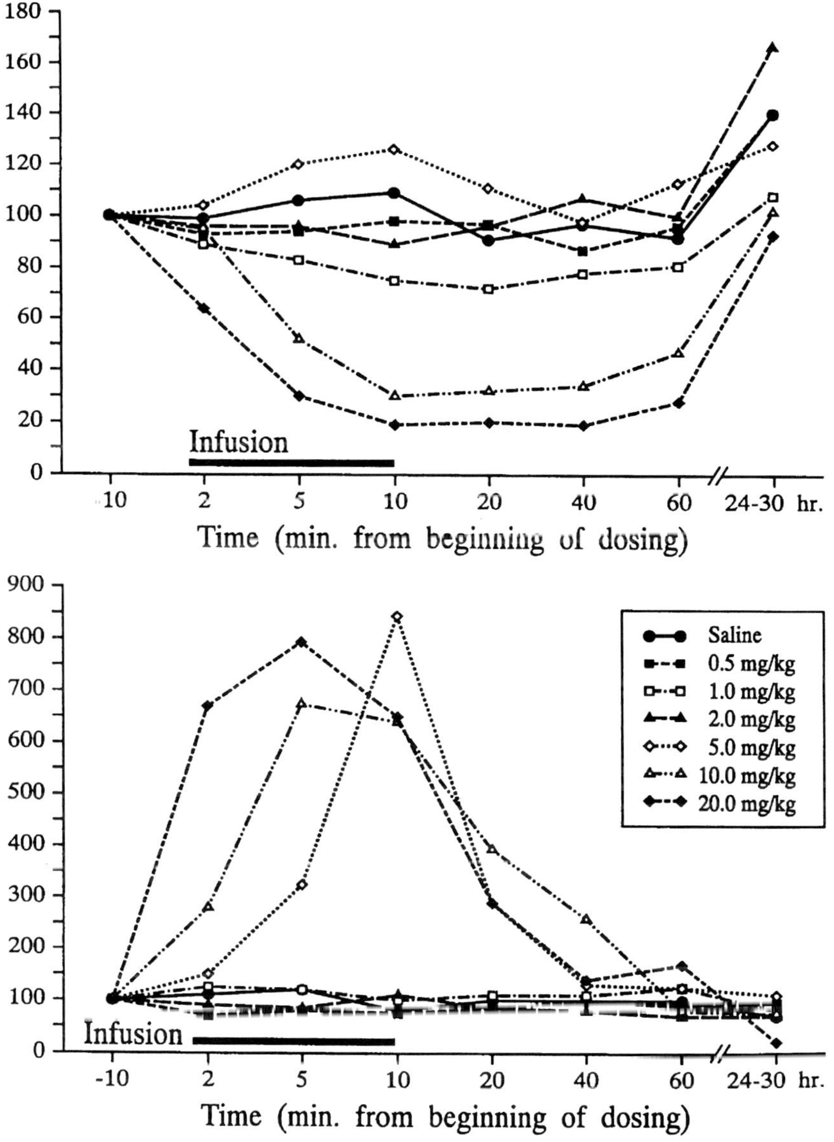

Figure 7. Level of serum complement (CH50) **(Top)** and serum complement (C5a) **(Bottom)** in monkeys, as percent of each individual's baseline following infusion of various doses of GEM®91 i.v. over a 10-min period. The blood samples were drawn at time indicated and analyzed for level of (CH50) complement. *(reprinted, with permission, from Galbraith et al., 1994).*

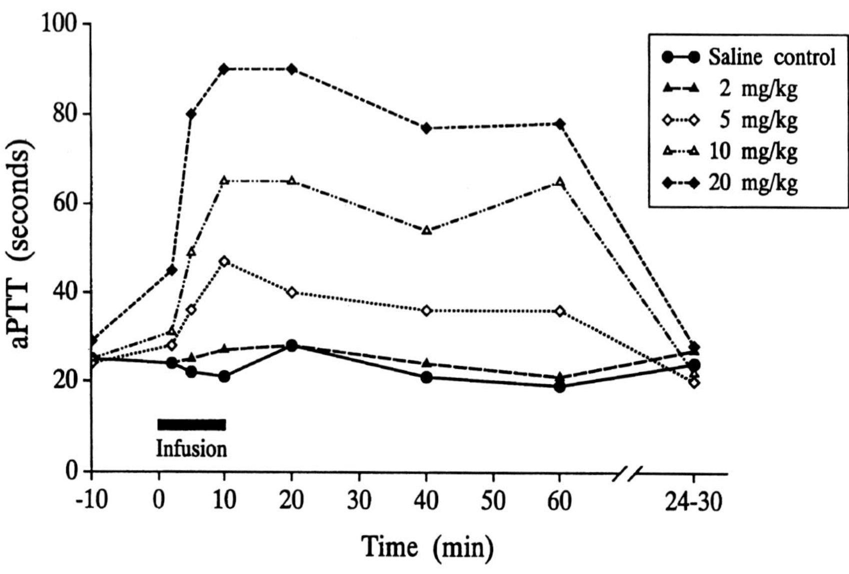

Figure 8. Activated partial thromboplastin tissue in monkeys following infusion of various doses of GEM®91 i.v. over a 10 min. period. The blood samples were drawn at times indicated and analyzed for aPTT. *(reprinted, with permission, from Galbraith et al., 1994).*

Complete urine collection over a 24-hour period was carried out on day-1 (before administration of GEM®91) for each patient to determine the urinary function, including total volume, total protein, and total creatinine clearance. Normal function was evident in each subject.

Table 2. Characteristics of the Human Subjects in the Phase I Trial

Pt. No.	Age Yr.	Sex	Body Weight (kg)	Height (in)	HIV	HBV	Total Dose (µg)
1	31	M	66.7	71	+	-	6670
2	33	M	78.5	68	+	+	7850
3	28	M	68.1	68	+	-	6810
4	28	M	69.5	70	+	+	6950
5	41	M	102.6	73	+	-	10260
6	43	M	75.4	73	+	-	7540

VI.A. Pharmacokinetic Study in HIV-1 Infected Patients

Each patient received GEM®91 by 2-hour continuous i.v. infusion at a dose of 0.1 mg/kg, including approximately 500 μCi [^{35}S]GEM®91 (Zhang *et al.*, 1995). Actual radioactivity in the dosing solution for each patient was determined immediately before administration and specific activity was then calculated (38.6 to 64.7 μCi/mg). Blood samples (20 ml for each time point) were collected in heparinized tubes from each subject at various times: 0 (immediately after infusion), 5, 30, 60, and 90 minutes and 2, 4, 6, 8, 10, 24, 48, 72, and 96 hours after the end of the 2-hour infusion of radiolabeled oligonucleotide. Plasma was then separated by centrifugation. Complete collection of urine was carried out for each subject at intervals of 0 to 6, 6 to 12, 12 to 24, 24 to 48, 48 to 72, and 72 to 96 hours after the end of infusion of GEM®91. These biological samples were kept at ≤-70°C until analyzed. To inactivate the HIV, these biological samples were heated at 56°C for 2 hours.

In addition to the pharmacokinetic measurements, the safety of GEM®91 was assessed during infusion and for 30 days after oligonucleotide administration. Clinical observations included cardiac monitoring by telemetry, evaluation of electrocardiograms, and recording of vital signs for 4 days after infusion. Hematologic profile, clinical chemistry, and urinalysis were performed until discharge on postinfusion day 4. Physical examination and interim history to capture adverse events were repeated at 10 and 30 days after GEM®91 administration.

VI.A.1 Radioactivity measurements

The total radioactivity in plasma and urine samples was determined by liquid scintillation spectrometry. In brief, the biological fluids (100 to 200 μl plasma; 50 to 100 μl urine) were mixed with 6 ml scintillation solvent (Beckman Ready-Safe biodegradable liquid scintillation cocktail), and counted using a spectrometer (LS 6000TA, Beckman Instruments, Inc., Irvine, CA) to determine total radioactivity.

VI.A.2. Gel electrophoresis of [^{35}S]GEM®91 from biological samples

Plasma (200 to 400 μl) was incubated with proteinase K (2 mg/ml) in an extraction buffer (0.5% sodium dodecyl sulfate/10 mM sodium chloride/20 mM Tris-HCl, pH 7.6/10 mM ethylenediamine tetraacetic acid) for 2 hours at 37°C. The samples were then extracted twice with phenol/chloroform (1:1, vol/vol) and once with chloroform. After ethanol precipitation, the oligonucleotides were analyzed by electrophoresis in 20% polyacrylamide gels that contained 7 M urea. The gels were fixed in

10% acetic acid/10% methanol solution and then dried before autoradiography.

VI.A.3 Kinetics of [³⁵S]GEM®91 in the plasma of HIV-infected subjects

Figure 9A illustrates the plasma concentration-time course of GEM®91 equivalents after 2-hour continuous i.v. infusion of radiolabeled oligonucleotide. Mean concentrations (±SD) at various times are presented in Table 3. Pharmacokinetic analysis revealed that plasma disappearance curves for GEM®91-derived radioactivity could be described by the sum of two exponentials, with $t_{1/2}$ values (mean±SE) of 0.18±0.04 and 26.71±1.67 hours, respectively (Table 3). Other estimated pharmacokinetic parameters are shown in Table 3, including apparent area under the curve (AUC), MRT, V_{88}, and clearance. The chemical forms of radioactivity in plasma were further evaluated by PAGE, showing the presence of both intact GEM®91 and lower molecular weight oligonucleotide fragments.

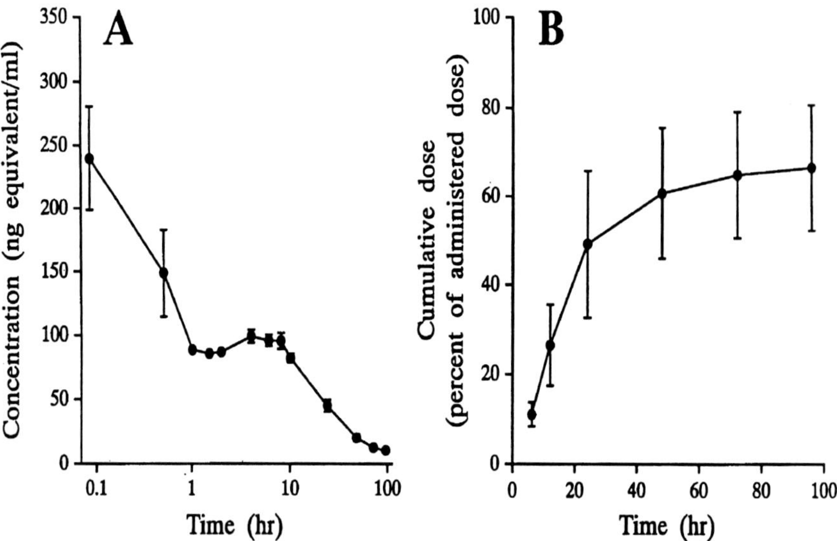

Figure 9. **A:** Plasma concentration over time of GEM®91 derived radioactivity in HIV-infected subjects, expressed as nanograms of oligonucleotide equivalents per milliliter after a 2-hour intravenous infusion of [³⁵S]GEM®91 of a dose of 0.1 mg/kg body weight. The mean plasma concentration is based on the radioactivity levels from six HIV-1 infected patients (See Table 2). **B:** Cumulative urinary excretion of GEM®91-derived radioactivity, expressed as the cumulative percentage of administered dose excreted over time. *(reprinted, with permission, from Zhang, et al., 1994).*

Table 3. Pharmacokinetic Analysis of GEM®91 in Humans

Pt. No.	Cmax (ng/ml)	t1/2α (hr.)	t1/2β (hr.)	AUC (hr•ng/ml)	MRT (hr.)	VDss (l/kg)	CL (l/hr/kg)
1	284.38	0.10	19.25	3131.19	27.56	0.880	0.032
2	418.47	0.36	30.93	5272.57	43.32	0.822	0.019
3	180.81	0.16	26.82	3479.41	38.46	1.105	0.029
4	160.95	0.06	25.80	3284.42	37.17	1.132	0.030
5	385.83	0.23	27.98	3712.12	39.34	1.060	0.027
6	344.20	0.192	29.46	4374.33	41.86	0.957	0.023
Mean	295.77	0.18	26.71	3875.67	37.95	0.993	0.027
S.E.	43.60	0.04	1.67	331.04	2.27	0.052	0.002

GEM®91-derived radioactivity in plasma following i.v. continuous infusion of radiolabeled GEM®91 at a dose of 0.1 mg/kg.

For the initial 2 hours after the end of infusion, the major radioactivity in plasma was present as intact GEM®91, which could be detected up to 6 hours after infusion. Separate experimental data indicated that, in this study, heat treatment (up to 2 hours) of plasma samples had no effect on the stability of the oligonucleotide in plasma. The shorter oligonucleotide metabolites in plasma represent, therefore, the *in vivo* enzymatic degradation products of GEM®91.

VI.A.4. Cumulative urinary excretion

Figure 9B shows the cumulative excretion of urinary activity over 96 hours after the end of administration of radiolabeled GEM®91. Urinary excretion represented the major pathway of elimination of the oligonucleotide. Rapid excretion of radioactivity was observed for the first 48 hours after GEM®91 administration. Of the administered dose, 49.15%± 6.80% (mean±SE, range 35.57% to 81.18%) was excreted within 24 hours, 60.71%±6.00% (range 46.90% to 89.54%) within 48 hours, and 70.37%± 6.72% (range 50.42% to 93.31%) over 96 hours. The majority of radioactivity in urine was present as degradation fragments of the oligonucleotide.

VI.B. Safety Evaluation

Systematic vital sign monitoring data were collected for 30 days after GEM®91 was given. During this period, three of six subjects reported no adverse events. One subject reported headache, and also experienced nausea and vomiting. Self-limiting diarrhea was reported by one subject.

One subject on outpatient status had blurred speech and an abnormal gait, after returning with alcohol intoxication. Three subjects were classified as having sinus bradycardia, with resting heart rates ranging from 50 to 59 beats/min on isolated occasions over 4 days after GEM®91 administration. There were no associated symptoms or blood pressure abnormalities. All adverse events noted were mild in nature and resolved without intervention. Extensive evaluation of hemotology and clinical chemistry tests did not reveal any treatment-associated abnormalities. GEM®91 was safe and well tolerated at the dose and administration schedule used.

VI.C. Ongoing Clinical Studies

In further clinical studies, escalating doses of GEM®91 (0.3, 0.5, 1 and 2.5 mg/kg) were administered in a two hour infusion, which resulted in higher levels of the oligonucleotide in plasma. Also, administration of multiple doses of GEM®91 in humans every other day for 27 days (14 doses) by two hour infusion at doses of 0.5, 1 and 2 mg/kg showed that oligonucleotide did not accumulate in plasma following multiple doses. Studies are also ongoing in which GEM®91 is being administered to HIV-1 infected patients by continuous intravenous infusion for 8 or 14 days at the doses of 0.2 mg/kg/day to 3.2 mg/kg/day.

VII. CONCLUSIONS

GEM®91 is an oligonucleotide phosphorothioate which is complementary to the HIV-1 *gag* mRNA initiation site. In various cell culture assays, GEM®91 inhibited virus replication at concentrations of 0.2 μM to 2.0 μM in a dose-dependent fashion. In rats, pharmacokinetics of the oligonucleotide showed rapid plasma elimination and disposition to various tissues. The highest concentrations of GEM®91 were observed in kidney, liver, spleen and bone marrow. At 24 hours post-administration, both intact and partially degraded GEM®91 were observed in the tissues. The major route of elimination of the oligonucleotide was by urinary excretion, and to a lesser extent by fecal excretion. Safety studies of GEM®91 in rats showed dose-dependent side effects, including thrombocytopenia, elevation of liver enzymes (ALT, AST), splenomegaly and some pathological changes in liver, kidney and spleen. In monkeys, the side effects observed were dependent on the dose and rate of administration. These side effects included activation of complement and prolongation of aPTT. These side effects were significantly minimized in monkeys following administration of GEM®91 by slow (2 hour) infusion. Based on the experience in monkeys, to avoid side effects in ongoing human clinical studies, the oligonucleotide is being administered to patients by 2 hour intravenous infusion or by continuous infusion. While present studies are ongoing by intravenous infusion, the subcutaneous or intramuscular administration of GEM®91 may also be viable routes for long term therapy.

VIII. REFERENCES

Agrawal, S., Goodchild, J., Civeira, M.P., Thornton, A.H., Sarin, P.S., and Zamecnik, P.C. (1988) Oligodeoxynucleoside phosphoramidates and phosphorothioates as inhibitors of human immunodeficiency virus. *Proc. Natl. Acad. Sci. USA* **85**:7079-7083.

Agrawal, S., Ikeuchi, T., Sun, D., Sarin, P.S., Konopka, A., Maizel, J., and Zamecnik, P.C. (1989) Inhibition of human immunodeficiency virus in early infected and chronically infected cells by antisense oligodeoxynucleotides and their phosphorothioate analogs. *Proc. Natl. Acad. Sci. USA* **85**:7790-7794.

Agrawal, S. (1991) A possible approach for chemotherapy of AIDS. In Wickstrom, E., ed., *Prospects for Antisense Nucleic Acid Therapy of Cancer and AIDS*, Wiley-Liss, New York, 143-158.

Agrawal, S., and Tang, J.-Y. (1992) GEM®91 - An antisense oligonucleotide phosphorothioate as a therapeutic agent for AIDS. *Antisense Res. Dev.* **2**:261-266.

Agrawal, S., Sarin, P.S., and Zamecnik, P.C. (1992a) Cellular uptake and anti-HIV activity of oligonucleotides and their analogs. In Erickson, R.P., and Izant, J.G., eds., *Gene Regulation: Biology of Antisense and DNA*, Raven Press, New York, 273-283.

Agrawal, S., Tang, J.-Y., Sun, D., Sarin, P.S., and Zamecnik, P.C. (1992b) Synthesis and anti-HIV activity of oligoribonucleotides and their phosphorothioate analogs. In Baserga, R., and Denhardt, D.T., eds., *Antisense Strategies, Ann. New York Acad. Sci.*, 660-662.

Agrawal, S., Tang, J.-Y., Groopman, J., and Byrn, R. (1993) GEM®91 - An antisense oligonucleotide as a therapeutic agent for AIDS. National Cooperative Drug Development Group Conference, NIAID, July 11-16.

Agrawal, S., and Lisziewicz, J. (1994) Potential for HIV-1 treatment with antisense oligonucleotides. *J. Biotech. Healthcare* **1**:167-182.

Agrawal, S. (1995) Antisense oligonucleotide approach for therapy of AIDS. In Mohan, P., and Baba, M., eds., *Anti-AIDS Drug Development: Challenges, Strategies and Prospects*, Harwood Academic Publishers, London, 143-162.

Agrawal, S., Temsamani, J., Galbraith, W., and Tang, J.-Y. (1995) Pharmacokinetics of antisense oligonucleotides. *Clin. Pharmacokin.* **28**:7-16.

Aldovini, A., and Young, R.A. (1990) Mutations of RNA and protein sequences involved in human immunodeficiency virus type 1 packaging result in production of noninfectious virus. *J. Virol.* **64**:1920-1926.

Balotta, C., Lusso, P., Crowley, R., Gallo, R.C., and Franchini, G. (1993) Antisense phosphorothioate oligonucleotides targeted to the *vpr* gene inhibit human immunodeficiency virus type 1 replication in primary human macrophages. *J. Virol.* **67**:4409-4414.

Baudin, F., Marquet, R., Isel, C., Darlix, J.-L., Ehresmann, B., and Ehresmann, C. (1993) Functional sites in the 5' region of human immunodeficiency virus type 1 RNA form defined structural domains. *J. Mol. Biol.* **229**:382-383.

Byrn, R.A., Chay, H., and Agrawal, S. (1995) *In vitro* antiviral activity of GEM®91 against diverse clinical isolates of HIV-1. 2nd Natl. Conf. Human Retroviruses Relat. Infect. (Jan 29 - Feb 2, Washington, DC) (Abst 469).

DeBouck, C. (1992) The HIV-1 protease as therapeutic target for AIDS. *AIDS Res. Human Retroviruses* **8**:153-164.

DeClercq, E. (1992) HIV inhibitors targeted at the reverse transcriptase. *AIDS Res. Human Retroviruses* **8**:119-134.

Degols, G., Devaux, C., and Lebleu, B. (1994) Oligonucleotide-poly-(L-lysine)-heparin complexes: Potent sequence specific inhibitors of HIV-1 infection. *Bioconjugate Chem.* **5**:8-13.

Galbraith, W.M., Hobson, W.C., Giclas, P.C., Schechter, P.J., and Agrawal, S. (1994) Complement activation and hemodynamic changes following intravenous administration of phosphorothioate oligonucleotides in the monkey. *Antisense Res. Dev.* **4**:201-206.

Harrison, G.P., and Lever, A.M.L. (1992) The human immunodeficiency virus type 1 packaging signal and major splice donor region have a conserved stable secondary structure. *J. Virol.* **66**:4144-4153.

Hostomsky, Z., Hostomska, Z., and Mathews, D.A. (1993) Ribonuclease H. In Linn, S.M., Lloyd, R.S., and Roberto, R.J., eds., *Nucleases*, Cold Spring Harbor Press, 341-376.

Letsinger, R.L., Zhang, R.L., Sun, D.K., Ikeuchi, T., and Sarin, P.S. (1989) Cholesterol-conjugated oligonucleotides: Synthesis, properties and activity as inhibitors of replication of human immunodeficiency virus in cell culture. *Proc. Natl. Acad. Sci. USA* **86**:6553-6556.

Lever, A., Gottlinger, H., Haseltine, W., and Sodroski, J. (1989) Identification of a sequence required for efficient packaging of human immunodeficiency virus type 1 RNA into virions. *J. Virol.* **63**:4085-4087.

Li, G., Lisziewicz, J., Sun, D., Zon, G., Daefter, S., Wong-Staal, F., Gallo, R.C., and Klotman, M.E. (1993) Inhibition of *rev* activity and human immunodeficiency virus type 1, replication by antisense oligodeoxynucleotide phosphorothioate analogs directed against the *rev*-responsive element. *J. Virol.* **67**:6882-6888.

Lisziewicz, J., Sun, D., Klotman, M., Agrawal, S., Zamecnik, P.C., and Gallo, R.C. (1992) Specific inhibition of human immunodeficiency virus type 1 replication by antisense oligonucleotides: An *in vitro* model for treatment. *Proc. Natl. Acad. Sci. USA* **89**:11209-11213.

Lisziewicz, J., Sun, D., Metelev, V., Zamecnik, P.C., Gallo, R.C., and Agrawal, S. (1993) Long-term treatment of human immunodeficiency virus-infected cells with antisense oligonucleotide phosphorothioates. *Proc. Natl. Acad. Sci. USA*

90:3860-3864.

Lisziewicz, J., Sun, D., Weichold, F.F., Thierry, A.R., Lusso, P., Tang, J.-Y., Gallo, R.C., and Agrawal, S. (1994) Antisense oligonucleotide phosphorothioate complementary to *gag* mRNA blocks replication of human immunodeficiency virus in human peripheral blood cells. *Proc. Natl. Acad. Sci. USA* **91**:7942-7946.

Matsukura, M., Zon, G., Shinozuka, K., Robert-Guroff, M., Shimada, T., Stein, C.A., Mitsuya, H., Wong-Staal, F., Cohen, J.S., and Broder, S. (1989) Regulation of viral expression of human immunodeficiency virus *in vitro* by an antisense phosphorothioate oligodeoxynucleotide against *rev (art/trs)* in chronically infected cells. *Proc. Natl. Acad. Sci. USA* **86**:4244-4247.

Papp, B., Agrawal, S., and Byrn, R.A. (1995) Failure to generate HIV-1 resistance *in vitro* of the antisense therapeutic agent GEM®91. 2nd Natl. Conf. Human Retroviruses Relat. Infect., Washington, DC, Abst 477.

Sarin, P.S., Agrawal, S., Civeira, M.P., Goodchild, J., Ikeuchi, T., and Zamecnik, P.C. (1988) Inhibition of acquired immunodeficiency syndrome virus by oligodeoxynucleoside methylphosphonates. *Proc. Natl. Acad. Sci. USA* **85**:7448-7451.

Shibahara, S., Mukai, S., Morisasa, H., Nakashima, H., Kobayashi, S., and Yamamoto, N. (1989) Inhibition of human immunodeficiency virus (HIV-1) replication by synthetic oligo-RNA derivatives. *Nucleic Acids Res.* **17**:239-252.

Stein, C.A., Matsukura, M., Subasinghe, C., Broder, S., and Cohen, J.S. (1989) Phosphorothioate oligodeoxynucleotides are potent sequence nonspecific inhibitors of de novo infection by HIV. *AIDS Res. Hum. Retroviruses* **5**:639-646.

Stein, C., Neckers, L., Nair, B., Mumbauer, S., Hoke, G., and Pal, R. (1991) Phosphorothioate oligodeoxycytidine interferes with binding of HIV-1 gp120 to CD4. *J AIDS* **4**:686-693.

Stein, C.A., Cleary, A., and Yakubov, L. (1993) Phosphorothioate oligonucleotides bind to the third variable loop domain (V3) of human immunodeficiency virus type 1 gp 120. *Antisense Res. Dev.* **3**:19-31.

Weichold, F., Lisziewicz, J., Zeman, K.A., Nerkurkar, L.S., Agrawal, S., Reitz, M.S., Jr. and Gallo, R.C. (1995) Antisense phosphorothioate oligodeoxynucleotide alter HIV-1 replication in cultured human macrophages and peripheral blood mononuclear cells. *AIDS Res. Hum. Retroviruses* **11**:863-867.

Zamecnik, P.C., Goodchild, J., Taguchi, Y., and Sarin, P.S. (1986) Inhibition of replication and expression of human T-cell lymphotrophic virus type III in cultured cells by exogenous synthetic oligonucleotide. *Proc. Natl. Acad. Sci. USA* **83**:4143-4146.

Zamecnik, P.C. (1996) History of antisense oligonucleotides. In Agrawal, S., ed., *Antisense Therapeutics*. Humana Press, Totowa NJ, 1-11.

Zelphati, O., Imbach, J.-L., Signoret, N., Zon, G., Rayner, B., and Leserman, L. (1994) Antisense oligonucleotides in solution or encapsulated in immunoliposomes inhibit replication of HIV-1 by several different mechanisms. *Nucleic Acids Res.* **22**:4307-4314.

Zhang, R., Yan T., Shahinian, H., Amin, G., Lu, Z., Liu, T., Saas, M.S., Jiang, Z.,

Temsamani, J., Martin, R.R., Schechter, P.J., Agrawal, S., and Diasio, R.B. (1995) Pharmacokinetics of an anti-human immunodeficiency virus antisense oligodeoxynucleotide phosphorothioate (GEM®91) in HIV-infected subjects. *Clin. Pharm. Therap.* **58**:44-53.

Zhang, R., Diasio, R., Lu, Z., Lu, T., Jiang, Z., and Galbraith, W. (1995) Pharmacokinetics and tissue distribution in rats of an oligodeoxynucleotide phosphorothioate (GEM®91) developed as a therapeutic agent for human immunodeficiency virus type-1. *Biochem Pharmacol.* **49**:929-939.

18

Treatment of Retinitis Induced by Cytomegalovirus Using Intravitreal Fomivirsen (ISIS 2922)

Stanley T. Crooke

Isis Pharmaceuticals, Inc., Carlsbad, California

I. INTRODUCTION

Cytomegaloviruses (CMV) are ubiquitous, highly species-specific herpes viruses (Alford and Britt, 1985). As with other herpes viruses, primary CMV infection is often followed by persistent and/or recurrent infections. The molecular biology of the virus and the infectious cycle are well characterized. The infectious cycle is divided into immediate-early, early, and late phases, relative to the time of appearance of CMV-specific proteins and sensitivity to different pharmacological agents (Stinski *et al.*, 1981). The immediate early period is usually defined as up to 2-4 hours post-infection, and is characterized by the production of a number of proteins responsible for regulation of viral gene expression (Hermiston *et al.*, 1987). These proteins are encoded by the major immediate-early transcriptional unit of the virus, and are required, along with the viral DNA polymerase, for viral replication (Heilbronn *et al.*, 1987).

The epidemiology of CMV infections is also well characterized. The virus can be transmitted sexually, parenterally or vertically from mother to child (Ives, 1996). Antibodies to CMV are present in up to 90% of homosexual males and female prostitutes. More than 50% of women who attend sexually transmitted clinics have CMV antibodies, and the likelihood of infection is proportional to the number of sexual partners.

CMV disease in adults generally results from reactivation of latent infections in patients with advanced HIV disease (Ives, 1996). CMV disease is among the most common opportunistic diseases in patients with HIV, with the incidence of *Candida* esophagitis being the only opportunistic disease occurring at a significantly greater rate. The incidence of CMV disease is related to the $CD4^+$ cell count. Virtually no CMV disease is observed in patients with $CD4^+$ cell counts greater than 100, and the median $CD4^+$ cell count at diagnosis in one large series was 15 (Moore and Chaisson, 1996). The incidence of CMV infection increases with time post-diagnosis of HIV, resulting in an overall incidence of 15-30% of all HIV-infected patients.

The most frequent and problematic manifestation of CMV infection in HIV-infected patients is CMV retinitis. Retinitis typically presents unilaterally, although as many as 35% of patients may have bilateral disease at presentation, and most develop bilateral disease (Heinemann, 1992). The disease is slowly progressive, may involve any part or all of the retina, and is necrotizing. It is often associated with significant inflammation and/or hemorrhage. Retinal detachment is common (up to 25% of patients). Although a number of drugs are available to treat CVM retinitis, none have been shown to cure the disease or to provide long-term disease control. All of the available agents are associated with significant toxicities.

II. MOLECULAR PHARMACOLOGY

Evaluation of the effects of antisense drugs in *in vitro* antiviral assays is particularly difficult. This results from the fact that relatively large concentrations of antisense drugs are coincubated with large concentrations of viruses and cells. The problem in interpreting data with regard to molecular mechanism of action is compounded by the fact that in many assays the endpoint measured is survival of target cells. As this may be the consequence of many mechanisms of action, proof of an antisense mechanism is complex (Crooke, 1995).

The molecular mechanisms of the oligonucleotide drug fomivirsen, Isis 2922, 5'-dGCGTTGCTCTTCTTCTTGCG phosphorothioate, have been well characterized in a series of studies in which human cytomegalovirus was incubated with normal human dermal fibroblast (NHDF) cells. The choice of this cell line was fortuitous, because it is one of the few cell lines that have displayed significant uptake of phosphorothioate oligonucleotides in the absence of cationic lipids *in vitro* (Nestle *et al.*, 1994). In these studies, all phosphorothioate oligonucleotides tested at 5 µM resulted in some inhibition of viral infectivity. However, the IC_{50} for fomivirsen was approximately 0.1 µM, while the IC_{50} for non-complementary phosphorothioate oligonucleotides was typically 2 µM or greater (Azad et al., 1993). Fomivirsen was shown to reduce the levels of the immediate-early proteins in a concentration and time-dependent fashion consistent with antisense inhibition of immediate-early region of HCMV RNA (Azad et al., 1993). Again, control phosphorothioate oligonucleotides were less effective. Nevertheless, substitutions of several nucleotides in fomivirsen with non-complementary nucleotides resulted in only modest loss of antiviral potency in this assay. This result suggests that although fomivirsen has an antisense effect, in the antiviral assays used, its antiviral activity is also due to effects that cannot be obviously ascribed to an antisense mechanism.

To better understand the mechanisms of fomivirsen antiviral activities, additional studies were performed. To avoid non-antisense effects and to avoid binding to viral coat proteins in the antiviral assays, U373 cells were transformed with a cDNA that encoded only the fomivirsen target, a CMV immediate-early 55 kDa protein. The activities of fomivirsen and other phosphorothioate oligonucleotides in this assay were compared to those observed in a traditional antiviral assay to dissect out potential mechanisms of action. A direct assay of inhibition of production of the 55 kDa protein revealed that fomivirsen displayed potent, sequence-dependent effects consistent with an antisense mechanism. Treatment of these cells with fomivirsen also resulted in loss of the message for the 55 kDa protein, consistent with RNase H mediated degradation. Thus, in U373 cells transformed with the antisense target, the drug displayed potent antisense activity (Anderson et al., 1996).

In antiviral assays, both sequence dependent and independent antiviral effects were observed. The sequence dependent antisense effects were shown to result in loss of both target RNA and target protein. The sequence-independent effects were due primarily to inhibition of viral adsorption. Finally, as replacement of cytosines in fomivirsen with 5-methyl-cytosine had no effect on antiviral potency (Anderson et al., 1996), it is unlikely that CpG-related B-cell stimulation (Krieg et al., 1995) is a factor.

Thus, the mechanism of action of fomivirsen in vitro is complex. Perhaps more complex is the assessment of the relevance of these data to the activity of the drug in patients with CMV retinitis. It is generally believed that there is no free virus in the infected retina and viral transmission is typically cell-to-cell (Alford and Britt, 1985). Consequently, one might expect that in vivo activity may be more dependent on the antisense mechanism than non-antisense mechanisms, such as inhibition of viral adsorption, but as there are no useful animal models of HCMV retinitis, this is very difficult to evaluate.

III. ANTIVIRAL ACTIVITY

Fomivirsen is a potent selective inhibitor of HCMV replication and infection. The IC_{50} in an immunoassay is approximately 0.37 µM, at least 30 times more potent than ganciclovir. In an infectious virus yield reduction assay, fomivirsen reduced viral yield by 2 logs at a dose of 2.2 µM, while ganciclovir produced a 2 log decrease at a dose of 35 µM. The inhibitory effects *in vitro* of fomivirsen are reduced in the presence of fetal bovine serum, consistent with the known propensity of phosphorothioate oligonucleotides to bind to serum proteins (Azad *et al.*, 1993).

Fomivirsen and ganciclovir, or foscarnet, displayed additive anti-HCMV activity when coincubated. Additive activity was also observed with 2'-3'-dideoxycytidine (ddC), which displays slight anti-HCMV activity. The activity of fomivirsen was unaffected by 3'-azido-3'-deoxythymidine (AZT), which exhibits no anti-HCMV activity. Against HIV, however, fomivirsen is slightly active, and is additive with AZT at most concentrations and synergistic at others (Azad *et al.*, 1995). Fomivirsen is active against all laboratory induced and clinically isolated resistant strains of HCMV studied to date.

IV. PHARMACOKINETICS

As fomivirsen is administered intravitreally, the pharmacokinetics of fomivirsen were evaluated after intravitreal administration to rabbits and monkeys. In neither species, at any time or dose, was any intact fomivirsen detected in plasma after intravitreal administration.

In humans, doses used were 150 and 330 µg per eye to result in vitreal concentrations of 5 and 10 µM, respectively. To achieve comparable concentrations in the rabbit model, single doses of 66 µg per eye of [^{14}C]fomivirsen were administered. The drug and metabolites in vitreous humor and retinal tissue were extracted and evaluated using HPLC and polyacrylamide gel electrophoretic analyses. Four hours after dosing, the concentration of fomivirsen in vitreous was 3.9 µM. Elimination of intact fomivirsen from vitreous exhibited first-order kinetics with a $t_{1/2}$ of 60 hours. By 10 days after dosing, the vitreal concentration of intact fomivirsen was 0.15 µM, and the total concentration of drug and metabolites was approximately 0.75 µM. The metabolites were all shorter oligonucleotides, 6-19 nucleotides in length. Intact fomivirsen accumulated in the retina over the first three days after dosing, reaching a maximum concentration of 3.5 µM, then declined with $t_{1/2}$ of 96 hours. Ten days after dosing, the concentration of intact fomivirsen in the retina was 1.8 µM, approximately ten times the vitreal concentration. Approximately 28% of the total radioactivity in the retina at that time was intact fomivirsen (Leeds *et al.*, 1996).

In the monkey, both single and multiple intravitreal doses were evaluated. Doses of 11, 57 and 115 µg/eye were administered (to result in vitreal concentrations of 1, 5, and 10 µM, respectively). Drug and metabolites were evaluated by solid phase extraction and capillary gel electrophoresis. Ocular exposure to fomivirsen was clearly dose dependent, and three days after a single dose, the vitreal concentrations ranged from 80 nM to approximately 1.5 µM, dependent on the initial dose. The concentrations in retina ranged from 50 nM to 1.1 µM over the range of doses. The estimated $t_{1/2}$ of intact fomivirsen in vitreous and retina after a single dose was approximately 22 and 75-85 hours, respectively. After three every-other-week doses of 115 µg fomivirsen, slight accumulation in the retina was observed. Similarly, after three weekly doses of 57 µg/eye, slight accumulation in the retina was observed. Based on an evaluation of the data from both species, it was concluded that it was appropriate to investigate administration of fomivirsen to humans with treatment intervals ranging from weekly to every other week (Leeds *et al.*, 1997).

V. TOXICOLOGICAL EFFECTS

Toxicological studies in mice of one-month, repeat-dose intravenous fomivirsen at 0.2-50 µg/kg every other day produced a toxicological profile similar to those of other phosphorothioate oligonucleotides. Next it was appropriate to evaluate its potential toxicological effects after intravitreal administration. The effects of single-intravitreal doses ranging from 16.5 to 330 µg/eye yielding theoretical vitreal concentrations as high as 40 mM were evaluated in the pigmented rabbit. The drug induced a dose dependent inflammatory response in the eye. At lower doses and earlier time points, the ciliary body was the primary structure affected. At doses that resulted in theoretical concentrations of 20 and 40 µM, other ocular structures were affected. In animals treated with 40 µM drug, retinal degeneration was observed, probably secondary to inflammation (Fomivirsen Investigator's Brochure, 1996).

The effects of weekly and biweekly administration of fomivirsen for one month to monkeys were also evaluated. Doses ranged from 3.3 to 33 μg/eye, which resulted in theoretical vitreal concentrations of 0.4 to 8.0 μM. Although less sensitive than rabbits, monkeys also displayed evidence of ocular inflammation, including cyclitis and vasculitis. However, in monkeys the incidence of these inflammatory changes was variable, and not dose-dependent. Local steroid treatment prior to intravitreal injection of fomivirsen eliminated the ocular inflammation, and no other ocular abnormalities have been attributed to fomivirsen in repeat-dose studies. The sporadic incidence and severity of ocular inflammation of fomivirsen in monkeys makes it difficult to attribute these findings to the compound, and suggest the possible influence of injection procedure (Fomivirsen Investigator's Brochure, 1996). Based on these studies, rabbits were more sensitive to the inflammatory effects of fomivirsen than monkeys, and doses up to 330 μg/week per eye were considered to be safe enough to evaluate in patients with CMV retinitis.

VI. CLINICAL STUDIES

Fomivirsen has been evaluated in phase I/II studies and is now being evaluated in several phase III studies. The phase III studies in progress are evaluating the drug as a single agent and in combination with ganciclovir in patients with early and advanced CMV retinitis.

Table 1 shows the safety profile of fomivirsen observed in patients treated with 150 μg or 330 μg per eye every two weeks. Most of these patients had far advanced AIDS and very advanced CMV retinitis. As can be seen, the major side effect was inflammation. In general, the inflammation was mild to moderate, could be treated with topical steroids when necessary, and has not resulted in significant dose reductions or patients being withdrawn from studies. Thus, the incidence of retinal detachment is remarkably low. That the drug is well tolerated is further demonstrated by the fact that many patients have been treated for periods of six months to more than a year. The severity of inflammation observed in patients with HIV and CMV infections has been substantially lower than that observed in rabbits, and probably less than the minimal effects observed in monkeys.

The side effect incidence shown in Table 1 represents the occurrence of any event at any time in the course of therapy, emphasizing the tolerability of the drug (Kisner, 1996). In initial studies, patients with early-stage CMV retinitis were treated with 330 μg/eye every two weeks. Changes in peripheral vision were observed in a few patients. As a consequence, the dose in patients with early stage CMV retinitis was lowered to 150 μg/eye every two weeks. The total incidence of this effect at 330 μg was approximately 6%.

Table 1. Safety with Fomivirsen

Dose; No. of Patients	150 µg; N = 17		330 µg; N = 107	
Ocular Side Effect	No. Pts.	Incidence	No. Pts.	Incidence
Vitritis	2	12%	16	15%
AC Inflammation	1	6%	4	4%
KP	0	0%	1	1%
Iritis	0	0%	2	2%
Increased IOP	1	6%	4	4%
Transient blurred vision	1	5%	10	9%
Photopsia	0	0%	4	4%
Photophobia	4	24%	6	6%
Chromatopsia	0	0%	8	7%
DPV	0	0%	6	6%
RPE pigment Inc.	1	6%	13	12%
Retinal detachment	0	0%	3	3%

Safety profile of fomivirsen in patients tolerated with 150 µg/eye and 330/µg/eye every two weeks. Patients were initially treated weekly, then maintained with biweekly doses any event at any time in the course of therapy. No fomivirsen related systemic adverse experiences or ocular adverse experiences have been observed to date.

Some patients experienced both subjective and objective improvement in peripheral vision with time. The problems in evaluating patients in this area include imprecision in perimetry measurements and variability in the ability of patients to participate in perimetry because of their advanced AIDS. Additionally, variability in vision associated with progression in the patients' CMV retinitis is a major problem in assessing the potential toxicity of any drug used in this disease. In patients treated with 150 µg/eye, to date no changes in peripheral vision have been observed (Kisner, 1996).

Efficacy has been evaluated most extensively in patients with advanced CMV retinitis. These patients were quite ill with a median CD4' cell count of 4. Patients had all failed at least two prior courses of therapy with ganciclovir and/or foscarnet. Disease control in many patients was effected rapidly (usually after one or two doses), and maintained for prolonged periods in many patients. In fact, in these studies most patients withdrew from the study because they were too sick to be treated, they expired, or left the study for reasons unrelated to the drug or CMV retinitis (usually progression of AIDS). The median time to disease progression was approximately 100 days in this single arm study. Many patients were treated bilaterally.

The drug has also demonstrated efficacy in early-stage patients. Based on the safety and efficacy observed in the Phase I/II studies, Phase III studies were initiated. These are in progress and results are expected in the third quarter of 1997.

VII. CONCLUSION

Fomivirsen is the antisense oligonucleotide that is most advanced in clinical development. It has demonstrated safety and efficacy sufficient to warrant Phase III trials, and to support cautious enthusiasm that another drug to treat CMV retinitis may soon be available.

ACKNOWLEDGEMENTS

The author appreciates the helpful comments of Drs. Daniel Kisner, Janet Leeds, Lisa Grillone, and Scott Henry, and the excellent administrative assistance of Ms. Donna Musacchia.

VIII. REFERENCES

Alford, C. A., and Britt, W. J. (1985) Cytomegalovirus. In Fields, B.N., Knipe, D.M., Melnick, J.L., Chanock, R.M., Roizman, B., and Shope, R.E., eds., *Virology*. Raven Press, New York, 629-660.

Anderson, K. P., Fox, M. C., Brown-Driver, V., Martin, M. J., and Azad, R. F. (1996) Inhibition of human cytomegalovirus immediate-early gene expression by an antisense oligonucleotide complementary to immediate-early RNA. *Antimicrob. Agents & Chemotherapy* **40**:2004-2011.

Azad, R. F., Driver, V. B., Tanaka, K., Crooke, R. M., and Anderson, K. P. (1993) Antiviral activity of a phosphorothioate oligonucleotide complementary to RNA of the human cytomegalovirus major immediate-early region. *Antimicrob. Agents & Chemotherapy* **37**:1945-1954.

Azad, R. F., Brown-Driver, V., Buckheit, R. W., Jr., and Anderson, A. P. (1995) Antiviral Activity of phosphorothioate oligonucleotide complementary to human cytomegalovirus RNA when used in combination with antiviral nucleoside analogs. *Antiviral Res.* **28**:101-111.

Crooke, S. T. (1995) *Therapeutic Applications of Oligonucleotides*. G. Landes Company, Austin, TX.

Fomivirsen Investigator's Brochure (1996) Isis Pharmaceuticals, Carlsbad, CA.

Heilbronn, R., Jahn, G., Burkle, A., Freese, U. K., Fleckenstein, G., and zur Hausen, H. (1987) Genomic localization, sequence analysis, and transcription of the putative human cytomegalovirus DNA polymerase gene. *J. Virol.* **61**:119-124.

Heinemann, M. H. (1992) Characteristics of cytomegalovirus retinitis in patients with acquired immunodeficiency syndrome. *Am. Jour. Med.* **92**(2A):12S-16S.

Hermiston, T. W., Malone, C. L., Witte, P. R., and Stinski, M. F. (1987) Identification and characterization of the human cytomegalovirus immediately-early region 2 gene that stimulates gene expression from an inducible promoter. *J. Virol.* **61**:3214-3221.

Ives, D. (1996) Cytomegalovirus infection. In Libman, H., and Witzburg, R. A., eds., *HIV Infection, a Primary Care Manual* **3**(24):363-381.

Kisner, D. L. (1996) Effects of fomivirsen on patients with human cytomegalovirus. The Macrae Group, International Society for Antiviral Research, National Institute of Allergy and Infectious Diseases, New York, NY, Apr. 21-23.

Krieg, A., Yi, A. K., Matson, S., Waldschmidt, T. J., Bishop, G. A., Teasdale, R., Koretzky, G. A., and Klinman, D. M. (1995) CpG motifs in bacterial DNA trigger direct B-cell activation. *Nature* **374**:546-549.

Leeds, J. M., Henry, S., Truong, L., Zutsi, A., Levin, A., and Kornbrust, D. (1996) Pharmacokinetics of a potential human cytomegalovirus therapeutic, a phosphorothioate oligonucleotide, after intravitreal injection in the rabbit. *Drug Metab. Disp.* **25**.921-926.

Leeds, J. M., Williams, K., Bistner, S., Scherrill, S., Levin, A., and Henry, S. (1997) The pharmacokinetics of a novel anti-CMV retinitis therapeutic agent, ISIS 2922, injected intravitreally in monkeys. (Submitted)

Moore, R. D., and Chaisson, R. E. (1996) Natural history of opportunistic disease in an HIV-infected urban clinical cohort. *Ann. Intern. Med.* **124**:633-642.

Nestle, F., Mitra, R. S., Bennett, C. F., Chan, H., and Hickeloff, B. J. (1994) Cationic lipid is not required for uptake and selective inhibitory activity of ICAM-1 phosphorothioate antisense oligonucleotides in keratinocytes. *J. Invest. Dermatol.* **103**:569-575.

Stinski, M. F., Thompson, D. R., and Wathen, M. W. (1981) Structure and function of the cytomegalovirus genome. In Nahmias, A., Dowdle, A., and Schinaza, R., eds., *The Human Herpesviruses*. Elsevier, New York, 72-84.

19

Synthetic DNA-Based Compounds for the Prevention of Coronary Restenosis: Current Status and Future Challenges

Andrew Zalewski, Yi Shi, John D. Mannion
Thomas Jefferson University, Philadelphia, Pennsylvania

Fernando Roqué
Clinica Olivos, Buenos Aires, Argentina

I. INTRODUCTION

Coronary restenosis is defined as a repeat arterial following mechanical revascularization. Although the initial stenosis is usually due to advanced atherosclerosis, subsequent restenosis reflects an excessive repair response after a therapeutic injury, such as balloon angioplasty. Restenosis remains the "Achilles' heel" of interventional cardiology, notwithstanding significant procedural improvements that allow the achievement of initially successful coronary revascularization. Combined clinical and angiographic restenosis occurs in 30 to 50% of patients within the first 6 months after transcatheter coronary revascularization. This translates to several hundreds of thousands of patients worldwide afflicted by this condition. Likewise, similar vascular narrowing may occur after coronary/peripheral bypass surgery or other procedures that induce acute vascular injury, highlighting the need for an effective therapy to control the vascular repair process (Table 1).

Table 1. Vascular Procedures Associated with Vasculoproliferative Responses

Condition	Patients/Year	(Re)stenosis/ Patency Loss	Source
Coronary Interventions	~400,000		
balloon angioplasty		30-50%	Holmes *et al.*, 1984; Nobuyoshi *et al.*, 1988*
atherectomy		50%	Topol *et al.*, 1993*
stent		22-32%	Serruys *et al.*, 1994; Fischman *et al.*, 1994*
Coronary Bypass Surgery	~400,000		
saphenous vein grafts		28%	Bourassa *et al.*, 1985[†]
arterial grafts		<5%	Acinapura *et al.*, 1992
Other	~100,000		
hemodialysis access		16-45%	Swedberg *et al.*, 1989[¶]

*Coronary restenosis refers to angiographic lesion 50% at 6-7 months after coronary angioplasty.
[†]Patency loss refers to angiographic lesion 50% at 18 months after bypass surgery.
[¶]Patency loss due to venous stenosis within 16-22 months after the placement of Brescia-Cimino fistulas or PTFE grafts.

Several pharmacological approaches have failed to prevent restenosis in patients due to the multifactorial pathogenesis of restenosis, and the sometimes simplistic applications of preclinical data to the clinical arena. The prospect of therapeutic administration of synthetic DNA-based technologies (e.g., oligonucleotides) has stimulated both great expectations and healthy skepticism. The classification of these agents as "antisense" has often been a subject of controversy, inasmuch as that adjective too narrowly defines their mechanism of action. The purpose of this chapter is to review the scientific rationale underlying the use of oligonucleotides for the prevention of coronary restenosis and to outline the challenges unique for cardiovascular applications.

Vascular injury, associated with balloon dilatation or newer revascularization methods, induces several overlapping responses that represent potential targets for the development of novel therapies.

I.A. Thrombus

Platelet deposition followed by fibrin generation is the earliest manifestation of vascular injury (Heras *et al.*, 1989). Deep medial dissection after angioplasty can sometimes create conditions for extensive thrombus formation (Figure 1A).

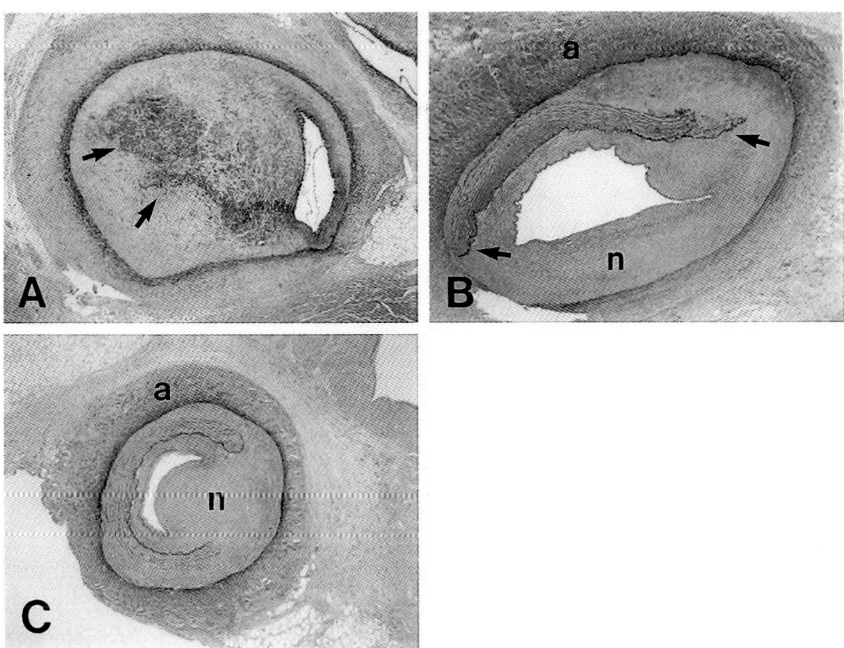

Figure 1. Vascular responses which contribute to (re)stenosis after balloon angioplasty. **A:** Thrombus *(arrows)* partially infiltrated by cells is present at the site of balloon injury. **B:** Neointima *(n)* occupies the space between the edges of broken media *(arrows)* and encroaches on the lumen. Note a profound scarring of the adventitia *(a)*. **C:** Constrictive remodeling, manifested by dramatically decreased overall vessel dimensions (compare with panels **A** and **B**), accentuates the effects of neointima *(n)* on lumen narrowing. Collagen-rich adventitia *(a)* surrounds the injured vessel. Panels **A-C** represent Verhoeff's stain of porcine coronary arteries at 2 weeks **(A)** and 3 months **(B,C)** after balloon injury. Magnification 17X.

Platelet degranulation and the release of several growth factors contribute to the activation of resident cells in the vessel wall and may induce a vasospastic response in injured vessels. In addition, the presence of a thrombus likely enhances cell migration, contributing to neointimal formation (Schwartz *et al.*, 1992).

I.B. Neointima

A rapid release of growth factors and cytokines from platelets and cell debris at the site of transcatheter interventions sets the conditions for paracrine and autocrine stimulation of resident cells in the arterial wall. This ultimately leads to the formation of neointima, which differs from atherosclerotic plaque or normal vessel wall (Ross *et al.*, 1984). Its primary function is to repair the injured tissue (e.g., fill dissection in the media), although often neointimal formation is excessive and may contribute to restenosis (Figure 1B).

I.C. Geometric remodeling

Recent experimental and clinical observations point to a poor correlation between neointimal formation and late vascular re-narrowing (i.e., restenosis) (Post *et al.*, 1994; Kakuta *et al.*, 1994; Lafont *et al.*, 1995; Mintz *et al.*, 1996). This is due to unfavorable remodeling of the entire vessel, manifested by luminal narrowing later in the process of vascular repair. The healing artery is either unable to dilate to accommodate the growing neointima or it may even constrict the vessel lumen, resembling wound contraction (Figure 1C) (Darby *et al.*, 1990; Welch *et al.*, 1990).

II. WHAT TO TARGET?

Although the prevention of coronary restenosis represents an important goal for cardiovascular applications of synthetic DNA-based compounds, it is surprising how little information exists regarding the repair process in the coronary vasculature. Based on the observations derived from a porcine model of coronary injury, however, the following critical steps associated with the repair of coronary arteries have been identified, which are relevant for the development of synthetic DNA-based therapy.

II.A Cell activation (hours - days)

Adventitial fibroblasts, rather than medial smooth muscle (SM) cells, are rapidly activated after balloon injury (Shi *et al.*, 1996a). This is manifested by their proliferation, which begins within 24 hours of injury and peaks at 3 days after vascular insult (Figure 2). The activation of adventitial fibroblasts is also accompanied by a dramatic upregulation of their synthetic function (Shi *et al.*, 1997b). In regard to proliferative and synthetic properties, however, medial SM cells remain largely indolent during the vascular response to injury. These findings suggest the need for an early delivery of therapeutic compounds, preferentially to the adventitia, in order to reduce fibroblast activation.

II.B. Myofibroblast formation (days - weeks):

Within days after balloon injury, activated fibroblasts begin to differentiate to myofibroblasts, which is reflected by the appearance of α-SM actin (Figure 3) (Shi *et al.*, 1996a). Myofibroblasts exhibit remarkable migratory characteristics, translocating from the adventitia to the gap in the dissected media, constituting the major cellular component of the

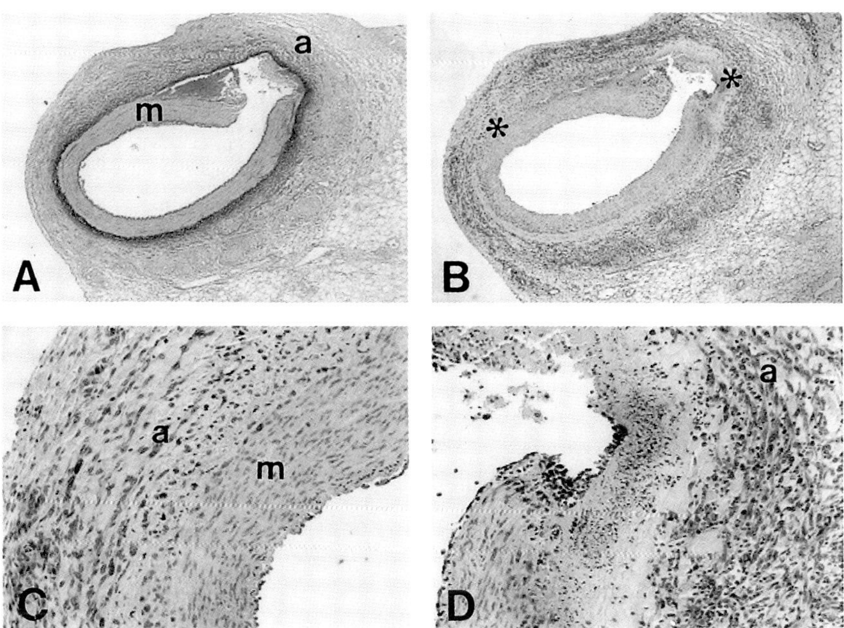

Figure 2. Cell proliferation in porcine coronary arteries at 3 days after balloon angioplasty. **A:** Verhoeff's stain shows disruption of the media *(m)*; which resulted in the exposure of the adventitia *(a)* to the vessel lumen. **B:** Proliferating cells (dark) exhibit circumferential distribution. Asterisks point to the regions shown in **C and D:** At higher magnification, the majority of proliferating cells (brown nuclear stain) are localized in the adventitia *(a)* with only infrequent staining present in the media *(m)*. Panels **B-D** represent proliferating cell nuclear antigen immunostaining. Magnification 25X (**A,B**) and 124X (**C,D**) *(reprinted, with permission, from Shi, et al., 1996a).*

neointima (Shi *et al.*, 1996b). Undoubtedly, similarities between myofibroblasts and SM cells (e.g., α-SM actin) have contributed to difficulties in the recognition of their respective roles in vascular repair. Myofibroblasts also appear to be involved in the repair of non-coronary arterial beds (Holifield *et al.*, 1996) and arterialized vein grafts (O'Brien *et al.*, 1997; Shi *et al.*, 1997b).

II.C. Remodeling (weeks - months)

Coronary myofibroblasts exhibit autocrine expression of the profibrotic cytokine, TGFß1, and are endowed with ultrastructural characteristics which reflect their enhanced synthetic functions (Shi *et al.*, 1996c). The interactions between these cells and the surrounding extra-

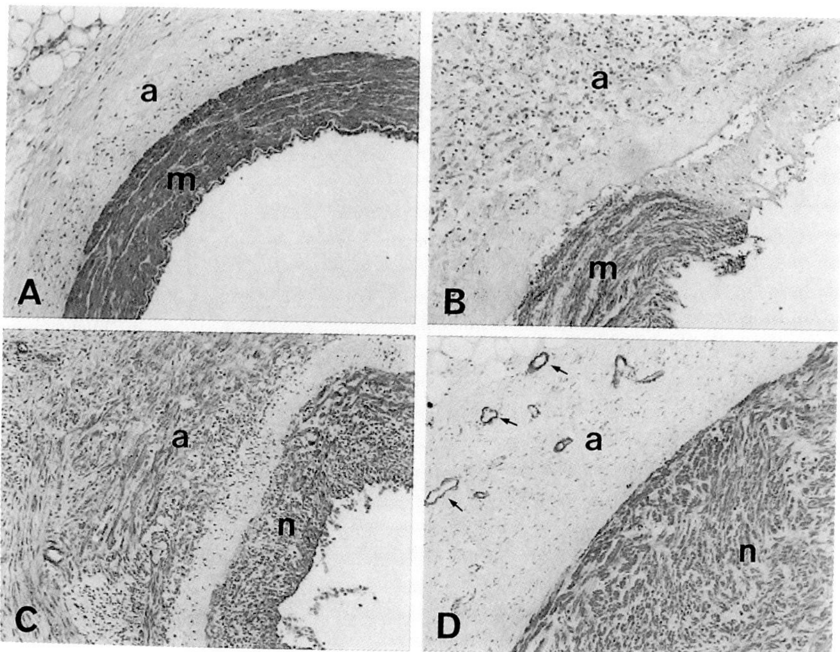

Figure 3. Myofibroblast formation in porcine coronary arteries after balloon injury. **A:** Normal coronary artery demonstrates α-SM actin immunostaining confined to the media *(m)*, whereas the adventitia *(a)* is negative. **B:** At 2 days after injury, hypercellular adventitia *(a)* remains negative for α-SM actin, the media *(m)* shows signs of dissection. **C:** At 8 days after injury, myofibroblasts (α-SM actin positive cells) are common in the adventitia; note the similarity between the adventitia and early neointima *(n)* regarding α-SM actin immunostaining. **D:** At 3 months, myofibroblasts disappear from the remodeled adventitia *(a)* that shows increased neovascularization *(arrows)*. Note the persistence of myofibroblasts in neointima *(n)*. Magnification 21X *(reprinted, with permission, from Shi, et al., 1996a)*.

cellular matrix lead to the repair and remodeling of injured tissue (Shi *et al.*, 1996a). Several studies in non-vascular tissues have identified the unique ability of myofibroblasts to reorganize the surrounding collagen matrix, resulting in constrictive remodeling (Arora *et al.*, 1994; Darby *et al.*, 1990; Welch *et al.*, 1990). The persistence of myofibroblasts in the neointima, coupled with their sustained synthetic activities, may provide a cellular basis for late vascular remodeling (Shi *et al.*, 1996c, 1997a).

The cellular responses of atherosclerotic coronary arteries to injury are far less understood than that of normal, non-atherosclerotic arteries. For example, the extent of cell proliferation in human coronary arteries after transcatheter interventions has been the subject of controversy (Pickering *et al.*, 1993; O'Brien *et al.*, 1994).

A short-time course of cell proliferation and the poor accessibility of the adventitia to retrieval may provide an explanation for their paucity in some atherectomy samples (O'Brien *et al.*, 1993). The modulating effects of underlying coronary atherosclerosis on other responses to injury remain unknown, although the presence of inflammatory cells and TGFß1, which promote myofibroblast formation, may accentuate these events in human atherosclerotic vessels (Nikol *et al.*, 1992).

Experimentally, adventitial cell proliferation, neointimal formation and remodeling changes are clearly linked to the severity of vascular injury (Shi *et al.*, 1996b), which raises the question whether such findings are applicable to the clinical setting. Although not appreciated angiographically, the adventitia is often exposed to the lumen after balloon angioplasty or other procedures in patients (e.g., stent implantation, atherectomy), setting conditions that promote fibroblast activation and the ensuing events leading to restenosis (Kohchi *et al.*, 1987; den Heijer *et al.*, 1994; Colombo *et al.*, 1995).

III. DOES SYNTHETIC DNA-BASED THERAPY AFFECT VASCULAR REPAIR?

The interest in synthetic DNA-based therapy has been based on a simple premise that oligomers complementary to mRNA coding for a cell-cycle regulating gene can block the proliferative response by means of antisense gene inhibition (Table 2). A short half-life of target transcripts and proteins, low levels after stimulation, as well as their rapid increase in injured vessels, all favor the antisense approach. A closer examination of preclinical studies, however, indicates several inconsistencies that suggest that the cardiovascular application of synthetic DNA is more complex than was initially envisioned.

III.A. "Immediate early" genes

These genes are rapidly expressed after growth factor and receptor interactions on the surface of cells. Their activation appears to be due to generation of a second messenger and is independent of protein synthesis (Gadeau *et al.*, 1991). The pattern of expression of individual genes varies considerably. For example, c-*fos* is induced within minutes of vessel injury making it an unsuitable target for antisense inhibition. Other "immediate early" genes like the c-*myc* nuclear proto-oncogene, peak after the first few hours, which is sufficient for the intracellular accumulation of oligonucleotides (Shi *et al.*, 1993).

The c-*myc* nuclear proto-oncogene is a ubiquitous phosphoprotein that regulates gene expression on transcriptional (Kaddurah-Daouk *et al.*,

Table 2. Vascular Responses After Treatments With Synthetic DNA-Based Compounds Targeting Cell Cycle Regulating Genes

Gene	Carotid (rat)	Coronary (pig)	Coronary (human)	Effects
c-*myc*	periadventitial*	transcatheter**	transcatheter[†]	↓cell proliferation ↓neointima ↑luminal size (Bennett, 1994; Shi, 1994) safe in humans (Roqué, 1997)
c-*myb*	periadventitial*	transcatheter**	ND	contradictory results (Simons, 1993; Villa, 1995; Gunn, 1997)
PCNA	periadventitial*	ND	ND	↓cell proliferation ↓neointima (Simons, 1994)
cdk2	periadventitial* endoluminal[‡]	ND	ND	↓cell proliferation ↓neointima (Abe, 1994; Morishita, 1994)
cdc2	endoluminal[‡]	ND	ND	↓cell proliferation ↓neointima (Morishita, 1993)
E2F	endoluminal[‡§]	ND	ND	↓cell proliferation ↓neointima (Morishita, 1993)

*Oligomers were delivered in pluronic gel after surgical exposure into the periadventitial space. **Oligomers were intramurally injected into balloon-injured coronary arteries mimicking clinical conditions. [†]Oligomers were injected through a guiding catheter at the completion of PTCA to assess the safety profile in humans. [‡]Oligomers were complexed with liposomes and inactivated hemagglutinating virus of Japan, and then instilled into the lumen of carotid arteries. [§]Double-stranded oligomers (phosphorothioate) binding E2F transcriptional factor downregulated c-*myc*.

1987) and post-transcriptional (Prendergast and Cole, 1989) levels, resulting in cell proliferation. Vascular cells derived from atherosclerotic lesions overexpress c-*myc* (Parkes *et al.*, 1991) and balloon injury itself provides a powerful stimulus for an increase in c-*myc* levels (Miano *et al.*, 1990; Bauters *et al.*, 1992). C-*myc* represents an important growth regulatory pathway in vascular cells. This is evidenced by the down-regulation of this proto-oncogene with several inhibitors of proliferation or the inability to maintain a growth arrest state when c-*myc* is constitutively overexpressed (Bennett *et al.*, 1994ab). The inhibition of c-*myc* expression by antisense oligonucleotides is associated with a decrease in cell proliferation (Wickstrom *et al.*, 1988; Biro *et al.*, 1993, Shi *et al.*, 1993; Bennett *et al.*, 1994ac).

The "anti-restenotic" efficacy of oligonucleotides targeting the c-*myc* proto-oncogene has been demonstrated in different models of vascular injury (Bennett *et al.*, 1994c; Shi *et al.*, 1994). A single transcatheter delivery of c-*myc* oligonucleotides has been shown to inhibit cell proliferation in porcine coronary arteries, with the most prominent effect in the adventitial layer (Figure 4). The formation of neointima was also reduced, demonstrating that the early inhibition of cell activation has an effect on later vascular events (Figure 5). The concomitant improvement in luminal diameter of treated coronary arteries suggests that unfavorable geometric remodeling could be limited (Shi *et al.*, 1994). These findings represented the first demonstration of the potential therapeutic effect of synthetic DNA-based compounds in the coronary vasculature using a clinically applicable protocol. In an independent study, c-*myc* antisense oligomers have also been shown to reduce c-*myc* expression and neointimal formation in denuded rat carotid arteries after periadventitial delivery in a pluronic gel (Bennett *et al.*, 1994c).

III.B. "Late G1" genes

These genes are required for DNA synthesis during the S phase of the cell cycle. They include the *c-myb* nuclear proto-oncogene, proliferat-

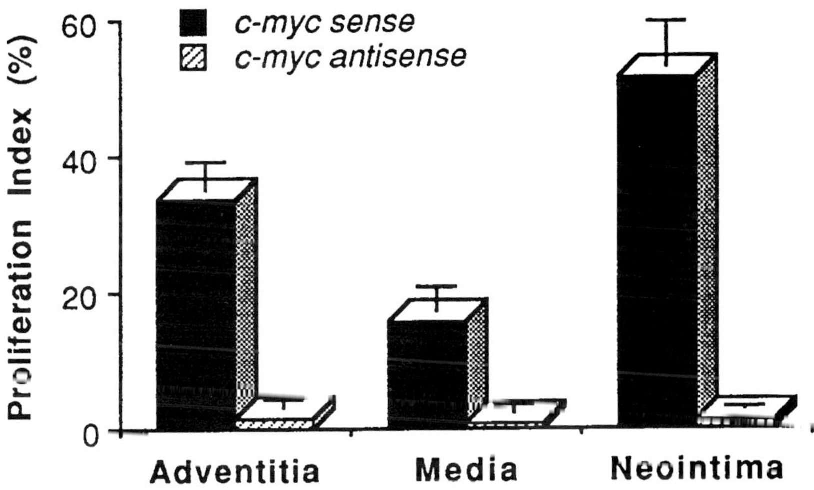

Figure 4. The inhibition of cell proliferation after transcatheter delivery of antisense oligomers directed against c-*myc* proto-oncogene. Oligomers (100 µM) were administered immediately after balloon injury using a porous balloon (delivery time 22±1 sec). Cell proliferation was assessed at 7 days by PCNA immunostaining. Note significant reduction of cell proliferation in all layers of porcine coronary arteries, as compared to control group receiving sense oligomers (p<0.01) *(reprinted, with permission, from Shi et al., 1994).*

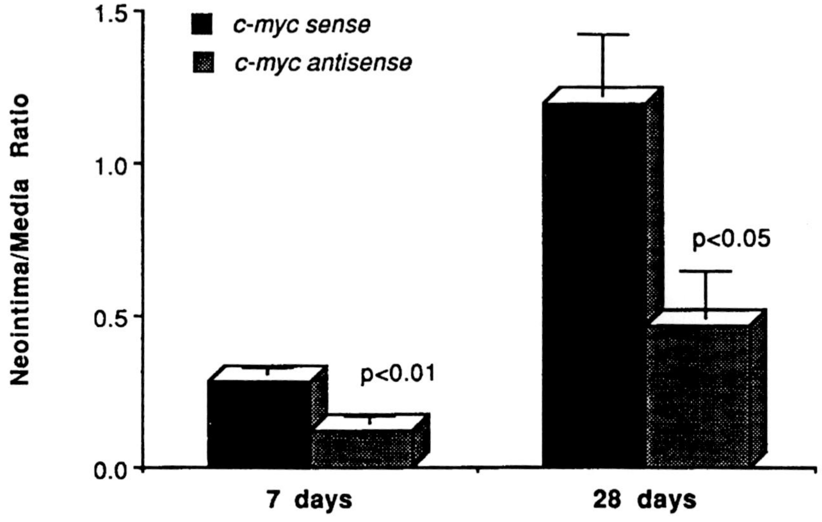

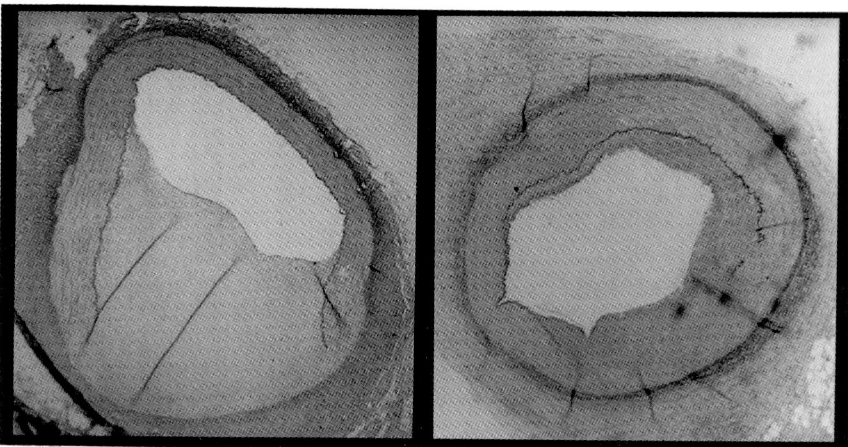

Figure 5. The inhibition of neointimal formation after transcatheter delivery of c-*myc* antisense oligomers. **Top:** Intima/media ratio is reduced at 7 and 28 days after coronary injury and antisense oligomer administration (100 µM). **Bottom:** Photomicrographs illustrating the effects of sense (**left**) and antisense oligomers (**right**) on neointimal formation at 28 days. Both vessels demonstrate similar degrees of initial injury; however, neointima is reduced only in the antisense-treated artery. Verhoeff's stain, magnification 15X *(reprinted, with permission, from Shi et al., 1994).*

ing cell nuclear antigen (PCNA), cyclin-dependent kinases and thymidine kinase genes (Gadeau *et al.*, 1991; Brown *et al.*, 1992). The *c-myb* proto-oncogene is indirectly involved in DNA synthesis, inducing PCNA, DNA polymerase, and histone H3 expression (Travali *et al.*, 1991). The *c-myb* expression level is low, and it peaks about 18 hours after the addition of growth factors (Brown *et al.*, 1992). These characteristics make c-*myb* an attractive target for antisense inhibition.

Initial observations in cell culture were suggestive of the sequence-dependent effect of phosphorothioate oligomers targeting c-*myb* (Simons and Rosenberg, 1992; Brown *et al.*, 1992), however, later studies have strongly suggested nonspecific effects at high concentrations *in vitro* (Villa *et al.*, 1995; Burgess *et al.*, 1995; Jarvis *et al.*, 1996). The controversy surrounding c-*myb* oligomers persisted when the reduction of neointimal formation *in vivo* could not be reproduced despite the use of the same sequences, methods of delivery, and animal model (Simons *et al.*, 1993; Villa *et al.*, 1995).

PCNA is a cofactor of DNA polymerase, both of which are required for DNA replication (Prelich and Stillman, 1988). The antiproliferative effect of PCNA antisense (phosphorothioate) in combination with oligo-nucleotides directed against cell division cycle 2 (*cdc2*) kinase has been demonstrated. This represents an example of combined oligomer application against multiple targets to enhance antisense effects and to reduce nonspecific cell inhibition (Morishita *et al.*, 1993). Antisense targeting PCNA alone has also been demonstrated to reduce PCNA expression and neointimal formation in a rat model after periadventitial delivery in a pluronic gel (Simons *et al.*, 1994).

The activation of *cyclin dependent kinase 2 (cdk2)* is required for the G1/S phase transition (Tsai *et al.*, 1993). Although the levels of the $p33^{cdk2}$ protein are constant throughout the cell cycle, *cdk2* activity rises just a few hours prior to DNA synthesis (S phase). Surprisingly, *cdk2* kinase activity is the highest at 14 days following vascular balloon injury (Abe *et al.*, 1994), lagging behind the expected burst of cell proliferation *in vivo* (Clowes *et al.*, 1983; Shi at al., 1996a). Notwithstanding the above, *cdk2* antisense delivered in pluronic gel to the adventitia, or instilled into the lumen of denuded rat carotid arteries, resulted in the reduction of both *cdk2* kinase activity and neointima (Abe *et al.*, 1994, Morishita *et al.*, 1994).

III.C. "G2/M boundary" genes

The $p34^{cdc2}$ protein kinase is critical for phosphorylation of key proteins required for cell entry into the M phase and the final act of the cell cycle, mitosis (Riabowol *et al.*, 1989; Nurse, 1990). *Cdc2* kinase activation requires $p34^{cdc2}$ dephosphorylation and association with cyclin, which signals cell readiness to divide (M phase). In contrast, *cdc2* kinase is inactivated as cells exit the M phase. This essential role of *cdc2* kinase and its downstream position in the transduction pathways provided the rationale for its inhibition. The combined use of an antisense "cocktail" against PCNA and *cdc2* kinase inhibited neointimal formation in a rat model of carotid denudation after endoluminal delivery (Morishita *et al.*, 1993). Periadventitial application of *cdc2* kinase antisense alone has also been demonstrated to be effective in the same model (Abe *et al.*, 1994).

III.D. Other molecular targets

The transcriptional factor E2F binds to double stranded DNA containing an 8 bp consensus sequence, which results in the activation of cell cycle regulating genes (c-*myc*, c-*myb*, *cdc2*, PCNA). An interesting approach of using double-stranded phosphorothioate oligomers containing the 8 bp *cis* element (E2F decoy) has been used to inhibit coordinated stimulation of several growth stimulatory genes and prevent neointimal formation in the rat model (Morishita *et al.*, 1995). This is particularly interesting, since it represents a novel application of synthetic DNA (double-stranded decoy oligomer) targeting a key transcriptional factor, as opposed to interacting with a single cell cycle regulating gene.

IV. NON-ANTISENSE AND ANTISENSE EFFECTS OR NO EFFECTS?

There is a considerable controversy surrounding the use of synthetic DNA-based compounds due to several non-antisense effects elicited in cultured cells (Stein and Cheng, 1993, Epstein *et al.*, 1993, Stein, 1995). It has become apparent that modified oligomers (e.g., phosphorothioate) are complex molecules able to interfere not only with the intended mRNA targets (Table 3). In fact, pure "antisense" gene inhibition becomes exceedingly difficult to discern even in the presence of sequence-dependent effects and multiple controls (Guvakova *et al.*, 1995; Stein, 1995). The relevance of non-antisense mechanisms remains to be carefully examined *in vivo*, as they may have an impact on the therapeutic efficacy and the safety profile of oligomers.

In particular, the interpretation of several studies regarding the potential cardiovascular application of synthetic DNA-based compounds is marred by a possible dichotomy between the findings *in vitro* and *in vivo*, the choice of animal models for preclinical studies, as well as difficulty related to oligomer delivery.

IV.A. Diverse mechanisms of action

Non-sequence-dependent effects are due to the avid binding of negatively charged oligonucleotides (e.g., phosphorothioate) to proteins. The increase in length and concentration of oligomers appears to enhance

Table 3. Effects of Synthetic DNA-Based Compounds Due to Non-Antisense Mechanisms in Biological Systems

Condition	Putative mechanism
Accentuates antiproliferative effects:	
phosphorothioate backbone	inactivation of growth factors[*]
higher concentration	
oligomer length	inhibition of DNA polymerases[†]
sequence-dependent	formation of large tetraplexes (G_4)
	induction of interferons[¶]
Attenuates antiproliferative effects:	
phosphodiester backbone	enzymatic degradation
higher concentration	inhibition of RNase H activity[†]
oligomer length	

[*]Phosphorothioate oligomers (25-mers and 28-mers) in concentrations >1 µM bound to basic fibroblast factor (bFGF) and inhibited proliferation *in vitro*. *In vivo*, neointimal formation was inhibited with 120 µM of dC_{28} phosphorothioate (Guvakova *et al.*, 1995, Wang *et al.*, 1996).
[†]The inhibitory effect requires a minimum phosphorothioate oligomer length of 15 deoxynucleotides, reaching a plateau at 28 deoxynucleotides (Gao *et al.*, 1992).
[¶]Palindromic sequences may induce interferons that possess growth-inhibitory effects (Yamamoto *et al.*, 1992).

nonspecific inhibition (Cazenave *et al.*, 1989; Gao *et al.*, 1992; Guvakova *et al.*, 1995). At higher doses oligonucleotides inhibit RNase H activity, thereby eliminating an important mechanism aiding the inhibition of gene expression by antisense. Longer sequences (28-mer) of phosphorothioate homopolymers of cytidine or thymidine have also demonstrated the ability to interact with extracellular growth factors (e.g., bFGF) and their binding with cell surface receptors (Guvakova *et al.*, 1995). This is consistent with their polyanion structure, thereby reducing cell proliferation in non-sequence-dependent manner *in vitro* and neointimal formation in the rat model *in vivo* (Guvakova *et al.*, 1995; Wang *et al.*, 1996).

Sequence-dependent nonspecific effects are related to the oligomer sequence producing biological changes by non-antisense mechanisms. For example, the presence of four contiguous deoxyguanosines (...dGGGG...) has been shown to produce the effects previously attributed to only an antisense mechanism (Ho *et al.*, 1991; Yaswen *et al.*, 1993; Burgess *et al.*, 1995). Oligonucleotides can function as aptamers interacting with a specific protein. A sequence-specific binding of phosphorothioate oligomers to thrombin inhibits its function *in vitro* and represents an example of such interactions (Bock *et al.*, 1992). Interestingly, *c-myb* antisense oligomers have been shown to bind bFGF and abrogated bFGF growth stimulatory response, whereas sense sequences had no such effect (Guvakova *et al.*, 1995). Some oligomers have also been reported to induce the expression of non-targeted genes

(Yamamoto *et al.*, 1992; McIntyre *et al.*, 1993). The mechanism of this phenomenon is unclear and it may involve oligomer interactions with a natural antisense RNA to a non-targeted gene.

IV.B. Dichotomy of *in vitro* and *in vivo* studies

Cell culture studies are used to screen synthetic DNA-based compounds to determine their potential value in the prevention of vascular restenosis. The experience with c-*myb* oligomers, however, clearly illustrates a limited predictive value of such approach to more clinically relevant properties *in vivo*. In some studies, high concentrations of c-*myb* oligomers containing a dG$_4$ motif had a strong antiproliferative effect in cell culture due to non-antisense mechanisms (Burgess *et al.*, 1995; Villa *et al.*, 1995; Guvakova *et al.*, 1995).

However, when c-*myb* antisense oligomers were applied *in vivo*, no reduction in neointimal formation could be appreciated, despite adequate vascular localization (Villa *et al.*, 1995). Likewise, control sequences exhibited no effects on neointimal formation in several animal studies, which questions the relevance of a wide spectrum of non-antisense effects observed in cell culture to the effects observed *in vivo*. Perhaps longer sequences of phosphorothioate oligomers (28-mer) are necessary to elicit nonspecific effects *in vivo* (Wang *et al.*, 1996). To specifically address this issue, we have examined the inhibition of cell proliferation after c-*myc* antisense oligomers in vascular grafts *in vivo*.

Although the antiproliferative effect of c-*myc* antisense in cell culture may depend, in part, on a dG$_4$ motif (Burgess *et al.*, 1995), a significant reduction in vascular cell proliferation *in vivo* was present only with antisense oligomer (20 µM), but not with scrambled sequences, regardless of four contiguous dGs (Figure 6). The additional growth-inhibitory effect due to synergistic non-antisense mechanisms, however, cannot be excluded at higher concentrations (Table 3). The dichotomy between the findings *in vitro* and *in vivo* may depend on several factors. These may include a higher concentration of oligomers in cell culture experiments than in the vicinity of cellular targets in the vessel wall and non-physiologic concentrations of growth stimulators in culture.

IV.C. Choice of animal model(s)

The rat model of carotid injury has been most often used to assess the efficacy of synthetic DNA-based compounds to reduce neointimal formation (Simons *et al.*, 1992; Morishita *et al.*, 1993, 1994; Bennett *et al* 1994; Abe *et al.*, 1994). The results of these studies, however, have to be assessed with caution, inasmuch as pharmacologic intervention(s) effective in the rat have generally been shown devoid of anti-restenosis properties in other species, including human trials (Powell *et al.*, 1989; Faxon *et al.*, 1995; Huckle *et al.*, 1996).

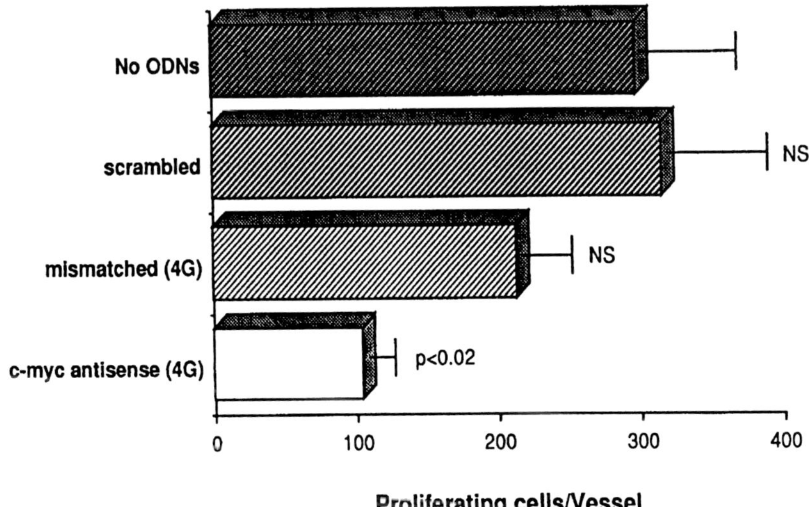

Figure 6. The effects of the dG_4 motif on cell proliferation *in vivo*. The graph represents cell proliferation in saphenous vein grafts interpositioned in porcine carotid arteries. Harvested saphenous veins were incubated with no oligomers, scrambled oligomers devoid of dG_4 motif, mismatched oligomers with dG_4 motif and c-*myc* antisense oligomers (20 µM) *ex vivo* (n = 6-8 vessels/group). After 30 min, grafts were surgical placed in the carotid artery. Cell proliferation was assessed using BrdU immunostaining at 3 days later. Note significant reduction in cell proliferation in the media of arterialized grafts after c-*myc* antisense oligomers, whereas control sequences exert no such effect.

The differences in vascular response to injury, and a likely different pattern of gene expression, have set the experiments in the rat carotid apart from the responses in the coronary vasculature in other models, possibly including humans. The porcine model may offer a better alternative for preclinical studies because of the similarity between porcine and human coronary vessels, as well as the ability to develop spontaneous atherosclerosis in pigs.

The size of the coronary vasculature is also comparable to that in humans, which makes possible the testing of transcatheter administration of oligomers into the injured arterial wall (Shi *et al.*, 1994). Coronary balloon overstretch in a porcine model induces clinically relevant responses to injury, including thrombus, neointimal formation and geometric remodeling (Figure 7) (Andersen *et al.*, 1996; Shi *et al.*, 1996a). Nevertheless, a significant variability exists in regard to oligomer delivery and arterial repair, requiring better standardization of this experimental approach for future comparative studies (Shi *et al.*, 1994, 1996a).

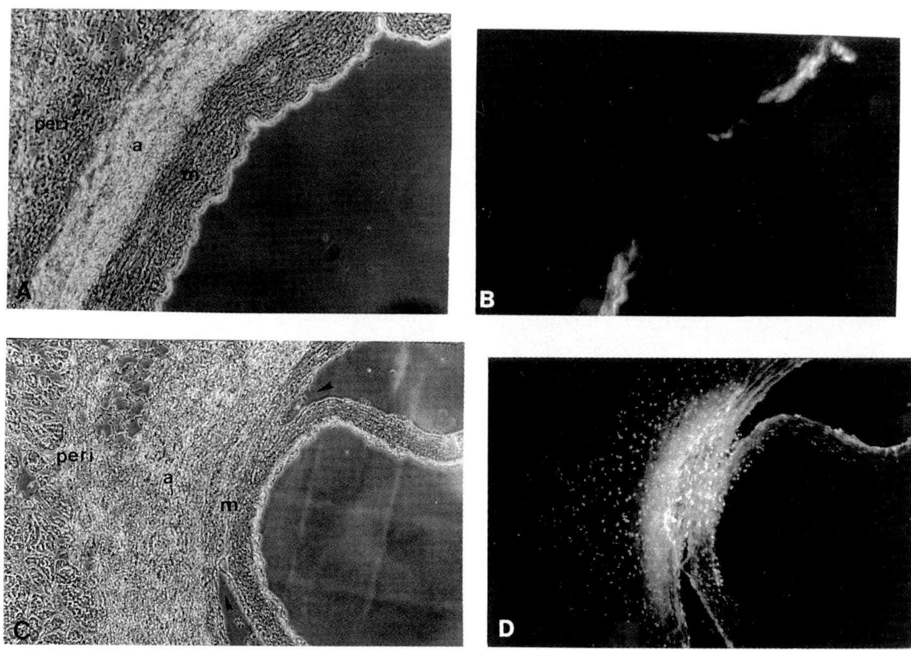

Figure 7. Intramural distribution of oligomers in balloon-injured coronary arteries is dependent on the underlying vascular injury. **A and B:** Mild injury: phase-contrast and fluorescence microscopy showing subintimal, non-transmural oligomers distribution. Note two areas of fluorescence, which correspond to the contact between side-holes on the infusion sleeve and the vessel wall. The absence of medial dissection prevents deeper oligomer penetration. **C and D:** Severe injury: phase-contrast and fluorescence microscopy demonstrating transmural distribution of fluorescein-labeled oligomers. Antisense oligomers are present in the media, adventitia, and perivascular tissues. Note medial dissection *(arrow)* facilitating oligomer entry to the vessel wall. m: media; a: adventitia; peri: perivascular tissues. Oligomers were delivered using infusion sleeve catheter (Infusasleeve, LocalMed) immediately after balloon injury and their vascular localization was examined at 30 min after transcatheter administration. Magnification 62.5X *(reprinted, with permission, from Shi et al., 1996a).*

V. WHEN, WHERE, AND HOW TO DELIVER SYNTHETIC DNA?

Focal changes induced by vascular injury have stimulated interest in the local administration of therapeutic agents in order to augment their concentration in target tissues and to minimize untoward effects associated with conventional therapy. The complexity of local drug delivery into injured coronary arteries, however, sets apart the use of synthetic DNA-based therapy in cardiac patients from other clinical applications of oligomers.

V.A. When to deliver?

Rapid release of growth factors and the induction of cell cycle regulating genes, which result in the activation of vascular fibroblasts within the first 24 hours, provide the rationale for the administration of synthetic DNA-based therapy immediately after coronary angioplasty. Furthermore, since the coronary circulation is inaccessible without invasive cardiac catheterization, this approach takes advantage of the coronary instrumentation required to perform transcatheter re-vascularization. Accordingly, it becomes apparent that a single delivery of synthetic DNA-based compounds becomes a critical step toward therapeutic success of this approach for the prevention of coronary restenosis.

V.B. Where to deliver?

Local activation of adventitial fibroblasts requires precise administration of therapeutic compounds into the site of injury and their deposition into the adventitial layer. To determine the feasibility of the intramural delivery of synthetic DNA-based compounds, we have studied the distribution of fluorescent-labeled phosphorothioate oligomers after transcatheter administration in injured porcine coronary arteries.

The presence of oligomers was highly variable in the arterial wall; it ranged from none to transmural or perivascular distribution. The examination of serial coronary sections revealed that oligomer distribution depends on the degree of vascular injury (Figure 7). The media of normal or minimally injured coronary artery (no medial dissection) forms an effective barrier, preventing oligomers from appearing in the vessel wall.

In contrast, in adjacent sections exhibiting medial dissection, oligomers were present in adventitial and perivascular layers. Nuclear localization of oligomers was evident within 30 min of transcatheter administration (Figure 8). These results illustrated the selectivity of oligomer deposition into the sites with maximal vascular injury and adventitial activation, which provide the nidus for the ensuing neointimal formation and remodeling changes (Figure 9) (Shi *et al.*, 1996c).

V.C. How to deliver?

Several attractive designs have been proposed for intramural drug delivery, although none of them have been thoroughly evaluated in the clinical setting (Table 4). Local intramural administration of therapeutic compounds in the coronary circulation is fraught by several challenges. These include the development of myocardial ischemia during even a short coronary occlusion (Zalewski *et al.*, 1985). This either shortens the time available for drug administration, or it requires more complicated designs that would allow distal coronary perfusion.

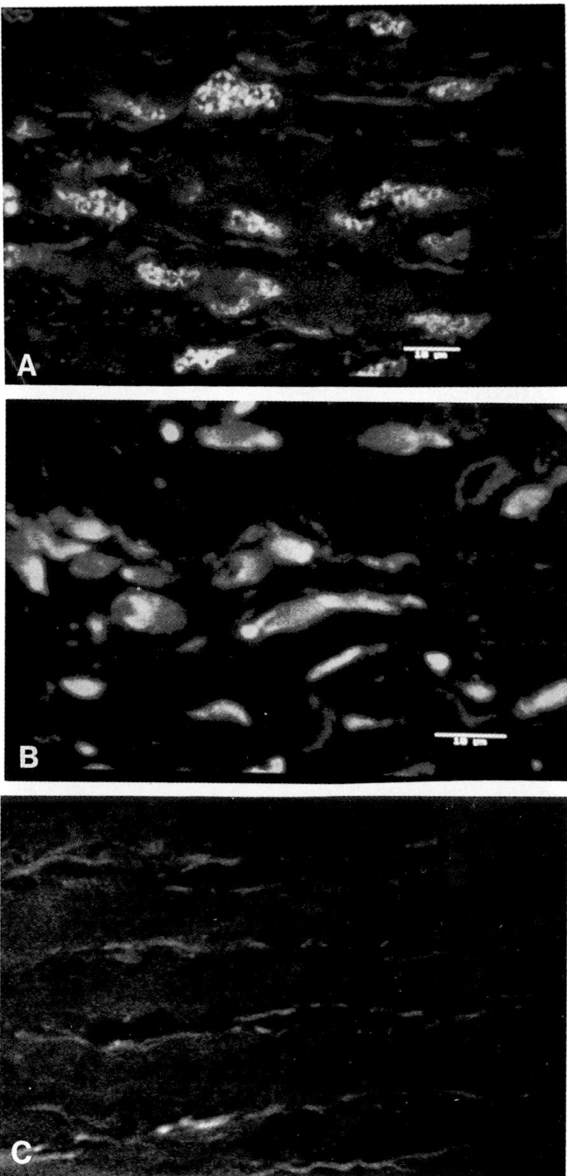

Figure 8. Nuclear localization of fluorescein-labeled c-*myc* antisense oligomers by confocal microscopy. At 30 min. after intramural delivery in porcine coronary artery (Infusasleeve, LocalMed), nuclear localization is visible in medial smooth muscle cells (**A**) and adventitial fibroblasts (**B**). In contrast, the contralateral (i.e., non-injected) coronary artery demonstrates no oligomers (**C**, negative control) *(reprinted, with permission, from Shi et al., 1996c).*

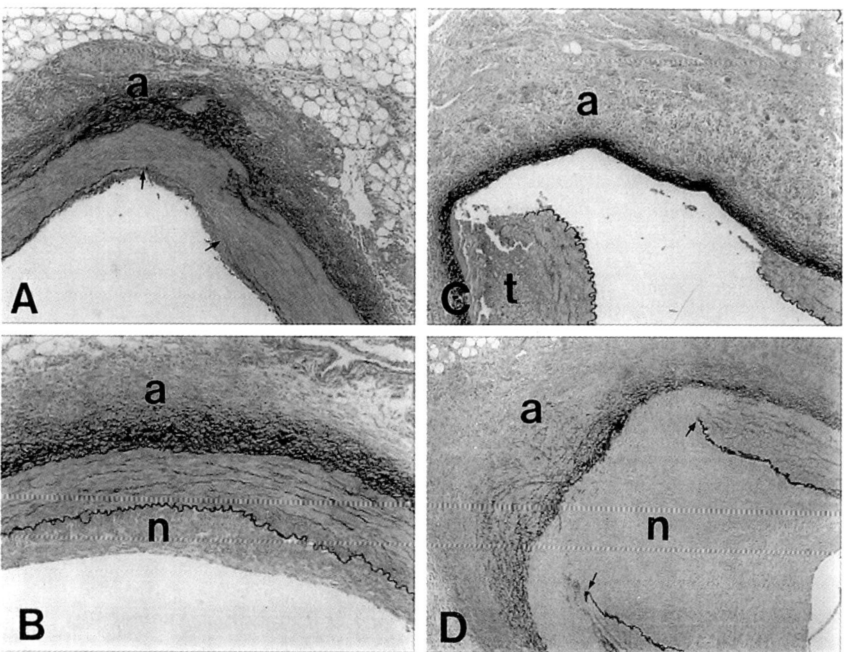

Figure 9. The severity of coronary injury determines the extent of vascular repair. **A:** Mild injury (2 days): breaks in the internal elastic lamina *(arrows)* without medial dissection. **B:** Mild injury (28 days): a well-preserved continuity of the media is associated with a thin layer of neointima. **C:** Severe injury (2 days): medial disruption results in the exposure of the adventitia to the vessel lumen. **D:** Severe injury (28 days): a large neointimal lesion is present filling the gap in the disrupted media *(arrows)*. The adventitia demonstrates a marked remodeling. a: adventitia; n: neointima; t: thrombus. Verhoeff's stain. **A, B, C:** magnification 58X; **D:** magnification 40X *(reprinted, with permission, from Shi et al., 1996c).*

There are several devices that may prove to be useful in the delivery of synthetic DNA-based therapy. Some of them reflect remarkable ingenuity, with various designs presenting a dilemma as to the best choice for the testing of clinical safety and efficacy (Table 4). An alternative approach, which deserves special attention, involves stents that may offer a platform for the sustained release of oligomers into injured coronary arteries.

Table 4. Devices for Site-Specific Delivery of Synthetic DNA-Based Compounds in the Coronary Vasculature

Type	Device	Advantages	Disadvantages
Pressure-driven	Porous membrane allowing intramural drug delivery (e.g., porous or dual balloons, infusion sleeve) Double balloon with central port for endoluminal drug instillation Protruding ports for a deep drug delivery (e.g., needle device, "nipple" balloon)	Short time of delivery Delivery dependent on design (e.g., size of sideholes) or conditions (e.g., pressure) Deep intramural delivery (e.g., to adventitia)	Risk of vascular trauma Low efficiency of delivery (≤1 %) Significant drug loss through side-branches.
Passive diffusion	Polymer serves as a drug repository (e.g., hydrogel-coated balloon) Coil catheter	Short time of delivery Atraumatic drug delivery Perfusion capabilities allow for longer drug instillation (30 min)	Not suitable for deep drug delivery Low efficiency and rapid wash-out
Facilitated diffusion	Transport of charged molecules using electrical current (e.g., iontophoretic catheter)	Deep intramural delivery (e.g., to adventitia) Atraumatic drug delivery	Not developed for coronary applications yet
Slow release	Drug eluting endovascular prosthesis (e.g., coated stent)	Platform for prolonged drug delivery Deep intramural delivery Stents are widely used	Stents alter vascular repair Drug eluting stents not developed yet

V.D. Efficiency of delivery

Local drug delivery is associated with a low overall efficiency. This does not necessarily reflect improper factors employed for the intramural administration (e.g., pressure, volume, and the contact of the device with the vessel wall). Most importantly, it represents a short time for oligomer delivery and limited diffusion of these compounds into normal or mildly injured media. When radiolabeled oligomers were delivered into injured porcine coronary arteries, vessel-associated radioactivity corresponded to ~1% of the injectate at 30 min after delivery. The remaining oligomers were rapidly cleared from the circulation ($t_{1/2}$ ~15 min) through renal and hepatic pathways (Table 5) (Shi *et al.*, 1994). A deep intramural oligomer delivery and their rapid intracellular localization were associated with their presence at the site of cell activation for 3 days (Figure 10).

Table 5. Distribution of Oligomers in the Peripheral Organs Following Transcatheter Delivery into the Coronary Vasculature (1 mg/vessel)

Organ	30 min	1 day	3 days
Coronary artery	23.40	18.20	51.00
Kidney	10.12	4.80	0.70
Liver	2.18	2.39	1.71
Cardiac muscle	0.45	0.33	1.14
Skeletal muscle	0.34	0.25	0.29
Small intestine	0.53	0.51	0.48
Bone marrow	0.37	0.38	0.46
Lung	0.49	0.42	0.43

All values are given in µg/g; 30 min, 1 day, and 3 days refer to time after oligomer delivery when the organs were harvested. *(Reprinted, with permission, from Shi et al., 1994).*

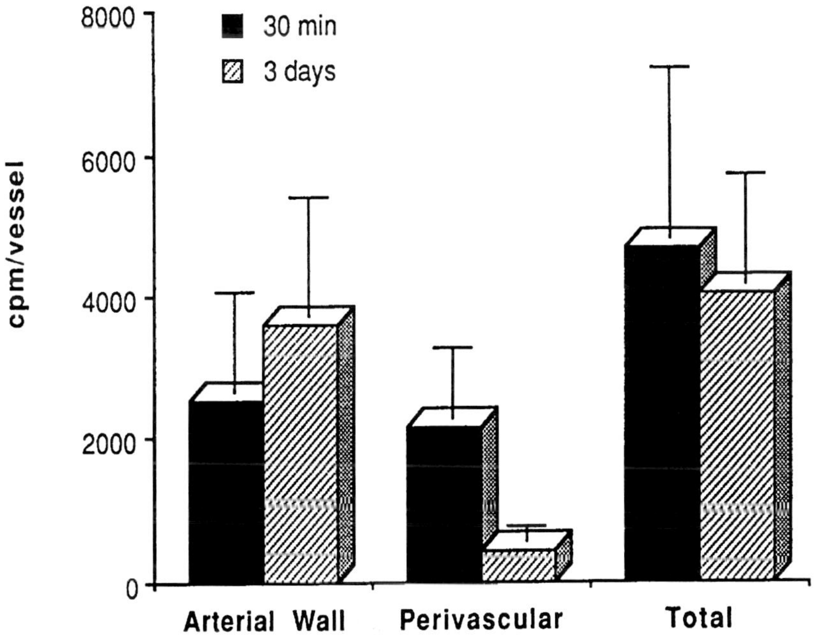

Figure 10. Bar graph depicting oligomer concentration in the vessel wall and in the perivascular tissues following transcatheter administration of ^{35}S-labeled c-*myc* antisense oligomers. Total oligomer content (cpm/vessel) was similar at 30 min (n = 6) and at 3 days (n = 4) after transcatheter delivery. Results are presented as mean±SEM.

Further improvements in delivery devices, or the development of eluting stents will be critical to enhance the efficiency of oligomer transfer into the arterial wall. The above observations regarding intramural administration of oligomers have been derived from normal, non-atherosclerotic coronary arteries that differ from diseased human arteries subjected to angioplasty.

The presence of atherosclerotic plaque may alter oligomer distribution and their stability in the vessel wall. Preliminary evidence, however, suggests that oligomers can penetrate atherosclerotic plaque and localize within cell nuclei (Pickering *et al.*, 1996).

In addition, newer formulations of oligomers, which are designed to favor cellular uptake, are being intensively pursued (Morishita *et al.*, 1993,1994). Several important issues related to local drug delivery await clinical trials. They include the potential impact of any intramural injection on the results of angioplasty in friable, atherosclerotic vessels and the development of more clinically applicable device(s).

VI. CLINICAL TESTING OF SYNTHETIC DNA-BASED COMPOUNDS

Clinical testing of synthetic DNA-based therapy for the prevention of restenosis requires taking into consideration several aspects that are unique to the use of oligomers in cardiac patients. They include the intracoronary delivery of oligomers, the choice of optimal patient population and endpoints to address the question of restenosis (Table 6).

These considerations are of paramount importance because of the novelty of these compounds, the multifactorial etiology of clinical restenosis, as well as the cost associated with large scale clinical trials. In addition, rapid changes affecting invasive cardiology (e.g., the use of stents or newer devices) should be taken into consideration to make the results applicable to clinical practice after these lengthy clinical investigations.

VI.A. Clinical safety

It is important to recognize that clinical testing of synthetic DNA-based therapy involves the assessment of two novel approaches, i.e., oligomers themselves, and the mode of their local delivery. To separate these issues, the clinical safety profile of c-*myc* phosphorothioate oligomers alone has been evaluated in patients after successful coronary angioplasty.

Several issues were addressed, including the changes in acute patency of dilated vessels, hemodynamic response after the intra-coronary injection of phosphorothioate oligomers, and the effects of this therapy on routine clinical chemistries. In a placebo controlled double-blinded

Table 6. Stepwise Approach to Clinical Development of Synthetic DNA-Based Compounds for the Prevention of Restenosis

1. Safety profile of intracoronary administration:
* no systemic toxicity (i.e., renal, hepatic, bone marrow, etc.);
* no local toxicity (i.e., coronary vasospasm, thrombosis);
* no hemodynamic response (i.e., hypotension);
* no adverse interactions with common cardiac medications.
2. Safety profile of intramural delivery:
* no adverse impact on acute results of angioplasty (e.g., abrupt closure);
* modalities suitable for coronary vasculature (e.g., device size, delivery time).
3. Impact on vascular repair at 3 - 6 months:
* reduction in neointima (intravascular ultrasound).
4. Impact on clinical and angiographic endpoints at 6 months:
* reduction in clinical events (i.e., angina, repeat revascularization,
* improved luminal patency (coronary angiography);
* benefit assessment in clinical subgroups (e.g., diabetes mellitus, multi-vessel disease etc.).

study, escalating doses of oligomers (1-24 mg) were injected via guiding catheter at 15 min after coronary angioplasty (Roqué *et al.*, 1997). Both the timing and a rapid injection (15-30 sec) of these compounds reflected the anticipated conditions to be employed in future efficacy trials. The quantitative coronary angiography revealed no acute changes in luminal dimensions, thereby excluding vasospastic or prothrombotic effects of c-*myc* oligomers in injured vessels. No significant hemodynamic changes were noted in response to oligomers.

This is particularly noteworthy, inasmuch as high doses of phosphorothioate oligomers have been reported to produce hypotension in non-human primates (Galbraith *et al.*, 1994). The administration of c-*myc* phosphorothioate oligomers resulted in no changes in routine clinical chemistries, as compared with control patients receiving placebo (Roqué *et al.*, 1997).

VI.B. Therapeutic efficacy

The fundamental question regarding the ability of synthetic DNA-based compounds to reduce both clinical ischemic events (e.g., recurrence of myocardial ischemia) and coronary restenosis over the 6 month period awaits clinical trials. Several issues remain unanswered, including the optimal method of local delivery of oligomers in human atherosclerotic vessels. Moreover, the effect of the intramural delivery of any compound on acute vessel patency (e.g., abrupt closure) is of concern, but is difficult to address clinically due to ethical reasons. The widening use of intracoronary stents, however, represents an opportunity to test the efficacy of transcatheter oligomer delivery in vessels supported by stents, thereby minimizing the potential adverse effects due to intramural drug administration.

In fact, a high pressure stent implantation may favor the accumulation of oligomers into the adventitia, inasmuch as the medial layer is often damaged in several points. This approach appears to be particularly important for the assessment of synthetic DNA-based therapy, since stents induce predominantly neointimal hyperplasia, thereby providing a clinical model for testing newer therapies that target cell activation, as opposed to other mechanisms. In addition, the ability to measure neointimal formation using intravascular ultrasound may provide a more sensitive index of oligomer efficacy than coronary angiography. The potential disadvantage of the above strategy is related to permanent placement of a foreign body (i.e., stent) in the arterial wall which may alter vascular repair (e.g., longer stimulation of adventitial fibroblasts, more intense formation of thrombus) making it less responsive to a single administration of oligomers.

VII. CONCLUSIONS

1. Coronary restenosis remains an important clinical problem that reflects an inappropriate vascular response to transcatheter coronary revascularization. A multifactorial etiology (i.e., formation of thrombus, neointima, and unfavorable remodeling) makes the development of effective therapeutic strategies quite challenging. Experimental studies indicate the activation of adventitial fibroblasts with their proliferation and differentiation to myofibroblasts after balloon injury. The latter not only contributes to the formation of neointima, rich in extracellular matrix proteins, but may also be involved in constrictive remodeling leading to restenosis. Accordingly, the inhibition of early cell activation may decrease the ensuing neointimal formation and geometric remodeling, thereby preventing restenosis.

2. Synthetic DNA-based compounds (oligomers) exert their biological effects through diverse mechanisms. The use of "antisense" oligomers to prevent restenosis has been based on the premise that targeting cell cycle regulating genes may inhibit the proliferative response in injured vessel wall. Synthetic DNA-based compounds (e.g., phosphorothioate oligomers), however, also demonstrate several non-antisense effects. Backbone modifications or enhanced cellular delivery of synthetic DNA-based compounds (e.g., oligomers conjugated with inactivated virus and liposomes) may increase their biological effects at lower concentrations and reduce non-antisense effects. It is unclear, however, whether the added complexity of such preparations is necessary to achieve therapeutic effects. Many non-antisense properties enhance antiproliferative actions of oligomers in cultured cells and their contribution to the effects *in vivo* remains to be determined. The diversity of actions exerted by synthetic DNA-based compounds does not

necessarily represent an impediment to their therapeutic utility, if restenosis can be reduced without untoward effects.

3. The effects of synthetic DNA-based compounds in cultured cells have a limited predictive value as to their efficacy in more complex conditions *in vivo*. Caution should also be advised against heavy reliance during preclinical testing on a single animal model utilizing non-coronary vascular injury (e.g., rat model) and clinically impractical methods of oligomer delivery (e.g., non-catheter administration). Although oligomers directed against several cell cycle regulating genes have been shown to reduce neointimal formation in rodents, the relevance of these findings to clinical restenosis remains uncertain due to species differences. C-*myc* oligomers have been shown to reduce neointimal formation in two animal species, including the porcine coronary model. The transcatheter delivery of c-*myc* oligomers, which resulted in their intramural deposition, favorably affected neointimal formation and luminal dimensions.

4. Therapeutic application of synthetic DNA-based compounds requires the development of practical and safe methods of their delivery to the site of coronary injury. To maximize tissue concentration and to minimize potential side effects of oligomers, local transcatheter administration immediately after coronary injury is envisioned for this purpose. Devices for intramural delivery should be a suitable size for use in the coronary vasculature, should deposit oligomers deep into the vascular wall (e.g., adventitia), and should produce no adverse effects (e.g., myocardial ischemia or abrupt closure). All current catheters for intramural administration demonstrate an overall low efficiency of coronary delivery. Oligomer deposition is focal through medial dissections caused by preceding balloon dilatation. A short residence time of oligomers in injured vessel wall represents an additional challenge. The development of stents eluting synthetic DNA-based compounds from their surface may represent a future alternative to circumvent the above difficulties.

5. Clinical testing of synthetic DNA-based compounds for the prevention of restenosis requires careful assessment of two components of this approach, that is, oligomers themselves and the method of their delivery. The first clinical study indicates that rapid intracoronary administration of c-*myc* phosphorothioate oligomers produces no acute adverse effects regarding coronary luminal dimension, hemodynamic response, and clinical chemistries in patients after coronary angioplasty. These results make it possible to consider these compounds for future efficacy trials, notwithstanding the complexity of their mechanism of action.

ACKNOWLEDGMENTS

The part of this study was supported by the National Institutes of Health grant RO1 HL44150. The authors would like to thank the staff of the Cardiovascular Research Center and many postdoctoral fellows who contributed to this work.

VIII. REFERENCES

Abe, J., Zhou, W., Taguchi, J., Takuwa, N., Miki, K., Okazaki, H., Kurokawa, K., Kumada, M., and Takuwa, Y. (1994) Suppression of neointimal smooth muscle cell accumulation in vivo by antisense *cdc2* and *cdk2* oligonucleotides in rat carotid artery. *Biochem. Biophys. Res. Comm.* **198**:16-24.

Acinapura, A. J., Jacobovitz, I. J., Kramer, M. D., Zisbrod, Z., and Cunningham, J. N. (1992) Internal mammary artery bypass: thirteen years of experience. Influence of angina and survival in 5125 patients. *J. Cardiovasc. Surg.* **33**:554-559.

Andersen, H. R., Mæng, M., Thorwest, M., and Falk, E. (1996) Remodeling rather than neointimal formation explains luminal narrowing after deep vessel wall injury. Insights from a porcine coronary (re)stenosis model. *Circulation* **93**:1716-1724.

Arora, P. D., and McCulloch, C. A. G. (1994) Dependence of collagen remodelling on α-smooth muscle actin expression by fibroblasts. *J. Cell Physiol.* **159**:161-175.

Bauters, C., de Groote, P., Adamantidis, M., Delcayre, C., Hamon, M., Lablanche, J. M., Bertrand, M. E., Dupuis, B., and Swynghedauw, B. (1992) Proto-oncogene expression in rabbit aorta after wall injury. First marker of the cellular process leading to restenosis after angioplasty? *Eur. Heart J.* **13**:556-559.

Bennett, M. R., Littlewood, T. D., Hancock, D. C., Evan, G. I., and Newby, A. C. (1994a) Down-regulation of the c-*myc* proto-oncogene in inhibition of vascular smooth-muscle cell proliferation: a signal for growth arrest? *Biochem. J.* **302**:701-708.

Bennett, M. R., Evan, G. I., and Newby, A. C. (1994b) Deregulated expression of the c-*myc* oncogene abolishes inhibition of proliferation of rat vascular smooth muscle cells by serum reduction, interferon, heparin, and cyclic nucleotide analogues and induces apoptosis. *Circ. Res.* **74**:525-536.

Bennett, M. R., Anglin, S., McEwan, J., Jagoe, R., Newby, A. C., and Evan, G. I. (1994c) Inhibition of vascular smooth muscle cell proliferation *in vitro* and *in vivo* by c-*myc* antisense oligodeoxynucleotides. *J. Clin. Invest.* **93**:820-828.

Biro, S., Fu, Y.-M., Yu, Z.-X., and Epstein, S. E. (1993) Inhibitory effects of antisense oligodeoxynucleotides targeting c-*myc* mRNA on smooth muscle cell proliferation and migration. *Proc. Natl. Acad. Sci. USA* **90**:654-658.

Bock, L. C., Griffin, L. C., Latham, J. A., Vermaas, E. H., and Toole, J. J. (1992) Selection of single-stranded DNA molecules that bind and inhibit human thrombin. *Nature* **355**:564-566.

Bourassa, M. G., Fisher, L. D., Campeau, L., Gillespie, M. J., McConney, M., and Lesperance, J. (1985) Long-term fate of bypass grafts: the Coronary Artery

Surgery Study (CASS) and Montreal Heart Institute experiences. *Circulation* 72(suppl V):V71-V78.

Brown, K. E., Kindy, M. S., and Sonenshein, G. E. (1992) Expression of the *c-myb* proto-oncogene in bovine vascular smooth muscle cells. *J. Biol. Chem.* 267:4625-4630.

Burgess, T. L., Fisher, E. F., Ross, S. L., Bready, J. V., Qian, Y.-X., Bayewitch, L. A., Cohen, A. M., Herrera, C. J., Hu, S., Kramer, T. B., Lott, F. D., Martin, F. H., Pierce, G. F., Simonet, L., and Farrell, C. L. (1995) The antiproliferative activity of c-*myb* and c-*myc* antisense oligonucleotides in smooth muscle cells is caused by a nonantisense mechanism. *Proc. Natl. Acad. Sci. USA* 92:4051-4055.

Cazenave, C., Stein, C. A., Loreau, N., Thuong, N. T., Neckers, L. M., Subasinghe, C., Hélène, C., Cohen, J. S., and Toulmé, J.-J. (1989) Comparative inhibition of rabbit globin mRNA translation by modified antisense oligodeoxynucleotides. *Nucleic Acids Res.* 17:4255-4273.

Clowes, A. W., Reidy, M. A., and Clowes, M. M. (1983) Kinetics of cellular proliferation after arterial injury. I. Smooth muscle growth in the absence of endothelium. *Lab. Invest.* 49:327-333.

Colombo, A., Hall, P., Nakamura, S., Almagor, Y., Maiello, L., Martini, G., Gaglione, A., Goldberg, S. L., and Tobis, J. M. (1995) Intracoronary stenting without anticoagulation accomplished with intravascular ultrasound guidance. *Circulation* 91:1676-1688.

Darby, I., Skalli, O., and Gabbiani, G. (1990) α-Smooth muscle actin is transiently expressed in by myofibroblasts during experimental wound healing. *Lab. Invest.* 63:21-29.

Epstein, S. E., Speir, E., and Finkel, T. (1993) Do antisense approaches to the problem of restenosis make sense? *Circulation* 88:1351-1353.

Faxon, D. B. (1995) MERCATOR Study Group. Effects of high-dose angiotensin-converting enzyme inhibition on restenosis: final results of the MERCATOR study, a multicenter, double-blind, placebo controlled trial of cilazapril. *J. Am. Coll. Cardiol.* 25:362-369.

Gadeau, A.-P., Campan, M., and Desgranes, C. (1991) Induction of cell cycle-dependent genes during cell cycle progression of arterial smooth muscle cells in culture. *J. Cell Physiol.* 146:356-361.

Galbraith, W. M., Hobson, W. C., Giclas, P. C., Schechter, P. J., and Agrawal, S. (1994) Complement activation and hemodynamic changes following intravenous administration of phosphorothioate oligonucleotides in the monkey. *Antisense Rev. Dev.* 4:201-206.

Gao, W.-Y., Han, F.-S., Storm, C., Egan, W., and Cheng, Y.-C. (1992) Phosphorothioate oligonucleotides are inhibitors of human DNA polymerases and RNase H: Implications for antisense technology. *Mol. Pharmacol.* 41:223-229.

Gunn, J., Holt, C. M., Francis, S. E., Shepherd, L., Grohmann, M., Newman, C. M. H., Crossman, D. C., Cumberland, D. C. (1997) The effect of oligonucleotides to *c-myb* on vascular smooth muscle cell proliferation and neointima formation after porcine coronary angioplasty. *Circ. Res.* 80:520-531.

Guvakova, M. A., Yakubov, L. A., Vlodavsky, I., Tonkinson, J. L., and Stein, C. A. (1995) Phosphorothioate oligodeoxynucleotides bind to basic fibroblast growth factor, inhibit its binding to cell surface receptors, and remove it from low affinity binding sites on extracellular matrix. *J. Biol. Chem.* **270**:2620-2627.

den Heijer, P., Foley, D. P., Escaned, J., Hillege, H. L., van Dijk, R. B., Serruys, P. W., and Lie, K. I. (1994) Angioscopic versus angiographic detection of intimal dissection and intracoronary thrombus. *J. Am. Coll. Cardiol.* **24**:649-654.

Heras, M., Chesebro, J. H., Penny, W. J., Bailey, K. R., Badimon, L., and Fuster, V. (1989) Effects of thrombin inhibition on the development of acute platelet-thrombus deposition during angioplasty in pigs. Heparin vs. hirudin, a specific thrombin inhibitor. *Circulation* **79**:657-665.

Ho, P. T. C., Ishiguro, K., Wickstrom, E., and Sartorelli, A. C. (1991) Non-sequence-specific inhibition of transferrin receptor expression in HL-60 leukemia cells by phosphorothioate antisense oligonucleotides. *Antisense Res. Dev.* **1**:329-342.

Holifield, B., Helgason, T., Jemelka, S., Taylor, A., Navran, S., Allen, J., and Seidel, C. (1996) Differentiated vascular myocytes: are they involved in neointimal formation? *J. Clin. Invest.* **97**:814-825.

Holmes, D. R. Jr., Vlietstra, R. E., and Smith, H. C. (1984) Restenosis after percutaneous transluminal coronary angioplasty (PTCA): a report from the PTCA Registry of the National Heart, Lung, and Blood Institute. *Am. J. Cardiol.* **53**:77C-81C.

Huckle, W. R., Drag, M. D., Acker, W. R., Powers, M., McFall, R. C., Holder, D. J., Fujita, T., Stabilito, I. I., Kim, D., Ondeyka, D. L., Mantlo, N.B., Chang, R. S. L., Reilly, C. F., Schwartz, R. S., Greenlee, W. J., and Johnson, R. G. (1996) Effects of subtype-selective and balanced angiotensin II receptor antagonists in a porcine coronary artery model of vascular restenosis. *Circulation* **93**:1009-1019.

Jarvis, T. C., Alby, L. J., Beaudry, A. A., Wincott, F. E., Beigelman, L., McSwiggen, J. A., Usman, N., and Stinchcomb, D. T. (1996) Inhibition of vascular smooth muscle cell proliferation by ribozymes that cleave c-*myb* mRNA. *RNA* **2**:419-428.

Kaddurah-Daouk, R., Greene, J. M., Baldwin, A. S. Jr., and Kingston, R. E. (1987) Activation and repression of mammalian gene expression by the c-*myc* protein. *Genes Dev.* **1**:347-357.

Kakuta, T., Currier, J. W., Haudenschild, C. C., Ryan, T. J., and Faxon, D. P. (1994) Differences in compensatory vessel enlargement, not intimal formation, account for restenosis after angioplasty in the hypercholesterolemic rabbit model. *Circulation* **89**:2809-2815.

Kohchi, K., Takebayashi, S., Block, P. C., Hiroki, T., Nobuyoshi, M. (1987) Arterial changes after percutaneous coronary angioplasty: Results at autopsy. *J. Am. Coll. Cardiol.* **10**:592-599.

Lafont, A., Guzman, L. A., Whitlow, P. L., Goormastic, M., Cornhill, J. F., and Chisolm, G. M. (1995) Restenosis after experimental angioplasty. Intimal, medial and adventitial changes associated with constrictive remodelling. *Circ. Res.* **76**:996-1001.

Miano, J. M., Tota, R. R., Vlasic, N., Danishefsky, K. J., and Stemerman, M. B. (1990) Early proto-oncogene expression in rat aortic smooth muscle cells following endothelial removal. *Am. J. Pathol.* **137**:761-765.

Mintz, G. S., Popma, J. J., Pichard, A. D., Kent, K. M., Satler, L. F., Wong, S. C., Hong, M. K., Kovach, J. A., and Leon, M. B. (1996) Arterial remodeling after coronary angioplasty. A serial intravascular ultrasound study. *Circulation* **94**:35-43.

McIntyre, K. W., Lombard-Gillooly, K., Perez, J. R., Kunsch, C., Sarmiento, U. M., Larigan, J. D., Landreth, K. T., and Narayanan, R. (1993) A sense phosphorothioate oligonucleotide directed to the initiation codon of transcription factor NF-kB p65 causes sequence-specific immune stimulation. *Antisense Res. Dev.* **3**:309-322.

Morishita, R., Gibbons, G. H., Ellison, K. E., Nakajima, M., Zhang, L., Kaneda, Y., Ogihara, T., and Dzau, V. J. (1993) Single intraluminal delivery of antisense *cdc2* kinase and proliferating-cell nuclear antigen oligonucleotides results in chronic inhibition of neointimal hyperplasia. *Proc. Natl. Acad. Sci. USA* **90**:8474-8478.

Morishita, R., Gibbons, G. H., Ellison, K. E., Nakajima, M., von der Leyen, H., Zhang, L., Kaneda, Y., Ogihara, T., and Dzau, V. J. (1994) Intimal hyperplasia after vascular injury is inhibited by antisense *cdk2* kinase oligonucleotides. *J. Clin. Invest.* **93**:1458-1464.

Morishita, R., Gibbons, G. H., Horiuchi, M., Ellison, K. E., Nakajima, M., Zhang, L., Kaneda, Y., Ogihara, T., and Dzau, V. (1995) A gene therapy strategy using a transcription factor decoy of E2F binding site inhibits smooth muscle proliferation *in vivo*. *Proc. Natl. Acad. Sci. USA* **92**:5855-5859.

Nikol, S., Isner, J. M., Pickering, J. G., Kearney, M., Leclerc, G., and Weir, L. (1992) Expression of transforming growth factor 1 is increased in human vascular restenosis lesions. *J. Clin. Invest.* **90**:1582-1592.

Nobuyoshi, M., Kimura, T., and Nosaka, H. (1988) Restenosis after successful percutaneous transluminal coronary angioplasty: serial angiographic follow-up of 229 patients. *J. Am. Coll. Cardiol.* **12**:616-623.

Nurse, P. (1990) Universal control mechanism regulating onset of M-phase. *Nature* **344**:503-508.

O'Brien, E. R., Alpers, C. E., Stewart, D. K., Ferguson, M., Tran, N., Gordon, D., Benditt, E.P., Hinohara, T., Simpson, J. B., and Schwartz, S. M. (1993) Proliferation in primary and restenotic coronary atherectomy tissue. Implications for antiproliferative therapy. *Circ. Res.* **73**:223-231.

O'Brien, J. E. Jr., Shi, Y., Fard, A., Bauer, T., Zalewski, A., and Mannion, J. D. (1997) Wound healing around and within saphenous vein bypass grafts. *J. Thorac. Cardiovasc. Surg.* **114**:38-45.

Parkes, J. L., Cardell, R. R., Hubbard, F. C., Hubbard, D., Meltzer, A., and Penn, A. (1991) Cultured human atherosclerotic plaque smooth muscle cells retain transforming potential and display enhanced expression of the c-*myc* proto-oncogene. *Am. J. Pathol.* **138**:765-775.

Pickering, J. G., Weir, L., Jekanowski, J., Kearney, M. A., and Isner, J. M. (1993) Proliferative activity in peripheral and coronary atherosclerotic plaque among patients undergoing percutaneous revasculization. *J. Clin. Invest.* **91**:1469-1480.

Pickering, J. G., Isner, J. M., Ford, C. M., Weir, L., Lazarovits, A., Rocnik, E. F., and Chow, L. H. (1996) Processing of chimeric antisense oligonucleotides by human vascular smooth muscle cells and human atherosclerotic plaque. *Circulation* **93**:772-780.

Post, M. J., Borst, C., and Kuntz, R. E. (1994) The relative importance of arterial remodeling compared with intimal hyperplasia in lumen renarrowing after balloon angioplasty. *Circulation* **89**:2816-2821.

Powell, J. S., Clozel, J. P., Muller, R. K., Kuhn, H., Hefti, F., Hosang, M., and Baumgartner, H. R. (1989) Inhibitors of angiotensin-converting enzyme prevent myointimal proliferation after vascular injury. *Science* **245**:186-188.

Prendergast, G. C., and Cole, M. D. (1989) Post-transcriptional regulation of cellular gene expression by the c-*myc* oncogene. *Mol. Cell Biol.* **9**:124-134.

Prelich, G., and Stillman, B. (1988) Coordinated leading and lagging strand synthesis during SV40 DNA replication in vitro requires PCNA. *Cell* **53**:117-126.

Riabowol, K., Draetta, G., Brizuela, L., Vandre, D., and Beach, D. (1989) The *cdc2* kinase is a nuclear protein that is essential for mitosis in mammalian cells. *Cell* **57**:393-401.

Roqué, F., Mon, G., Belardi, J., Rodriguez, A., Grinfeld, L., Fischman, D., Shi, Y., and Zalewski, A. (1997) Safety of intracoronary administration of c-*myc* antisense after PTCA. *J. Am. Coll. Cardiol.* **29A**:317A.

Ross, R., Wight, T. N., Strandness, E., and Thiele, B. (1984) Human atherosclerosis. I. Cell constitution and characteristics of advanced lesions of the superficial femoral artery. *Am. J. Pathol.* **114**:79-93.

Schwartz, R. S., Holmes, D. R., and Topol E. J. (1992) The restenosis paradigm revisited: an alternative proposal for cellular mechanisms. *J. Am. Coll. Cardiol.* **20**:1284-1293.

Shi, Y., Hutchinson, H. G., Hall, D. J., and Zalewski, A. (1993) Downregulation of c-myc expression by antisense oligonucleotides inhibits proliferation of human smooth muscle cells. *Circulation* **88**:1190-1195.

Shi, Y., Fard, A., Galeo, A., Hutchinson, H., Vermani, P., Dodge, G. R., Hall, D. J., Sheehan, F., and Zalewski, A. (1994) Transcatheter delivery of c-myc antisense oligomers reduces neointimal formation in a porcine model of coronary balloon injury. *Circulation* **90**:944-951.

Shi, Y., Pieniek, M., Fard, A., O'Brien, J., Mannion, J. D., and Zalewski, A. (1996a) Adventitial remodeling following coronary arterial injury. *Circulation* **93**:340-348.

Shi, Y., O'Brien, J., Fard, A., Mannion, J. D., Wang, D., and Zalewski, A. (1996b) Adventitial myofibroblasts contribute to neointimal formation in injured porcine coronary arteries. *Circulation* **94**:1655-1664.

Shi, Y., O'Brien, J. E., Jr., Fard, A., and Zalewski, A. (1996c) Transforming growth factor ß1 expression and myofibroblast formation during arterial repair. *Arterioscl. Thromb. Vasc. Biol.* **16**:1298-1305.

Shi, Y., O'Brien, J., Fard, A., Ala-Kokko, L., Chung, W. S., Mannion, J. D., and Zalewski, A. (1997a) The origin of extracellular matrix synthesis during coronary repair. *Circulation* **95**:997-1006.

Shi, Y., O'Brien, J. E., Mannion, J. D., Morrison, R. C., Chung W., Fard, A., and Zalewski, A. (1997b) Remodeling of anutologous saphenous vein grafts. The role of perivascular myofibroblasts. *Circulation* **95**:2684-2693.

Simons, M., and Rosenberg, R. D. (1992) Antisense nonmuscle myosin heavy chain and c-*myb* oligonucleotides suppress smooth muscle cell proliferation *in vitro*. *Circulation Res.* **70**:8350843.

Simons, M., Edelman, E. R., DeKeyser, J.-L., Langer, R., and Rosenberg, R. D. (1992) Antisense c-*myb* oligonucleotides inhibit intimal arterial smooth muscle cell accumulation *in vivo*. *Nature* **359**:67-70.

Simons, M., Edelman, E. R., and Rosenberg, R. D. (1994) Antisense proliferating cell nuclear antigen oligonucleotides inhibit intimal hyperplasia in a rat carotid artery injury model. *J. Clin Invest.* **93**:2351-2356.

Stein, C. A., and Cheng, Y. C. (1993) Antisense oligonucleotides as therapeutic agents - Is the bullet really magical? *Science* **261**:1004-1012.

Stein, CA. (1995) Does antisense exist? *Nature Med.* **1**:1119-1121.

Swedberg, S. H., Brown, B. G., Sigley, R., Wight, T. N., Gordon, D., and Nicholls, S. C. (1989) Intimal fibromuscular hyperplasia at the venous anastomosis of PTFE grafts in hemodialysis patients. *Circulation* **80**:1726-1736.

Topol. E. J., Leya, F., Pinkerton, C. A., Whitlow, P. L., Hofling, B., Simonton, C. A., Masden, R. R., Serruys, P. W., Leon, M. B., Williams, D. O., King, S. B., Mark, D. B., Isner, J. M., Holmes, D. R., Ellis, S. G., Lee, K. L., Keeler, G. P., Berdan, L. G., Hinohara, T., and Califf, R. M. (1993) A comparison of directional atherectomy with coronary angioplasty in patients with coronary artery disease. *New Engl. J. Med.* **329**:221-227.

Travali, S., Ferber, A., Reiss, K., Sell, C., Koniecki, J., Calabretta, B., and Baserga, R. (1991) Effect of the *myb* gene product on the expression of the PCNA gene in fibroblasts. *Oncogene* **6**:887-894.

Tsai, L.-H., Lees, E., Faha, B., Harlow, E., and Riabowol, K. (1993) The *cdk2* kinase is required for the G1-to-S transition in mammalian cells. *Oncogene* **8**:1592-1602.

Villa, A. E., Guzman, L. A., Poptic, E. J., Labhasetwar, V., D'Souza, S., Farrell, C. L., Plow, E. F., Levy, R. J., DiCorleto, P. E., and Topol, E. J. (1995) Effect of antisense *c-myb* oligonucleotides on vascular smooth muscle cell proliferation and response to vessel wall injury. *Circ. Res.* **176**:505-513.

Wang, W., Chen, H. J., Schwartz, A., Cannon, P. J., Stein, C. A., and Rabbani, L. E. (1996) Sequence-independent inhibition of *in vitro* vascular smooth muscle cell proliferation, migration, and *in vivo* neointimal formation by phosphorothioate oligodeoxynucleotides. *J. Clin Invest.* **98**:443-450.

Welch, M. P., Odland, G. F., and Clark, R. A. F. (1990) Temporal relationship of F-actin bundle formation, collagen and fibronectin matrix assembly, fibronectin receptor expression to wound contraction. *J. Cell Biol.* **110**:133-145.

Wickstrom, E. L., Bacon, T. A., Gonzalez, A., Freeman, D. L., Lyman, G. H., and Wickstrom, E. (1988) Human promyelocytic leukemia HL-60 cell proliferation and *c-myc* protein expression are inhibited by an antisense pentadecadeoxynucleotide targeted against c-*myc* mRNA. *Proc. Natl. Acad. Sci. USA* **85**:1028-1032.

Yamamoto, S., Yamamoto, T., Kataoka, T., Kuramoto, E., Yano, O., and Tokunaga, T. (1992) Unique palindromic sequences in synthetic oligonucleotides are required to induce INF and augment INF-mediated natural killer activity. *J. Immunol.* **148**:4072-4076.

Yaswen, P., Stampfer, M. R., Ghosh, K., Cohen, J. S. (1993) Effects of sequence of thioated oligonucleotides on cultured human mammary epithelial cells. *Antisense Res. Dev.* **3**:67-77.

Zalewski, A., Goldberg, S., Dervan, J. P., Slysh, S., and Maroko, P. R. (1986) Myocardial protection during transient coronary occlusion in man: Beneficial effects of regional beta adrenergic blockade. *Circulation* **73**:734-739.

20

Prevention of Restenosis by Gene Targeting

Michael J. Mann and Victor J. Dzau
Harvard Medical School and Brigham and Women's Hospital, Boston, Massachusetts

Heiko E. von der Leyen
Abteilung Kardiologie, Medizinische Hochschule, Hannover, Germany

I. INTRODUCTION

Researchers seeking new therapeutic strategies for the treatment of cardiovascular disease have turned their attention to the genetic basis of both normal and pathologic cardiovascular growth and function (Berg and Singer, 1995; Dzau *et al.*, 1993). Various molecular and biochemical pathways are known to play critical roles in the control of cell cycle progression and cell division, and these pathways can be influenced by the expression of certain genes. Control of cell growth is, in turn, an important determinant in the natural history of proliferative disorders of the cardiovascular system, such as the phenomenon of restenosis after balloon angioplasty. It has therefore become possible to identify genes that are important in disease progression, and to target the expression of these genes for modification in an attempt to alter cardiovascular pathobiology.

Two strategies have emerged for *in vivo* gene therapy, and both have been applied successfully in animal models of cardiovascular disease: 1) down-regulation of gene expression using antisense DNA, ribozyme RNA or, more recently, transcription factor "decoy" double-stranded DNA (Morishita *et al.*, 1993a,1995) and 2) overexpression of genes or introduction of exogenous genes (Ohno *et al.*, 1994; von der Leyen *et al.*, 1995) using plasmid or viral vectors. The development of these two complementary strategies has enabled the investigation of genes that are either upregulated or downregulated during cardiovascular pathogenesis as potential targets for therapeutic intervention.

In addition to the identification of appropriate target genes, the application of gene therapy in both experimental and clinical settings continues to be hindered by the inefficiency and potential hazards of techniques for the transfer of DNA into living cells. The most commonly used method of experimental *in vivo* gene transfer is transfection with viral vectors, such as replication-deficient adenoviral vectors. Despite their successes, the construction of these viral vectors remains difficult and time consuming, and their use continues to pose the threat of viral mutation and oncogenesis. Furthermore, they do not provide a mechanism for oligodeoxynucleotide (ODN) delivery. Alternative, nonviral vectors for DNA transfer, such as cationic lipids, have not yet been shown to yield consistently efficient DNA delivery to cells *in vivo*.

Our laboratory has therefore focused on the development of both novel strategies for genetic intervention in cardiovascular proliferative disorders, as well as improvements in current DNA transfer technology. We have exploited both gene blockade and gene transfer approaches in inhibiting neointimal hyperplasia in arterial balloon injury models (Morishita *et al.*, 1993a,1995; von der Leyen *et al.*, 1995). We have also extended some of these observations to other examples of cardiovascular proliferative disease, such as intimal hyperplasia and atherogenesis in bypass vein grafts (Mann *et al.*, 1995).

These successful *in vivo* manipulations of the genetic machinery of cardiovascular tissue have been achieved using a highly efficient vector, free of live viral components, the HVJ-liposome complex. In this review we will briefly describe the development of this vector, as well as its use in delivering ODN and plasmid DNA into vascular smooth muscle cell (VSMC) proliferation models. Our studies may help provide insight into the prevention of restenosis after balloon angioplasty of atherosclerotic vessels.

II. THE HVJ-LIPOSOME COMPLEX

The hemagglutinating virus of Japan (Sendai virus; HVJ), a member of the paramyxovirus family, has long been known to possess two viral coat proteins that enhance membrane fusion (Figure 1A) (Okada *et al.*,

1961; Okada, 1993). Hemagglutinating neuroaminidase (HN) mediates viral particle binding to sialoglycoproteins or sialolipids on the cell surface. It then catalyzes the removal of sugar moieties from these surface molecules. After particle binding, the second protein, Fusion protein (F), induces membrane fusion via interaction with the lipid bilayer. The active form of F protein consists of two polypeptides, F1 and F2, produced by proteolytic cleavage of the inactive F0 form. Following cleavage, F1 and F2 remain associated, anchored by a disulfide bridge.

As its name implies, the HVJ particle binds avidly to red blood cells, and the use of HVJ particles to enhance the *in vivo* delivery of DNA was first described using erythrocyte ghosts to encapsulate the DNA (Sugawa *et al.*, 1985; Kaneda *et al.*, 1989). More recently, a combination of cholesterol, phosphatidyl serine and phosphatidyl choline has been used to more efficiently encapsulate DNA into 200-300 nm liposomes. Incubation of these liposomes with HVJ particles that have been inactivated by UV irradiation leads to formation of the fusogenic HVJ-liposome complex. Although the liposomes do not possess receptors for the HN protein, the interaction of cholesterol with a hydrophobic region of F1 is known to take place during cell membrane fusion, and is believed to facilitate the fusion of viral envelopes with these liposomes (Nakanishi *et al.*, 1985).

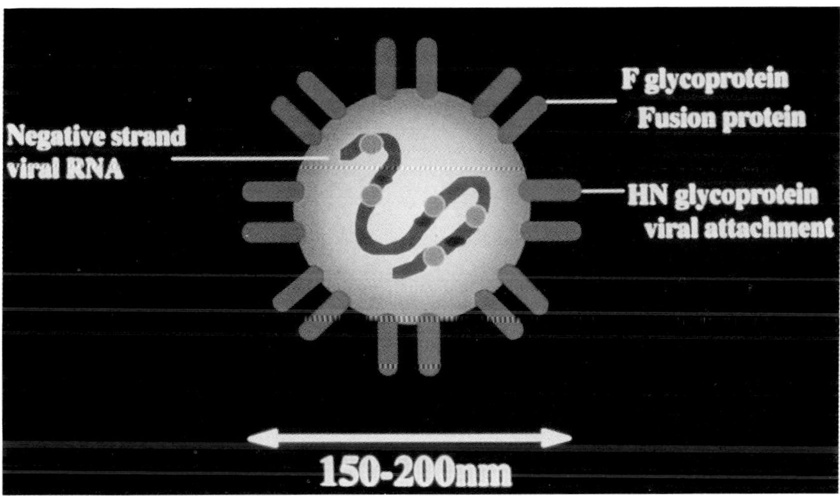

Figure 1. Hemagglutinating virus of Japan (HVJ). The viral envelope of this paramyxovirus contains two proteins, HN and F, that are essential for its membrane fusion properties.

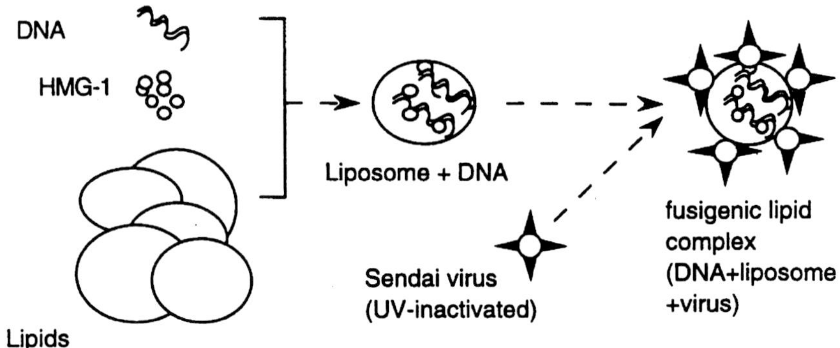

Figure 2. Process of HVJ-liposome complex formation.

The resulting complex possesses both HN and F proteins on its surface, and it can be used *in vivo* or *in vitro* for cellular delivery of a wide range of DNA, including single or double stranded ODN and plasmids up to 100 kb in size (Figure 2).

In addition to enhancing fusion of the HVJ-liposome complex with cell membranes, the viral coat proteins appear to influence the intracellular fate of DNA delivered by this fusion (Okada *et al.*, 1975). The passive uptake of ODN is believed to involve an endocytotic process that results in lysosomal degradation of most of the DNA.

In contrast, HVJ-liposome delivery triggers enhanced intracellular trafficking of ODN to the cell nucleus, where it can bind to target sequences of messenger RNA (mRNA) or to transcription factors. This improved intranuclear delivery of ODN can improve the efficiency of ODN transfection and improve the biologic effect of *in vivo* ODN treatment in intact tissues.

Transfection with plasmid vectors containing full gene constructs requires nuclear DNA delivery to achieve transgene expression. An important advantage of the HVJ-liposome technique is that it allows delivery of proteins together with DNA. This feature has been used to improve overexpression of target genes (Kaneda *et al.*, 1989). Kaneda and his associates studied the co-delivery of plasmid DNA with high mobility group 1 (HMG-1) protein, a non-histone nuclear protein that can influence transcription via effects on DNA bending or looping. The preincubation of plasmid DNA with HMG-1 prior to encapsulation into HVJ-liposomes has been found to yield higher levels of gene expression both *in vitro* and *in vivo*. This influence may be mediated both by improved intracellular trafficking as well as enhanced transcription of the transfected gene construct.

The HVJ virus has been associated with respiratory infection in rodent species, but human infection with HVJ has not been reported. Despite this limitation on viral pathogenicity, HVJ-liposome complexes have been used to enhance transfection in a number of different cell types *in vitro*, including VSMC from rat and rabbit, and both bovine aortic and human umbilical vein endothelial cells. This technique has also been reported to achieve efficient delivery of both plasmid DNA and ODN in a variety of tissues *in vivo*, including liver, kidney and the vascular wall (Kaneda *et al.*, 1989; Morishita *et al.*, 1993ab; Isaka *et al.*, 1993; von der Leyen *et al.*, 1995; Mann *et al.*, 1995). UV inactivation of the virus during complex formation has prevented infectious complications of HVJ-liposome transfection, even in rodent species.

Furthermore, despite the presence of viral coat proteins on the surface of HVJ-liposome complexes, immunologic reaction to the complex does not seem to limit transfection efficiency, as it does in transfection with live adenoviral vectors, even with repeated HVJ-liposome trans-fections *in vivo*. Investigations are currently underway to improve the standardization and safety of the HVJ-liposome technique by utilization of purified or recombinant HN and F proteins rather than inactivated viral particles during complex formation.

The potential advantages of HVJ-liposome mediated transfection compared to traditional liposome or viral mediated transfection *in vivo* and *in vitro* include: 1) *in vivo* transfection efficiency in replicating and non-replicating cells similar to that reported with adenoviral gene transfer, 2) avoidance of the hazards associated with live viral gene transfer, since HVJ particles are inactivated during complex preparation, 3) simplicity of preparation when compared to adenoviral vector construction, 4) ability to transfect cells with either ODN or gene constructs, and 5) no limitation on transgene size.

III. RESTENOSIS AND CELL CYCLE INHIBITION

Restenosis after balloon angioplasty is now known to be a complex event involving a number of different vascular remodeling pathways. None of the current animal models has succeeded in fully recapitulating the pathologic process in humans. One important component of resten-osis, however, is the reaction of VSMC to injury and the subsequent development of neointimal hyperplasia. This hyperplastic reaction begins with the activation of VSMC into a secretory and proliferative state. Elaboration of matrix metalloproteinases by proliferating VSMC in the medial layer of the vessel wall facilitates their migration across the internal elastic lamina to form the intimal lesion where cell division continues (Bendeck *et al.*, 1996). In addition to proliferation and migration, these cells also express an array of cytokines, chemoattractants and adhesion molecules that not only enhance the inflammatory response to injury but also act in a paracrine and autocrine fashion to stimulate further vascular cell abnormality (Loppnow *et al.*, 1990; Wenzel *et al.*, 1995). Carotid artery balloon injury in the rat provides a model of neointimal hyperplasia that we have utilized to investigate gene targeting strategies to reduce or eliminate this component of the restenosis phenomenon.

A number of molecular species have been implicated in the stimulation of VSMC to begin the intimal hyperplastic process, including platelet derived growth factor, basic fibroblast growth factor, transforming growth factor ß and angiotensin II (Gibbons *et al.*, 1994; Glagov, 1994). The redundant nature of these stimuli prompted our laboratory to look beyond the pharmacologic or even genetic blockade of these factors or their secondary messengers to a final common pathway in the cellular response to vascular injury. VSMC proliferation depends on the increased expression of certain cell cycle regulatory proteins at critical points during the cell cycle, including proliferating cell nuclear antigen (PCNA), cell division cycle 2 (*cdc2*) kinase, the cyclins and cyclin dependent kinases (*cdks*) (Figure 3). We hypothesized that by blocking expression of the genes that encode one or more of these proteins, we could prevent the progression of VSMC through the cell cycle and inhibit neointimal hyperplasia.

To achieve *in vivo* modification of cell cycle regulatory gene expression after carotid balloon injury, we first employed transfection of antisense ODN. These short chains of DNA are designed to bear a complementary base sequence to a portion of the target gene's mRNA, and are generally 15-20 bases in length to correspond to a unique sequence in the human genome (Wickstrom, 1991). Antisense mRNAs have been found to occur naturally as regulators of gene expression (Inouye, 1988; Krystal *et al.*, 1990). They may prevent gene expression by a number of different mechanisms after Watson-Crick basepairing to the target mRNA, including interference with ribosomal binding and alterations in mRNA conformation or splicing. In addition to these mechanisms, antisense ODN may act primarily through degradation of mRNA-DNA hybrids by RNase H (Colman, 1990; Cohen, 1992).

Although traditional phosphodiester ODN are susceptible to exonuclease and endonuclease digestion, ODN with phosphodiesterase-resistant modifications of the phosphate backbone, such as phosphoramidate and phosphorothioate ODN, can be used for more prolonged inhibition of gene expression (Wickstrom, 1991; Crooke, 1993).

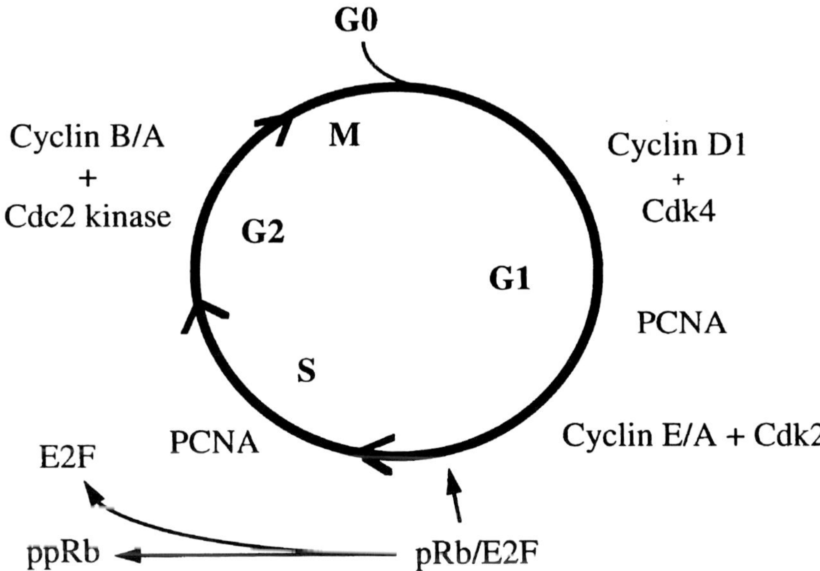

Figure 3. Cell cycle regulation. Progression through the early growth (G1), DNA synthesis (S), late growth (G2), and mitosis (M) phases of the cell cycle to cell division depends on the activity of various cell cycle regulatory proteins, including PCNA and *cdc2* kinase. The release of transcription factor E2F from its complex with hypophosphorylated retinoblastoma gene product (pRb) allows this molecule to further drive cell cycle progression *(reprinted, with permission, from Mann et al., 1995).*

Finally, to control for potential nonspecific, or "aptameric," effects on cell biology and protein expression that have been reported with ODN transfection (Matsukura *et al.*, 1987; Yaswen *et al.*, 1993; Villa *et al.*, 1995; Burgess *et al.*, 1995), it is important to design experiments with adequate controls, including sense, mismatch and reverse antisense sequence ODN.

Initial experiments in cell culture indicated that transfection of VSMC with antisense ODN directed against the translation initiation site of cell cycle regulatory genes such as PCNA could, in fact, partially inhibit cellular proliferation in response to serum stimulation. However, we found that this incomplete inhibition could be augmented by simultaneous transfection with two ODN species designed to block a combination of cell cycle genes (Morishita *et al.*, 1993a). We therefore undertook the transfection of rat carotid artery at the time of balloon injury with antisense ODN against cell cycle regulatory genes as a model treatment for the neointimal hyperplastic component of restenosis after balloon angioplasty.

Unlike some strategies involving gene transfer, in which successful transgene product secretion by a small percentage of cells can exert an effect on the surrounding tissue, *in vivo* gene therapy using antisense ODN requires a high transfection efficiency to influence the behavior of a large percentage of cells in an intact tissue. We therefore utilized HVJ-liposome technology to enable us to influence VSMC proliferation in the injured vessel wall. To further enhance HVJ-liposome activity, and to localize the effect of our treatment, we infused HVJ-liposome suspensions containing ODN into isolated segments of injured carotid. These segments were first flushed free of residual blood to reduce HVJ-liposome-erythrocyte interaction.

As in cell culture, we found that neointimal formation could be reduced by transfection of VSMC in the injured vessel wall with antisense ODN against a single cell cycle gene such as PCNA, but that near complete inhibition of neointimal hyperplasia could be achieved via transfection with a combination of antisense ODN against PCNA and *cdc2* kinase (Morishita *et al.*, 1993a,1994a). This effect on VSMC growth was not limited to blockade of this particular combination of genes, as similar inhibition was detected after treatment with combinations of antisense ODN against *cdc2* kinase and *cdk2*, against *cdc2* kinase and cyclin B1, and against *cdk2* and cyclin B1 (Morishita *et al.*, 1994ab). The ODN treatment was both locally selective and sequence specific, as neointimal inhibition was limited to the transfected segment of injured carotid; adjacent untransfected segments developed neointimal lesions similar to untreated carotids and to vessels transfected with control ODN that were subjected to balloon injury.

Although antisense inhibition of cell cycle regulatory gene expression after this single transfection is likely to last less than 2 weeks, we observed a persistent inhibition of neointimal formation up to 8 weeks after injury. These results suggested that blockade of cell cycle progression after vascular injury has a stabilizing effect on VSMC phenotype. We further hypothesized that cell cycle regulatory gene blockade could prevent other elements of VSMC activation. This hypothesis has subsequently been supported by observations made in a related example of vascular proliferative disease, the fibrointimal hyperplasia and accelerated atherosclerosis seen after surgery in bypass vein grafts.

The failure rate of 30-50% seen in longterm autologous vein grafts is due primarily to neointimal disease (Cox *et al.*, 1991). We therefore examined the effect of intraoperative transfection of rabbit vein grafts with a combination of antisense ODN against PCNA and *cdc2* kinase (Mann *et al.*, 1995). In this model we demonstrated a sequence-specific reduction of >90% in the increase of target protein levels seen four days after surgery. We also found a similar reduction in VSMC proliferation at 1 week, when cell division is normally at its peak in this model (Zwolak *et al.*, 1987), and a near complete inhibition of neointimal hyperplasia at 2 weeks after operation.

In contrast to arterial balloon injury, however, vein grafts are not only subjected to a single injury at the time of operation, but are also exposed to chronic hemodynamic stimuli for remodeling. In addition, some adaptive remodeling and wall thickening is necessary to reduce high levels of tangential wall stress in these previously thin-walled, low pressure conduits (Zwolak *et al.*, 1987). Despite these chronic stimuli, we found that our single intraoperative ODN treatment of vein grafts endowed them with a resistance to neointimal hyperplasia that persisted for at least six months after operation. During that time period, however, the antisense ODN treated grafts did exhibit the capacity for adaptive remodeling via hypertrophy of the medial layer. The VSMC of that layer retained their normal circumferential architecture, in sharp contrast to the chaotic disarray of neointimal VSM. The medial thickening observed in antisense treated grafts, without significant neointimal hyperplasia, successfully reduced graft wall tension levels to those of normal arteries. Furthermore, these genetically engineered conduits proved resistant to diet-induced graft atherosclerosis in our animal model (Figure 4).

We hypothesize that cell cycle regulatory gene blockade prevents phenotypic changes associated with the activation of neointimal VSMC. As described above, this activation involves the expression of cytokines and adhesion molecules that are likely to contribute to the pro-atherogenic environment of the untreated graft wall. Stabilization of the normal contractile VSMC phenotype would therefore help to explain the antisense ODN treated graft's resistance to graft atherosclerosis. In addition, the vein graft endothelium is known to exhibit abnormalities similar to those associated with native arterial atherogenesis.

We therefore studied the influence of PCNA antisense and *cdc2* kinase antisense treatment on vein graft endothelial function (Mann *et al.*, 1997). Endothelial cell nitric oxide synthase (ecNOS) activity, responsible for endothelial NO generation, was preserved in antisense treated grafts compared to sharp reductions seen in untreated and control ODN treated grafts. This preservation of ecNOS activity level correlated with a preservation of vascular reactivity in these genetically engineered grafts as well. Endothelial cell phenotypic stabilization in antisense treated grafts was also reflected in normally low levels of vascular cell adhesion molecule 1 (VCAM-1) expression. VCAM-1 is believed to play a role in the infiltration of macrophages into atherosclerotic lesions, and was found to be upregulated in control grafts for at least 4 weeks after surgery.

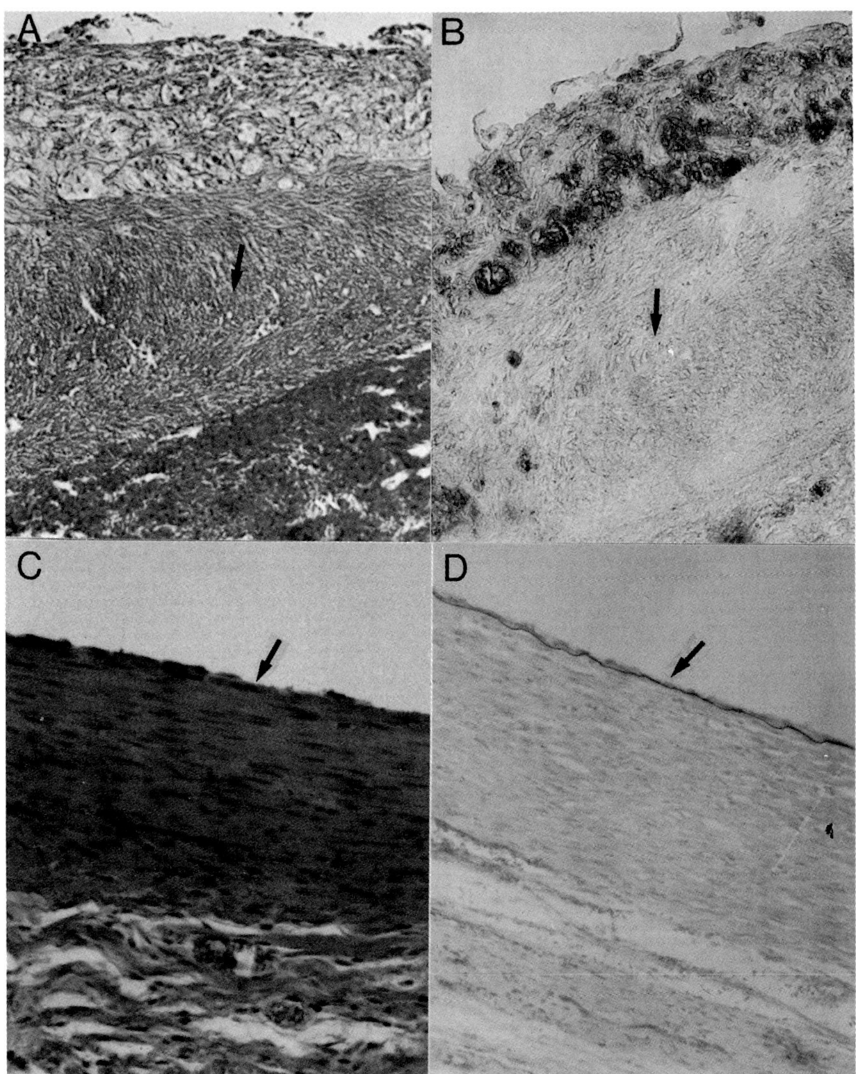

Figure 4. Light micrographs of vein grafts placed in cholesterol-fed rabbits after treatment with control ODN (**A,B**) or antisense (AS) ODN against cell cycle regulatory genes (**C,D**). Neointimal hyperplasia and foam cell lesions are prevented after AS ODN treatment, as is invasion of the vessel wall by macrophages as indicated by immunohistochemical staining of these cells (**B,D**) *(reprinted, with permission, from Morishita, et al., 1995).*

Prevention of VCAM-1 upregulation may be related to a reduction in monocyte adhesion that was also measured in these genetically engineered graft conduits. Either of two mechanisms may explain the prevention of endothelial dysfunction observed after PCNA/*cdc2* kinase antisense treatment of vein grafts: 1) a direct influence of cell cycle regulatory gene blockade on endothelial cell phenotype and/or 2) an indirect influence on endothelial cell behavior via the reduced elaboration of cytokines and other bioactive species by the underlying VSMC.

IV. THE TRANSCRIPTION FACTOR "DECOY" APPROACH

Our success at achieving long term modification in vascular cell biology through a single intervention in cell cycle regulatory gene expression led us to explore other, potentially more efficient approaches to genetic control of the cell cycle. Among the events now known to take place early in the transition of cells from G0 phase to progression through the cell cycle is the release of transcription factor E2F. During cell quiescence, this molecule is bound to a complex of cyclin A, *cdk2*, and the retinoblastoma gene product. Phosphorylation of the retinoblastoma gene product after cell stimulation leads to release of E2F from this complex. The unsequestered molecule is then free to bind with its target *cis* elements in the promoter regions of a number of genes that must be expressed during cell cycle progression. This list of genes upregulated by E2F binding includes both PCNA and *cdc2* kinase, as well as c-*myb* and c-*myc*, proto-oncogenes that have also been shown to play a critical role during VSMC proliferation and neointimal hyperplasia (Simons *et al.*, 1992; Shi *et al.*, 1994).

The activity of a transcription factor can be blocked via the introduction into cells of double stranded ODN containing the sequence to which the factor normally binds on chromosomal DNA (Bielinska *et al.*, 1990; Sullenger *et al.*, 1990). Such *cis* element "decoys" bind to free transcription factors and prevent their interaction with the promoter regions of target genes. We therefore tested the ability of *in vivo* transfection with E2F decoy ODN to inhibit *cis*-activation of cell cycle genes and prevent neointimal hyperplasia after rat carotid balloon injury (Morishita *et al.*, 1995).

Preliminary *in vitro* studies demonstrated that the serum stimulated upregulation of PCNA, *cdc2* kinase and c-*myc* could all be blocked at the mRNA and protein levels via E2F decoy ODN transfection of VSMC. *In vitro* proliferation of these cells was also inhibited, demonstrating the dependence of VSMC cell cycle progression on E2F release. *In vivo* HVJ-liposome transfection of rat carotid arteries with E2F decoy at the time of

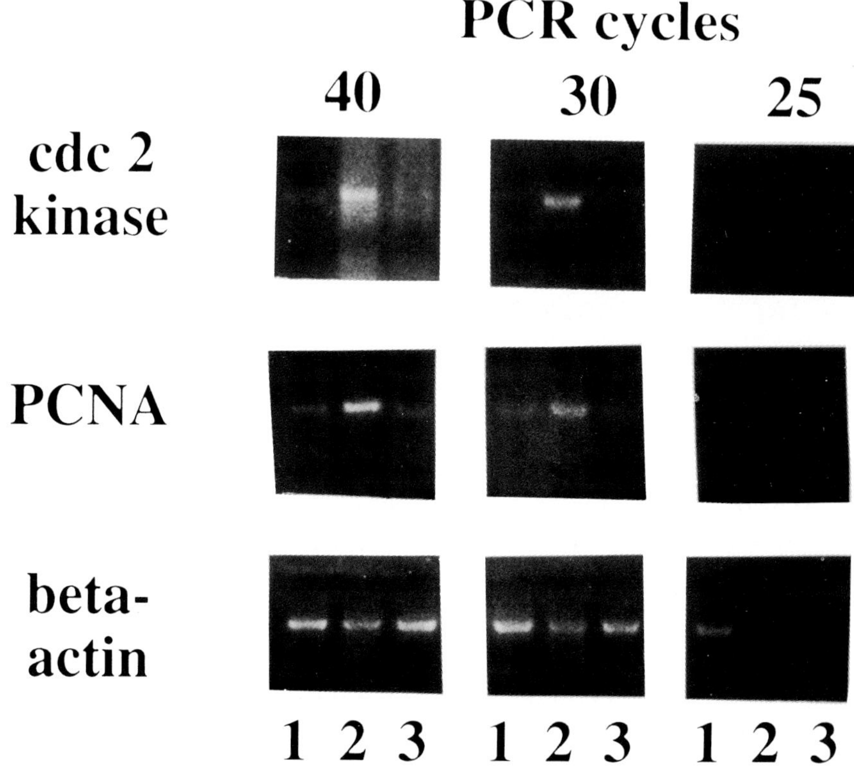

Figure 5. Reverse transcriptase-polymerase chain reaction (RT-PCR) analysis of *cdc2* kinase and PCNA gene expression after *in vivo* transfection of rat carotids with E2F decoy oligonucleotide, compared to expression of the control gene, ß-actin. Lane 1 = normal carotid, lane 2 = injured carotid, lane 3 = injured carotid treated with E2F decoy. The lower intensity of the bands in lane 3 indicates a lower level of target gene expression *(reprinted, with permission, from von der Leyen, et al., 1995)*.

balloon injury yielded a similar, sequence specific inhibition of expression of these three genes as documented by mRNA analysis (Figure 5). As a result of this blockade of E2F mediated gene upregulation, a significant reduction in VSMC replication four days after injury was exhibited through a decrease in bromodeoxyuridine incorporation. Furthermore, neointimal hyperplasia was again inhibited for at least 8 weeks after injury and transfection, providing further support to the hypothesis that interventions in gene expression at critical points during pathogenesis can yield long term effects on vascular biology and remodeling.

The transcription factor decoy approach has certain advantages over transfection with antisense ODN in the potential treatment of proliferative disorders such as restenosis. Multiple genes that are all regulated by a single transcription factor can be targeted simultaneously through transfection with decoy ODN. This can obviate the need for transfection of two or more antisense ODN species into a single cell, and may therefore increase the efficiency of this gene targeting approach. In addition, it has yet to be determined whether transfection with double stranded ODN produces the same nonspecific effects on cell biology and protein expression as have been observed with single stranded ODN.

V. GENE OVEREXPRESSION AND RESTENOSIS: ecNOS

Endothelial dysfunction plays an important role in vascular prolif erative disease (Ross, 1993) The correlation between endothelial function and neointimal disease was further underscored by our observations of endothelial preservation and neointimal inhibition in genetically engineered autologous vein grafts. Denudation of the endothelium during balloon angioplasty is therefore believed to contribute to the pathogenesis of restenosis. Endothelial loss results in vasomotor abnormalities and in increased adhesion of platelets and blood borne inflammatory cells. These phenomena may be partially related to the loss of nitric oxide (NO) generation by endothelial cell nitric oxide synthase (ecNOS). Endothelial cell production of NO has other important homeostatic effects on the vessel wall (Vane *et al.*, 1990), including inhibition of VSMC proliferation and migration, and moderation of superoxide radical generation and oxidative state. We therefore hypothesized that expression of ecNOS in VSMC after balloon injury could replace this important source of NO production and ameliorate the effects of endothelial denudation on arterial remodeling after balloon injury

To achieve this overexpression of ecNOS, HVJ-liposomes were used to deliver a plasmid containing ecNOS cDNA driven by a ß-actin promoter and CMV enhancer to medial VSMC at the time of balloon injury in rat carotid arteries (von der Leyen *et al.*, 1995). Successful gene transfer in this *in vivo* model, in which no significant re endothelialization is present for up to 2 weeks after injury, was documented in several ways: Western blot demonstration of ecNOS protein in vessel homogenates, localization of ecNOS enzyme expression in medial VSMC via *in situ* histochemical staining using the NADPH-diaphorase assay, measurement of increased nitric oxide generation from transfected vessel segments using the chemi-

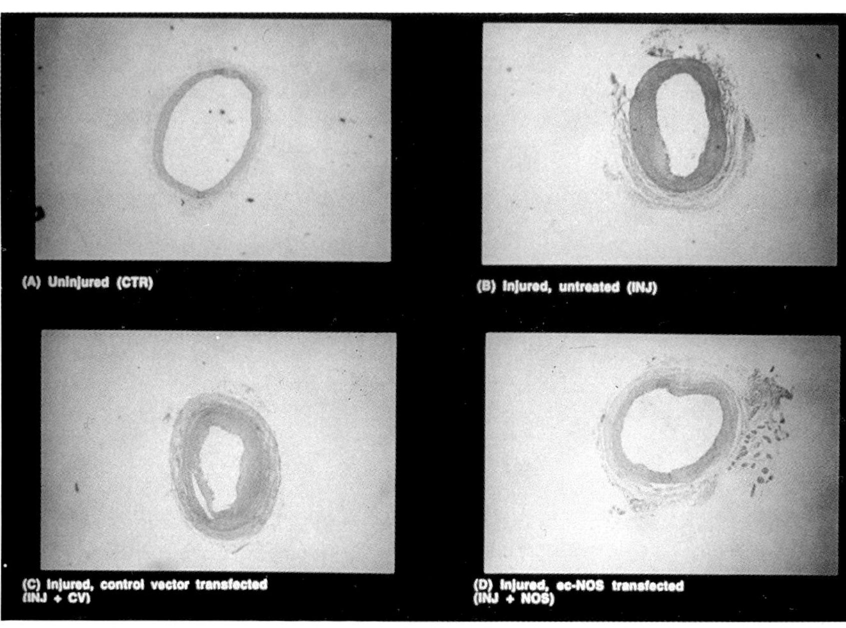

Figure 6. Inhibition of neointimal hyperplasia after rat carotid balloon injury via ecNOS gene transfer.

luminescence assay, and verification of the biological effectiveness of the ecNOS transgene product through analysis of ecNOS mediated changes in vascular reactivity.

Endothelial cell NOS overexpression after gene transfer therapy succeeded in reducing neointimal formation after carotid balloon injury by approximately 70% (Figure 6). These data reflected the important role a single gene product can play even in a complex disease process such as neointimal hyperplasia, and demonstrated the feasibility of targeting a physiologically important gene for overexpression as a therapeutic strategy in cardiovascular disease. Transgene expression after HVJ-liposome transfection is likely to last only two weeks, but this study demonstrates that transient overexpression of a critical gene product during certain phases of proliferative disease can have an effect on vascular remodeling. Furthermore, it may be possible to augment the efficacy of gene therapy approaches by combining gene overexpression and gene blockade strategies via cotransfection of plasmid and oligonucleotide DNA.

VI. SUMMARY

The study of molecular genetics and cell biology has yielded important insights into the function of cells and tissues during the development of cardiovascular disease. Technologies that allow the manipulation of gene expression in intact tissues in live animals have allowed investigators to extend these observations and to design preliminary strategies for genetic intervention in the disease process. A first critical step in developing such gene therapies is the targeting of individual genes or combinations of genes, the down-regulation or overexpression of which are likely to influence vascular pathobiology. Our group and others are currently investigating both the methods of *in vivo* genetic manipulation as well as the identification of such target genes to bring this exciting arena of molecular science out of the laboratory and into the clinic.

VII. REFERENCES

Bendeck, M. P., Irvin, C., and Reidy, M. A. (1996) Inhibition of matrix metalloproteinase activity inhibits smooth muscle cell migration but not neointimal thickening after arterial injury. *Circ. Res.* **78**:38-43.

Berg P., and Singer M. F. (1995) The recombinant DNA controversy: twenty years later. *Proc. Natl. Acad. Sci. USA* **92**:9011-9013

Bielinska, A., Schivdasani, R. A., Zhang, L., and Nabel, G. (1990) Regulation of gene expression with double-stranded phosphothioate oligonucleotides. *Science* **250**:997-1000.

Burgess, T. L., Fisher, E. F., Ross, S. L., Bready, J. V., Qian, Y. X., Bayewitch, L. A., Cohen, A. M., Herrera, C. J., Hu, S. S., Kramer, T. B., Lott, F. D., Martin, F. H., Pierce, G. F., Simonet, L., and Farrell, C. (1995) The antiproliferative activity of c-*myb* and c-*myc* antisense oligonucleotides in smooth muscle cells is caused by a nonantisense mechanism. *Proc. Natl. Acad. Sci. USA* **92**:4051-4055.

Cohen, J. S. (1992) Oligonucleotide therapeutics. *Trends Biotechnol.* **10**:87-91.

Colman, A. (1990) Antisense strategies in cell and developmental biology. *J. Cell Sci.* **97**:399-409.

Crooke, S. T. (1993) Progress toward oligonucleotide therapeutics: pharmacodynamic properties. *FASEB J.* **7**:533-539.

Cox, J. L., Chiasson, D. A. and Gotlieb, A. I. (1991) Stranger in a strange land: the pathogenesis of saphenous vein graft stenosis with emphasis on structural and functional differences between veins and arteries. *Prog. Cardiovasc. Dis.* **34**:45-68.

Dzau V. J., Gibbons G. H., Cooke J. P., and Omoigui, N. (1993) Vascular biology and medicine in the 1990s: Scope, concepts, potentials, and perspectives. *Circulation* **87**:705-719

Gibbons, G. H., and Dzau, V. J. (1994) The emerging concept of vascular remodeling. *New Engl. J. Med.* **330**:1431-1438.

Glagov, S. (1994) Intimal hyperplasia, vascular remodeling and the restenosis problem. *Circulation* **89**:2888-2891.

Inouye, M. (1988) Antisense RNA: its functions and applications in gene regulation--review. *Gene* **72**:25-34.

Isaka, Y., Fujiwara, Y., Ueda, N., Kaneda, Y., Kamada, T., and Imai, E. (1993) Glomerulosclerosis induced by *in vivo* transfection of transforming growth factor-ß or platelet-derived growth factor gene into the rat kidney. *J. Clin. Invest.* **92**:2597-2601.

Kaneda, Y., Iwai, K., and Uchida, T. (1989) Increased expression of DNA cointroduced with nuclear protein in adult rat liver. *Science* **243**:375-378.

Krystal, G. W., Armstrong, B. C., and Battey, J. F. (1990) N-*myc* mRNA forms an RNA-RNA duplex with endogenous antisense transcripts. *Mol. Cell Biol.* **10**:4180-4191.

Loppnow, H., and Libby, P. (1990) Proliferating or interleukin 1-activated human vascular smooth muscle cells secrete copious interleukin 6. *J Clin. Invest.* **85**:731-738.

Mann, M. J., Gibbons, G. H., Kernoff, R. S., Diet, F. P., Tsao, P., Cooke, J. P., Kaneda, Y., and Dzau, V. J. (1995) Genetic engineering of vein grafts resistant to atherosclerosis. *Proc. Natl. Acad. Sci. USA* **92**:4502-4506.

Mann, M. J., Gibbons, G. H., Tsao, P. S., von der Leyen, H. E., Buitrago, R., Kernoff, R., Cooke, J. P., and Dzau, V. J. (1997) Cell cycle inhibition preserves endothelial function in genetically engineered vein grafts. *J. Clin. Invest.* **99**:1295-1301

Matsukura, M., Shinozuka, K., Zonm, G., Mitsuya, H., Reitz, M., Cohen, J. S., and Broder, S. (1987) Phosphorothioate analogs of oligodeoxynucleotides: inhibitors of replication and cytopathic effects of human immunode-ficiency virus. *Proc. Natl. Acad. Sci. USA* **84**:7706-7710.

Morishita R., Gibbons G.H. , Ellison K. E., Nakajima, M., Zhang, L., Kaneda, Y., Ogihara, T., and Dzau, V. J. (1993a) Single intraluminal delivery of antisense *cdc2* kinase and proliferating-cell nuclear antigen oligo-nucleotides results in chronic inhibition of neointimal hyperplasia. *Proc. Natl. Acad. Sci. USA* **90**:8474-8478.

Morishita, R., Gibbons, G. H., Kaneda, Y., Ogihara, T., and Dzau, V. J. (1993b) Novel and effective gene transfer technique for study of vascular renin angiotensin system. *J. Clin. Invest.* **91**:2580-2585.

Morishita, R., Gibbons, G. H., Ellison, K. E., Nakajima, M., von der Leyen, H., Zhang, L., Kaneda, Y., Ogihara, T., and Dzau, V. J. (1994a) Intimal hyperplasia after vascular injury is inhibited by antisense *cdk2* kinase oligonucleotides. *J. Clin. Invest.* **93**:1458-1464.

Morishita, R., Gibbons, G. H., Kaneda, Y., Ogihara, T., and Dzau, V. J. (1994b) Pharmacokinetics of antisense oligodeoxynucleotides (cyclin B1 and *cdc2* kinase) in the vessel wall *in vivo*: enhanced therapeutic utility for restenosis by HVJ-liposome delivery. *Gene* **149**:13-19.

Morishita, R., Gibbons, G. H. , Horiuchi, M., Ellison, K. E., Nakama, M., Zhang, L., Kaneda, Y., Ogihara, T., and Dzau, V. J. (1995) A novel molecular strategy using *cis* element "decoy" of E2F binding site inhibits smooth muscle proliferation *in vivo. Proc. Natl. Acad. Sci. USA* **92**:5855-5859.

Nakanishi, M., Uchida, T., Sugawa, H., Ishiura, M., and Okada, Y. (1985) Efficient introduction of contents of liposomes into cells using HVJ (Sendai virus). *Exp. Cell Res.* **159**:399-409.

Okada, Y. (1993) Sendai Virus-Induced Cell Fusion. In Düzgünes, N., ed., *Methods in Enzymology*, Vol. 221, *Membrane Fusion*, Part B. Academic Press, New York, 18-23.

Okada, Y., Koseki, I., Kim, J., Hashimoto, T., Kanno, Y., and Matsui, Y. (1975) Modification of cell membranes with viral envelopes during fusion of cells with HVJ (Sendai virus). *Exp. Cell Res.* **93**:368-378.

Okada, Y., and Tadokoro, J. (1961) Analysis of giant polynuclear cell formation caused by HVJ virus from Ehrlich's ascites tumor cells. *Exp. Cell Res.* **26**:108-118.

Ohno, T., Gordon, D., San, H., Pompili, V. J., Imperiale, M. J., Nabel, G. J., and Nabel, E. G. (1994) Gene therapy for vascular smooth muscle cell proliferation after arterial injury. *Science* **265**:781-784.

Ross, R. (1993) The pathogenesis of atherosclerosis: A perspective for the 1990s *Nature* **362**:801-809.

Shi, Y., Fard, A., Galeo, A., Hutchinson, H. G., Vermani, P., Dodge, G. R., Hall, D. J., Shaheen, F., and Zalewski, A. (1994) Transcatheter delivery of c-myc antisense oligomers reduce neointimal formation in a porcine model of coronary artery balloon injury. *Circulation* **90**:944-951.

Simons, M., Edelman, E. R., DeKeyser, J. L., Langer, R., and Rosenberg, R. D. (1992) Antisense c-*myb* oligonucleotides inhibit intimal arterial smooth muscle cell accumulation *in vivo. Nature* **359**:67-70.

Sugawa, H., Uchida, T., Yoneda,Y., Ishiura, M., and Okada, Y. (1985) Large macromolecules can be introduced into cultured mammalian cells using erythrocyte membrane vesicles *Exp. Cell Res.* **159**:410-418.

Sullenger, B. A., Gallardo, H. F., Ungers, G. E., and Gilboa, E. (1990) Overexpression of TAR sequence renders cells resistant to human immunodeficiency virus replication. *Cell* **63**:601-608.

Vane, J. R., Änggård, E. E. and Botting, R. M. (1990) Regulatory functions of the endothelium. *New Engl. J. Med.* **323**:27-36

Villa, A. E., Guzman, L. A., Poptic, E. J., Labhasetwar, V., D'Souza, S., Farrell, C. L., Plow, E. F., Levy, R. J., Dicorleto, P. E., and Topol, E. J. (1995) Effects of antisense c-*myb* oligonucleotides on vascular smooth muscle cell proliferation and response to vessel wall injury. *Circ. Res.* **76**:505-513.

von der Leyen, H. E., Gibbons, G. H., Morishita, R., Lewis, N. P., Zhang, L., Nakajima, M., Kaneda, Cooke, J. P., and Dzau, V. J. (1995) Gene therapy inhibiting neointimal vascular lesion: *In vivo* gene transfer of endothelial cell nitric oxide synthase gene. *Proc. Natl. Acad. Sci. USA* **92**:1137-1141.

Wenzel, V. O., Fouqueray, B., Grandaliano, G., Kim, Y. S., Karamitsos, C., Valente, A. J., and Abboud, H. E. (1995) Thrombin regulates expression of monocyte chemoattractant protein-1 in vascular smooth muscle cells. *Circ. Res.* **77**:503-509.

Wickstrom, E. (ed.) (1991) *Prospects for Antisense Nucleic Acid Therapy of Cancer and AIDS*. Wiley-Liss, New York.

Yaswen, P., Stampfer, M. R., Ghosh, K., and Cohen, J. S. (1993) Effects of sequence of thioated oligonucleotides on cultured human mammary epithelial cells. *Antisense Res. Dev.* **3**:67-77.

Zwolak, R. M., Adams, M. C., and Clowes, A. W. (1987) Kinetics of vein graft hyperplasia: association with tangential stress. *J. Vasc. Surg.* **5**:126-136.

Index

*Page numbers in italic indicate figures. Page numbers followed by
"t" indicate tables.*